Hydrocarbon Nation

THE JOHNS HOPKINS UNIVERSITY STUDIES
IN HISTORICAL AND POLITICAL SCIENCE
133rd SERIES (2018)

1. *Hydrocarbon Nation: How Energy Security Made Our Nation Great and Climate Security Will Save Us*   Thor Hogan

Hydrocarbon Nation

How Energy Security Made Our Nation Great
and Climate Security Will Save Us

THOR HOGAN

Johns Hopkins University Press
Baltimore

Johns Hopkins University Press
2715 North Charles Street
Baltimore, Maryland 21218-4363
www.press.jhu.edu

Library of Congress Cataloging-in-Publication Data

Names: Hogan, Thor, author.
Title: Hydrocarbon nation : how energy security made our nation great and climate
    security will save us / Thor Hogan.
Description: Baltimore : Johns Hopkins University Press, 2018. | Includes
    bibliographical references and index.
Identifiers: LCCN 2017033309 | ISBN 9781421425061 (hardcover : alk. paper) | ISBN
    9781421425078 (electronic) | ISBN 1421425068 (hardcover : alk. paper) | ISBN
    1421425076 (electronic)
Subjects: LCSH: Petroleum industry and trade—Government policy—United States. |
    Hydrocarbons—United States—History.
Classification: LCC HD9566 .H62 2018 | DDC 333.8/20973—dc23
LC record available at https://lccn.loc.gov/2017033309

A catalog record for this book is available from the British Library.

*Special discounts are available for bulk purchases of this book. For more information,
please contact Special Sales at 410-516-6936 or specialsales@press.jhu.edu.*

Johns Hopkins University Press uses environmentally friendly book materials,
including recycled text paper that is composed of at least 30 percent post-consumer
waste, whenever possible.

For Sam

CONTENTS

Acknowledgments   *ix*

Prologue: Prince William Sound   1

Introduction   9

PART I   HYDROCARBONS AND THE AMERICAN RISE

1 Steam, National Security, and the First Political Age   23

2 Coal, Macroeconomic Security, and the Second Political Age   56

3 Oil, Microeconomic Security, and the Third Political Age   104

PART II   SUSTAINABILITY AND AN AMERICAN REBIRTH

4 Energy Insecurity and the American Decline   181

5 Gas and National Renewal in the Fourth Political Age   268

6 Climate Security and a Sustainability Revolution   334

Epilogue: Napatree Point   381

Notes   *387*
Index   *409*

This book would not have been possible without financial assistance from Earlham College. I would like to thank Greg Mahler in particular for championing scholarly research at the college. I am especially indebted to several students who have engaged in student-faculty collaborative research projects with me over the past five years. David Goldenberg provided invaluable assistance by investigating and writing about the political history of highway developments in the United States. Regan Lowring and Truman McGee both drafted valuable notes for several key books on energy policy and energy history. Eliza Hudson and Ben Smith prepared very helpful annotated bibliographies for various aspects of the agricultural revolution. I have also benefitted from scores of invigorating class discussions with my students over the years, which were crucial in the development of many of the ideas shared in this book.

I am grateful to everyone who provided feedback on early drafts of the book, including Kevin Armitage, Matt Lloyd, and Friederike Sundaram. I would especially like to thank Brian Black for his valuable reviews of the manuscript, which without question greatly improved its quality. I have been fortunate to work with amazing editors at Johns Hopkins University Press. I will be forever grateful to Elizabeth Demers for her unwavering support for this book. Many thanks to Joanne Allen for her copyediting, which made this book far more readable. I am also indebted to Juliana McCarthy, Lauren Straley, Kyle Kretzer, Gene Taft, Meagan M. Szekely, and everyone else at Johns Hopkins University Press who worked so hard to produce this book.

Nobody has been more important to me in the pursuit of this project than the members of my family. Anna, Stefan, Linda, Diane, and Mark have offered unstinting encouragement and helped keep me sane. Most importantly, I wish to thank my loving and supportive partner, Kate, and my wonderful son, Sam, for their love and understanding and for providing me with boundless inspiration.

Hydrocarbon Nation

Prince William Sound

I was born and raised in Anchorage, Alaska. I grew up in a small house on the city's outskirts that had a mind-blowing view of the Chugach Mountains. Although Anchorage was a city of nearly a quarter million people, the last frontier was then, and still remains, a world apart. I survived the lengthy winters attending some wonderful schools and playing ice hockey. During the exceptionally long summer days, my family spent as much time as possible at a lake an hour north, which was where I learned to love the outdoors—hiking, biking, and swimming with my brothers and sister until being forced inside for the night. It was glorious. It was the place that taught me about the wonders of the natural world.

I remember the exact date of my environmental awakening. It was 24 March 1989, the day the *Exxon Valdez* ran aground on Bligh Reef in Prince William Sound. While this was a transformational moment for many in my generation, it had special significance for me because the disaster unfolded in my home state. Just a few years before I was born, the Atlantic Richfield Company (ARCO) had discovered the Prudhoe Bay oil field, which would prove to be the largest petroleum reserve on the continent. Shortly after the field was located, Americans were shocked when an offshore oil platform blowout released 3.4 million gallons of crude into the pristine waters of California's Santa Barbara Channel and ultimately contaminated more than forty miles of coastline. Three thousand miles up the coast, this awakened many Alaskans to the dangers the oil industry posed to critical ecosystems. There was enormous political pressure, however, to develop the North Slope. The endeavor gained significant public support when the state auctioned leases to 179 parcels of land for $900 million. The big question: how was the oil to be transported to the continental United States?

Alaska governor Wally Hickel and the Nixon administration supported the construction of an 800-mile pipeline south to Valdez, the northernmost ice-free port in North America. Alaskan commercial fishermen, however, were pushing for an all-land route through Canada, which would not endanger Prince

William Sound. These efforts were supported by a US Department of the Interior report that suggested the danger of marine pollution was a greater environmental threat than the pipeline itself. The only Alaskan politician who shared these concerns was Representative Nick Begich, but he was tragically killed in a plane crash. Senator Ted Stevens vociferously supported the pipeline, famously promising commercial fishermen that "not one drop of oil would touch the waters of Prince William Sound." In November 1973, after a series of lawsuits had delayed construction of the pipeline for a half decade, Congress passed the Trans-Alaskan Pipeline Authorization Act—one response to the OPEC oil embargo. Within a few years, the oil industry successfully eliminated most of the safety measures they had promised Congress they would employ, including double-bottom tankers, state-of-the-art navigation systems, and up-to-date oil spill contingency plans.[1]

While I was growing up, most Alaskans rightly believed oil production had been a boon for the state. Taxes on oil production funded nearly the entire state government budget. A share of oil revenues was redirected to a sovereign wealth fund established to benefit current and future residents. Starting in the early eighties, each resident received a cash payment from the fund that generally ranged from $600 to $1,500 annually. (By 2012 the fund was valued at more than $40 billion.) While there had been some land- and sea-based environmental damage resulting from the pipeline development, to that point these were widely seen as relatively minor. Global warming had yet to reach the national agenda, so concerns regarding the role Alaska's oil played in contributing to this growing crisis barely registered. In 1988, the Alaskan economy had a banner year. Commercial fisherman had their best year ever, and Prudhoe Bay saw its production peak (at 2.1 million barrels a day, which satisfied 12 percent of American demand).[2] The following year, however, events would cast serious doubts on the benefits of oil production in the last frontier.

The fully loaded *Exxon Valdez* departed the tanker terminal at the far end of the 800-mile pipeline at 21:12 on 23 March 1989. With a Coast Guard pilot on board, it successfully navigated what was considered the most dangerous portion of its route, the passage through the Valdez Narrows. About ninety minutes into the trip to Long Beach, after the pilot had disembarked, Captain Joseph Hazelwood requested and received permission to move from the outbound to the inbound tanker lane to avoid icebergs coming off Columbia Glacier. Hazelwood, who many suspect was drunk at the time, then retired to his cabin and left his third mate in charge. Without the captain on hand, the ship passed entirely out of the tanker lanes. At 00:04 on 24 March the massive ship smashed into Bligh Reef, damaging eight of the single-hulled vessel's eleven cargo holds. Within six hours, 10.1 million gallons of crude had poured into the chilly waters of Prince William Sound.[3]

I remember being frustrated as I watched the news in the first few days after the accident. It seemed that no one had a game plan. Like many others, I repeatedly asked myself what the Aleyeska Pipeline Service Company, Exxon, and the Coast Guard were actually doing? In those initial days, thanks to the surprisingly good weather, most of the oil was still massed around Bligh Island. But no one was taking action. Aleyeska was supposed to be in charge of the response, but its superintendent actually went back to bed after hearing about the tanker grounding. The company's cleanup barge was on shore for repairs, with no backup available, so an insufficient number of oil booms could be deployed to contain the spill. Then came what seemed to be an endless debate about whether dispersants should be used to break up the petroleum before it reached shore. The reality was that there weren't enough of these chemicals in the entire world to treat such an enormous spill. Some experts argued that in situ burning was the correct solution, but this path wasn't taken because of concerns about the tanker's safety, as the *Exxon Valdez* still had more than forty million gallons of oil on board. After a few days the weather turned foul and a huge storm pounded Prince William Sound. The slick would ultimately extend nearly five hundred miles, impacting more than thirteen hundred miles of shoreline.[4]

Although the *Exxon Valdez* oil spill was nowhere near the largest in history, it is still considered by many to have had the largest environmental impact: "The timing of the spill, the remote and spectacular location, the thousands of miles of rugged and wild shoreline, and the abundance of wildlife in the region combined to make it an environmental disaster well beyond the scope of other spills."[5] Once Aleyeska and Exxon lost control of the spill, their public-relations machine rolled into action. Their key spokesperson actually claimed that Alaskans had been lucky, because Exxon was such a solid corporate citizen. "We will consider whatever it takes to make you whole."[6] He was correct that Exxon would *consider* all claims; he just failed to honestly relate the strong likelihood that after careful consideration the company would decide to spend as much time in court as was necessary to duck its financial responsibility.

I was finishing my junior year in high school when the spill happened. While I grew up in a state shaped by oil, I had not thought much about that fact. Almost overnight that changed. Suddenly I was doing a lot of thinking about oil. After the efforts in Prince William Sound had shifted from containment to cleanup, I attended Boys State, a summer leadership and citizenship program sponsored by the American Legion. I spent a week at Fort Richardson Army Base with several dozen of my peers participating in a legislative simulation. The topic: protecting Alaskan ecosystems from future *Exxon Valdez*-scale catastrophes. That experience stimulated a growing interest on my part in public affairs and environmental issues.

In the years to come, my life would continue to be intertwined with the Alaskan oil industry. The summer after high school, I had an opportunity to work on the *Exxon Valdez* cleanup effort. This was the second of four expensive and ultimately futile summers dedicated to cleaning up oiled beaches in Prince William Sound and Cook Inlet. Despite the nearly $2 billion spent and the ten thousand workers employed, "it is widely believed . . . that wave action from winter storms did more to clean the beaches than all the human effort involved."[7] Fortunately, I didn't take the job working on the cleanup: two-thirds of those toiling on the beaches experienced respiratory problems because of inadequate safety equipment.[8] Instead, I found a job working at a cannery on the Kenai Peninsula. Although my lungs weren't injured, I didn't make much money that summer because of the catastrophic impacts of the spill on the fishing industry. Fortunately, I found a job that paid well for the month before I started college as a ticket agent for ARCO's charter service, which flew oil workers to the North Slope. I was from a struggling working-class family, and a job was a job. It never occurred to me to turn down the work just because of my growing unease about the role oil played in our society. This was the first of many summers that I would wind up employed by the oil industry.

The following summer, after my first year of college, I returned home once again and found a job working as a file clerk for the Trans Alaska Pipeline Liability Fund (TAPL). In the months after the *Exxon Valdez* spill, scores of lawyers had descended on Southcentral Alaska. Eventually, more than thirty thousand claims would be filed against the oil companies. Most of these were pushed into the federal court system and assigned to Judge Hezekiah Russel Holland. Exxon attorneys believed that Holland, a former law partner of Ted Stevens and a Reagan appointee, would be sympathetic to their arguments. Initially these hopes seemed justified when Judge Holland ruled that no plaintiffs' suits could go forward until all TAPL remedies had been exhausted. The problem was that TAPL only had about $86 million available, while the damages claimed were in the billions. So for about twelve weeks that summer, I found myself assisting several dozen insurance adjusters who had been given the thankless task of sifting through the thousands of claims and determining which should be paid.[9] It took two years to complete this process, which fit into the Exxon lawyers' "procedural war to confuse, consolidate, delay, draw out, and exhaust the plaintiff constituency."[10]

The next two summers, the pattern continued, and once again I was working directly for the oil industry. Every summer hundreds of college students are employed as "stick-pickers," which means that they walk around the North Slope collecting the trash that accumulated during the previous winter. The hours were long (I worked twelve hours a day, seven days a week, for seven weeks), and the work was pretty boring, but the pay was very good. Since I was paying my own way through college, this was impossible to pass up. It was also a fantastic experience. Very few people ever have an opportunity to spend time on the North Slope, largely because it is so completely isolated. I remember thinking that it was the closest I was ever likely to come to living on a moon base. For as far as the eye can see, lush, beautiful tundra flanks the Arctic Ocean. You cannot really build anything of significance on tundra, however, so an enormous amount of earth was moved to build up the immense gravel platforms that housed the oil industry infrastructure, from drilling sites to residential camps to service facilities. Connecting all these platforms were gravel roads, which were similarly raised half a dozen feet off the tundra. I vividly remember watching the porcupine caribou migration, with nearly 125,000 animals moving between their winter range and their summer calving grounds. It was surreal to watch thousands upon thousands of these majestic creatures passing through an environment that had been so fully scarred by humanity's lust for petroleum. I still wonder how many Americans would support drilling for oil in the National Arctic Wildlife Refuge if they had an opportunity to see this splendid landscape. Seeing the enormity of the Prudhoe Bay operation had a huge impact on me.

Meanwhile, the *Exxon Valdez* legal drama was just getting started in Judge Holland's courtroom. In 1994, five years after the spill and the year I graduated from college, a federal jury awarded the plaintiffs $287 million in actual damages and $5 billion in punitive damages, the latter equaling one year of Exxon's profits (for which the company had insurance covering nearly the entire amount). Seven years later, the US Court of Appeals for the Ninth Circuit, in San Francisco, concurred with the jury but ordered Judge Holland to reduce the punitive-damages award. The following year, he shocked everyone when he reduced the amount by only $1 billion, saying that he could not "by any principled means" reduce it further. A year later the Ninth Circuit vacated his decision and instructed Holland to reconsider the award based on a recent

Supreme Court case that suggested punitive damages should amount to no more than ten times the actual damages. Yet another year later, an apparently irritated Holland increased the punitive-damages award to $4.5 billion. Two more years and the Ninth Circuit finally took matters into its own hands, reducing the amount to $2.5 billion. Finally, in 2008 the Supreme Court reduced the amount to be paid to the plaintiffs to just over $500 million. In an opinion written by Associate Justice David Souter, the court found that in maritime cases a ratio of no more than one to one between actual and punitive damages is generally appropriate. Since Exxon had paid about $507 million to compensate landowners, commercial fisherman, and Alaska Natives, it should have to pay no more than that amount in punitive damages.[11]

The day the opinion was made public, the editorial board of the *Anchorage Daily News*, which is not exactly a bastion of liberal ideology, excoriated the court majority for taking "it upon themselves to write new law that shields corporate wrongdoers from appropriate penalties. Their ruling makes it much harder to prevent businesses from engaging in profitable but dangerous corner-cutting."[12] The newspaper was outraged that the court had overridden a duly constituted jury and a Reagan-appointed judge. The editorial pointed out that the company would be able to pay off the newly defined penalty with five days of profits (a 98.6 percent reduction in the financial pain the judge and jury deemed appropriate). It concluded with this pronouncement: "BOTTOM LINE: Exxon was the big winner."[13]

Nineteen years. That's how long it was from the time of the spill until the final adjudication. Writing this book, I found myself thinking a lot about how much happened to those directly impacted by the disaster during that long, painful, and strange interlude. The financial stress and mental anxiety they experienced, waiting year after year for the tragedy to finally end. The emotional seesaw they were on as one court after another changed direction. The anger they must have felt when they recalled Exxon executives telling them how lucky they had been because the company would make them whole again. The fury they surely experienced as they watched that same corporation, the largest in the world, walk all over them for nearly two decades, employing thousands of lawyers and spending tens of millions to avoid keeping that promise.

I also think a lot about the magnitude of the environmental disaster. Within days 250,000 shorebirds, 2,000 sea otters, and 300 harbor seals died. According to researchers at the University of North Carolina at Chapel Hill, "Oil . . . persisted in surprisingly large quantities for years after the *Exxon Valdez* spill in subsurface reservoirs under coarse intertidal sediments. This oil was sequestered in conditions where weathering by wave action, light, and bacteria was inhibited, and toxicity remained for a decade or more."[14] Salmon, duck, and large mammal populations were adversely impacted for many years. It has been estimated that mussel beds won't recover for three decades. Some researchers believe the spill was the direct cause of the collapse of the Prince William Sound herring fishery, which has not yet recovered.[15]

Thinking about all of this, it is hard not to feel angry. A couple of years ago I returned to Prince William Sound and spent a day on the water with my wife and young son. It is still a place of astounding beauty, with towering mountains and vast glaciers hanging above the icy waters. Although much damage remains that cannot be readily seen with the naked eye, much of the wildlife has returned. During a half day cruising around we spotted scores of sea otters, several porpoise pods, a large colony of harbor seals, several humpback whales, bald eagles, and gorgeous black puffins. It is hard to imagine anyone knowingly jeopardizing this natural gift. But as I've learned up close, the undeniable reality of our Hydrocarbon Nation is that we are now endangering innumerable species, and human existence itself, on a global scale. Why? Because hydrocarbons represent the most useful natural resources in human history. They have driven economic and social growth for several centuries.

The *Exxon Valdez* oil spill was the event that made me a committed environmentalist. Over the years, as I tracked the progress of the court case against Exxon, it also contributed to my evolving belief that our entire political system is badly broken. I found myself asking how business interests had come to so fully dominate our national government. If things can go so wrong at this relatively small scale, what hope is there of solving the vastly more significant environmental challenges associated with global warming? If it takes nearly two decades to provide redress to the victims of a regional calamity, what hope is there of taking real action to slow the concentration of greenhouse gases in the planetary atmosphere?

As I've struggled to answer these questions in the past few years, I haven't given in to despair. Quite the contrary, I've found hope. I've come to believe that solving the climate crisis could be the catalyst for solving many other seemingly intransigent national problems, ranging from our dysfunctional republican institutions to our deteriorating international reputation to our growing income insecurity to our failed energy policy. The reason we didn't start this process earlier is that the electorate was dominated by generations who had become too accustomed to a world of plenty and failed to recognize or take seriously our internal institutional decay. The reality, however, is that fundamental changes began to come about as Generation X and the Millennials came to compose a much larger share of the electorate. This was the result of the resolve within those generations to both redefine our governmental framework and halt global warming. I believe that by working in concert with ever more powerful minority voting blocks, these generational groups could continue to push forward a progressive agenda for a new political age. To do so, however, we must elect political leaders who are willing to envision a future beyond our Hydrocarbon Nation.

Introduction

Hydrocarbons are bad; the only people who argue otherwise are clearly evildoers who want to destroy the planet while getting filthy rich. Within the modern environmental movement, this is considered to be a universal truth. Supporting this uncomplicated belief is the equivalent of an environmentalist's shibboleth; it is sacrilegious to suggest otherwise. The reality, however, is far more complex and nuanced. The extraordinarily compelling reasons why the use of hydrocarbons should be dramatically decreased do not diminish their historical value and contemporary importance. More than any other factor, hydrocarbons created the modern world. The United States especially benefitted from the use of hydrocarbons; they were the driving force behind American economic, political, and social progress for nearly two centuries. Yet, the emissions from burning hydrocarbons quietly birthed a climate crisis that has come to threaten humanity's ability to survive on this planet. For most, this contradiction is hard to reconcile. Hydrocarbons should be either good or bad. But it is critical for everyday Americans and policymakers alike to chart a path forward that acknowledges the benefits hydrocarbons have provided while also embracing the need to reduce their use to safeguard the advances that they have delivered.

When I began this project five years ago, I had a number of different questions in mind. Was there a link between the expanded use of hydrocarbons and America's economic, political, and social advance? What role did hydrocarbons play in its geopolitical rise? What were the impacts of the peaking of domestic conventional petroleum production in 1970? And what are the prospects for revitalizing the Republic in a low-carbon future? Many scholars have provided answers to these questions. For the most part, political historians have taken too little notice of the part hydrocarbons performed in the American Rise (economic historians have been somewhat more successful). Environmental historians have largely chosen to malign the importance of hydrocarbons in creating the modern world. These omissions are troubling, because it usually proves difficult to find solutions to contemporary problems without first understanding their historical roots. I decided to write a book that would attempt to close these gaps by blending several different literatures into a comprehensive narrative that seeks to explain the American Rise and the recent relative decline, by describing the overwhelming but often ignored significance of hydrocarbons in the country's history.

The United States has long been blessed with enormous deposits of these valuable resources, spread over large portions of its territory. Although these highly prized reserves have been aggressively extracted for more than a century and a half, the nation still enjoys a top-three ranking in the annual production of coal, oil, and gas. America was not simply fortunate to possess large hydrocarbon supplies; it developed a standard model of economic advancement that allowed it to leverage them better than any other nation. This approach included a strong central bank, protective tariffs, transportation infrastructure, and mass education. When this economic model was combined with a geopolitical strategy that led to the creation of a continental nation nearly impervious to outside threats, the United States was uniquely positioned to take advantage of its hydrocarbon largesse. By the late nineteenth century it had exploited this reality to become the most prosperous nation on the planet. It became a country with literally unparalleled economic and military potential, which would lead it in the twentieth century to become a reluctant global hegemon. By the final quarter of the century, however, its reliance on fossil fuels had also resulted in an extended economic malaise caused by the peaking of domestic supplies of conventional petroleum.

This book does not seek to make an argument regarding the insidious role that coal, oil, and gas companies have had on American economics and politics. There are many fine books that pursue this thesis, in particular Timothy Mitchell's superlative *Carbon Democracy: Political Power in the Age of Oil* (but also Barbara Freese's *Coal: A Human History* and Robert Engler's *The Politics of Oil: A Study of Private Power and Democratic Directions*). While I deeply admire Mitchell's accomplishment in scrutinizing the effect corporations engaged in the various hydrocarbon industries have had on democracy and frequently find myself agreeing with specific arguments he makes, overall I come to a largely different conclusion regarding the direction of those impacts. Mitchell's overarching thesis is that hydrocarbon wealth leads inexorably to political poverty, to the erosion of the highest ideals and norms of democracy. Although I don't disagree that our use of hydrocarbons has had negative impacts, my take is that the positive impacts have, until relatively recently, outweighed them. Mitchell argues that oil corrupts, while I am making the case that for most of our history hydrocarbons have provided the basis for an expansion of economic security that was unprecedented in human history.[1]

Nor does this book examine the longstanding (and generally localized) environmental impacts of hydrocarbon extraction and use. Once again, there are many exceptional books that follow this path. I would point in particular to J. R. McNeill's *Something New Under the Sun: An Environmental History of the Twentieth Century*, Ted Steinberg's *Down to Earth: Nature's Role in American History*, and Carolyn Merchant's *American Environmental History*. The difference between those works and this one has less to do with the eventual conclusions, which are actually quite similar, than with the narrative and analytical approach. While these are primarily environmental histories that critique humanity's use of hydrocarbons, I focus much more on the economic and geopolitical benefits of fossil fuels before turning my attention to the climate crisis. My objective is most certainly not to challenge the conclusions of these authors or to defend the actions of corporate barons, who have often acted without taking the public interest to heart. Instead, this book seeks to chart a middle ground, in some respects similar to the approach of authors like Tyler Priest and Brian Black, by describing the political and economic benefits that have accrued to society as a result of our exploitation of hydrocarbons, while wholeheartedly embracing the need to reduce or eliminate their use to preserve the habitability of our planet.

* * *

So what does this book do? It seeks to synthesize selected scholarly work from several fields—primarily political and economic history, political science, and environmental policy—to tell the story of how hydrocarbons contributed to the American Rise and came to endanger its economic and geopolitical future. Part I, "Hydrocarbons and the American Rise," comprising chapters 1–3, builds a historical foundation by examining the role hydrocarbon-based energy security played in the development of the country. In particular, it provides a detailed account of how Americans used fossil fuels in a wide variety of endeavors and how government leaders took advantage of the resultant growth to provide national and economic security. In part II, "Sustainability and an American Rebirth," comprising chapters 4–6, this foundation is utilized as a tool to better understand contemporary political developments. Perhaps more importantly, it is used to consider paths that national security, economic security, and climate security policy might take in the coming years and decades.

I use two literatures to help guide the narrative. The first, which examines political-party realignments, was first developed by V. O. Key Jr. and later expanded upon by the likes of E. E. Schattschneider, Paul Kleppner, Walter Dean Burnham, and James Sundquist. In his seminal 1955 article "A Theory of Critical Elections," Key argued that certain elections have a far more important impact on the nation. Key wrote: "Even the most fleeting inspection of American elections suggests the existence of a category of elections in which voters are, at least from impressionistic evidence, unusually deeply concerned, in which the extent of electoral involvement is relatively quite high, and in which the decisive results of the voting reveal a sharp alteration of the preexisting cleavage within the electorate. Moreover, and perhaps this is the truly differentiating characteristic of this sort of election, the realignment made manifest in the voting in such elections seems to persist for several succeeding elections."[2] In 1960 E. E. Schattschneider expanded upon this idea by contending that not only does voting behavior shift but there is a marked change in policy outputs and institutional arrangements.[3] Thus we arrive at a workable definition of an electoral realignment: one that not only brings to power a new electoral coalition but also results in significant changes in how the nation is governed. Using this definition, one might consider the following to have been realigning elections: 1800 (Jefferson), 1828 (Jackson), 1860 (Lincoln), 1896 (McKinley), 1920 (Harding), 1932 (Roosevelt), 1968 (Nixon), 1980 (Reagan), and 2008 (Obama).

David Mayhew's *Electoral Realignments: A Critique of an American Genre*, published in 2004, provides an excellent assessment of the theory of electoral realignments. While I believe he goes too far, essentially discarding the entire theory, he does raise two important points that can help us make better use of the model. First, he discusses the importance of sectionalism and racial politics in our electoral history. There is no doubt that interests engaged in protecting slavery and southern apartheid had an enormous impact on both party strategy and election outcomes. The three-fifths compromise and the electoral college provided outsized power to the American South, and the latter institution still affords too much electoral influence to small rural states. Racial politics in the modern age has a completely different connotation, as the nation approaches a majority-minority status and racial minorities have an ever-larger impact on national politics. Second, Mayhew questions the idea that realignments lead to major policy changes. He argues that this seems to have been true in 1860 and 1932 but not in 1896. This is a point echoed by James Reichley, who argues that there have been only three realignments in American history, in 1800, 1860, and 1932. In his view, what have been commonly recognized as realignments in 1828 and 1896 were actually just restorations of previous critical elections. While both Mayhew and Reichley have struck upon an important point, I believe there is a far more compelling interpretation.[4]

I contend that American political history can be divided into four ages, each of which began with a *major realignment election*—in 1788, 1860, 1932, and 2008. Each of these contests saw a new political coalition, characterized by its own generational, sectional, and racial arrangements, come to power. These political ages began with a progressive cycle that saw significant changes in the types of policy interventions the federal government pursued. Each age was underpinned by an economic expansion made possible by the leveraging of a new type of nonanimal energy that provided the fiscal resources for the federal government to augment national, economic, and ultimately environmental security in ways that had never been seen in human history. In these larger political ages there were also *minor realignment elections*—1800, 1828, 1896, 1920, 1968, and 1980. These contests saw either a fundamental shift in the nature of the existing progressive cycle or the beginning of a conservative cycle. The conservative cycles were particularly important, because they saw a structural backlash against the provision of economic and environmental security. These cycles created the preconditions for the next major realignment election and the inauguration of a new political age.

I employ Frank Baumgartner and Bryan Jones's punctuated-equilibrium model to make the case that American political history can be separated into these four political ages. These two scholars suggest that the policy process has long periods of equilibrium that are periodically disrupted by some instability that results in dramatic policy change. Baumgartner and Jones describe "a political system that displays considerable stability with regard to the manner in which it processes issues, but the stability is punctuated with periods of volatile change." Only in times of unique crisis and instability do issues rise to the top of a government's agenda. At a fundamental level, the punctuated-equilibrium model seeks to explain why the policy process is largely incremental and conservative but also subject to periods of radical change.[5] I suggest that major party realignments are associated with important policy punctuations but also that the increased use of specific hydrocarbons helps explain why the realignments and punctuations happened at specific points in American history.

The second literature that helps guide this narrative examines the impact of generational dynamics and cycles on American politics, which turns out to be a valuable supplement to the electoral-realignment scholarship. In his 1927 article "The Problem of Generations," Karl Manheim argued that generations provide "one of the indispensible guides to an understanding of the structure of social and intellectual movements."[6] He went further to suggest that "the social phenomenon of 'generations' represents nothing more than a particular kind of identity of location, embracing related 'age-groups' embedded in a historical-social process."[7] Manheim and other developers of this theory, such as Françoise Mentré and José Ortega y Gasset, moved away from the idea of a generation as merely a sequential biological progression, instead focusing on the idea of the generation as a group of people of roughly the same age whose shared experience differentiates them from other such cohorts.[8] Efforts to determine the starting and ending points for a given generation, however, proved controversial either because they often seemed arbitrary or because the term did not apply across all human activities; for example, a cultural cohort and a political cohort might not overlap perfectly. The best attempt to answer such criticisms was offered by Julian Marías, who submitted that generations could be identified empirically.[9]

The theory of generational dynamics is based on scholarship considering the notion that American history is cyclical. As Mark Twain is purported to have said, "History doesn't repeat itself, but it does rhyme." The Pulitzer Prize–winning author Arthur M. Schlesinger Jr. was the first to rigorously analyze this idea (although he had inherited it from Henry Adams and from his eminent father, Arthur M. Schlesinger Sr.). In 1986 Schlesinger published *The Cycles of American History*, in which he presented his cyclical hypothesis. He found "a pattern of alteration in American history between negative and affirmative government—that is, between times when voters see private action as the best way of meeting our troubles and times in which voters call for a larger measure of public action."[10] The transitions from one cycle to another are rooted in the public mood, with political leaders elected specifically because they best represent this mood. One critically important idea within Schlesinger's work, however, is that shifts from a progressive (public-action) to a conservative (private-action) government do not often lead to the dissolution of the accomplishments of the progressive cycle, thus there is an accumulation of change over time.[11]

In 1991 William Strauss and Neil Howe published *Generations: The History of America's Future, 1584 to 2069*, their first of three books that explicate a model of generational cycles.[12] Their detailed historical and sociological research revealed a recurring cycle of four distinct types of generations, arriving in a repeating sequence, that correspond with specific types of political events. Each generation has a persona determined by common beliefs, behaviors, and age location. The model asserts that the four types of generations recur in a fixed order. First come the Prophets [e.g., Baby Boomers], who grow up as indulged youths after a crisis, come of age inspiring a spiritual awakening, fragment into narcissistic rising adults, cultivate principles as moralistic midlifers, and emerge as visionary elders guiding the next crisis. Next come the Nomads [e.g., Generation X'ers], who grow up as criticized youths during an awakening, mature into risk-taking and alienated rising adults, mellow into pragmatic midlife leaders during a crisis, and maintain respect as reclusive elders. Then come the Heroes [e.g., Millennials], who grow up as protected youths after an awakening, come of age overcoming a crisis, unite into a heroic and achieving cadre of rising adults, sustain that image while building institutions as powerful midlifers, and emerge as busy elders attacked by the next awakening. Finally come the Artists [e.g., Generation Z'ers], who grow up as overprotected and suffocated youths during a crisis; mature into risk-averse, conformist rising adults; produce indecisive midlife arbitrator-leaders during an awakening; and maintain influence as sensitive elders.[13]

Strauss and Howe conclude that generational change has a critical impact on how history unfolds. As the authors write, "Just as history produces generations, so too do generations produce history."[14] The practical outcome of this statement is that generational personalities lead to repeating *turnings*, periods when people perceive that historic events are radically altering the lived reality. There are four types of turnings: crises, highs, awakenings, and unravelings. *Crises* are periods when society focuses on reordering the outer world of institutions and public behavior (e.g., the Great Depression). *Highs* are periods of strong institutions and weak individualism (e.g., the Great Compression), which allow for *awakenings*, when society focuses on changing the inner world of values and private behavior (e.g., the Counterculture Revolution). Finally, an *unraveling* is a period when institutions are weak and distrusted, which leads full circle to another crisis (e.g., the Great Recession). Turnings normally occur about every twenty years and alternate in type. Strauss and Howe suggest that since the founding of the nation in 1776, there have been three crises (the American Revolution, the American Civil War, and the Great Depression–Second World War) and three awakenings (the Transcendental Awakening, the Reform Awakening, and the Counterculture Revolution). If this pattern is holding, as they argue it has since the sixteenth century, America is very likely in the middle of a crisis period right now—although I have fundamental disagreements with the authors regarding the nature of the current crisis and the response to it. Leveraging the generational-cycle theory, I argue in the coming chapters that specific generational turnings promoted the four major realignments in American political history—in 1788, 1860, 1932, and 2008. Generational turnings also fostered six minor realignments. Three of these signaled a fundamental expansion of an existing progressive cycle—in 1800, 1896, and 1968—and three signaled the beginning of a conservative cycle—in 1828, 1920, and 1980.

The final piece of the puzzle returns us to the role of hydrocarbons. One of the central contentions of this book is that providing security is the primary function of government and that the ever-growing use of hydrocarbon energy has been *the* crucial factor shaping politics and institutions to expand the scope of human security. For most of history, humans were focused almost entirely on securing adequate sustenance and personal safety. The creation of nation-states added the requirement to guarantee territorial integrity and domestic sovereignty. Security has been at the heart of the American experience for more than two centuries. During each of four political ages, the exploitation of a largely untapped energy source (e.g., water/steam, coal, oil, and gas) provided the federal government with the ability to expand human security. This began with national and macroeconomic security but ultimately extended to include microeconomic security and finally environmental security.

For millennia, individual economic security had essentially been beyond the realm of possibility for almost everyone. As the economist John Kenneth Galbraith suggests, there were two reasons. One was that the conventional wisdom (a term he coined) had long viewed insecurity as an economic positive. Without the possibility of failure, the argument went, capital and labor wouldn't struggle to discover new efficiencies. Economists like Adam Smith, David Ricardo, and Thomas Malthus all believed it was normal for the average farmer or worker to live on the brink of starvation. As Galbraith wrote, however, most everyone craves individual economic security.[15] Regardless, this had seemed impossible before the twentieth century. But during the American Century the idea that the nation could play a role in providing both macroeconomic *and* microeconomic security became central to notions of an American Dream. Why did this happen? Galbraith correctly argues that an interest in economic security is a luxury that is only available to farmers and workers who have something to lose, which was rare before the creation of an affluent society. He suggests that the real change heading into the twentieth century was that Americans had chanced upon the ultimate equalizer, ever-increasing industrial production. As the economy increased the output of food and goods, most Americans became wealthier. As the fate of average citizens improved, they had an increased desire for economic security.[16] While this analysis is compelling, it doesn't explain how the economy was able to increase output so significantly. The answer, in one word, is *hydrocarbons*.

In his fascinating book *Why The West Rules—For Now*, Ian Morris introduced a "social development index" that attempted to explain the shape of human history. He defined an index score as "the bundle of technological, subsistence, organizational, and cultural accomplishments through which people feed, clothe, house, and reproduce themselves, explain the world around them, resolve disputes within their communities, extend their power at the expense of other communities, and defend themselves against others' attempts to extend power." Social development is measured using four metrics: energy capture, organization and urbanization, war-making, and information technology. Morris's exhaustive research revealed that social development barely advanced at all from about 14000 BCE until the First Industrial Revolution. Both Western and Eastern civilizations would go through centuries-long periods in which they would surge ahead before hitting a "hard ceiling" caused by climate change, migration, famine, epidemics, or state failure—which he calls the "five horsemen of the apocalypse." Using a scale that reached 1,000 points in 2010 CE, Morris found that early tribes from 14000 BCE to 5000 BCE had never enjoyed a social development score that exceeded 10 points. By the height of the Roman Empire, about 100 CE, civilization had nearly reached 45 points, which was as high as it would reach until about 1700 CE. At that point, the opening of the First Industrial Revolution led to the skyrocketing growth in social development that has continued unabated.[17]

Why did things change after the First Industrial Revolution? As Morris discusses, of the four metrics he uses to measure social progress, energy capture is by far the most important. Prior to the Industrial Revolution, energy capture accounted for 75 to 90 percent of total social development scores. Harnessing fire and domesticating animals were the primary reasons for civilizational advance until a few hundred years ago, with better organization and war-making skills playing a secondary role (urbanization and information technology barely registered). Things changed spectacularly after the First Industrial Revolution, primarily because of the discovery of how to utilize hydrocarbons to generate energy. The energy resulting from combusting these highly concentrated forms of ancient biomass far surpassed the energy it had previously been possible to capture. While these new riches themselves provided tremendous growth in social development, they also formed the foundation for advances in war-making, organization and urbanization, and information technology.

So while energy accounts for about 28 percent of Western civilization's social development points circa 2010, most of the remaining points would not have been possible without the hydrocarbon foundation.[18]

Returning to Galbraith's thesis, the notion that high rates of energy capture provide resources for expanded social development meant that leveraging the nation's vast coal and oil reserves gave Americans the luxury of pursuing both macroeconomic and microeconomic security. It also supplied the nation with the income necessary to pursue a globe-straddling security policy, which formed the impetus for the transition from the American Republic to the American Empire. Ultimately the nation was also able to pursue environmental security. In the final third of the twentieth century, however, after the peaking of conventional oil production at home, it became far more difficult to provide security in these areas. Rather than turning to renewable energy sources to fill the gap, America engaged in a policy aimed at securing the global oil supply. This endeavor began to seriously erode institutions that had long held the nation together and has endangered the future.

In this book, we examine this story to better understand the role hydrocarbons played in America's two centuries of unparalleled progress and why our continued reliance upon them now threatens to tear the country apart. The good news is that there is a clear pathway that avoids this outcome, but it will require an abandonment of the apathy and complacency that have come to characterize the electorate in recent decades. We the People must work to reclaim and reinvigorate the values and ideals that have long guided the American story. We must restore energy security to the nation as we work toward a new energy economy capable of protecting the progressive gains that have been made since the founding of our country. The current state of affairs is dire, with our very security imperiled at every level. The time for action is upon us.

* * *

In part I, three interweaving storylines build upon one another to develop the case that without hydrocarbons the United States would never have become the most important global superpower. The first narrative thread argues that the American Rise has been defined by four energy-related revolutions that were almost entirely fueled by hydrocarbons: the Industrial Revolution, the Transportation Revolution, the Electrification Revolution, and the Agricultural Revolution. In this book, I refer to them collectively as the INNATE (INdustrial-N-Agricultural-Transportation-Electrification) revolutions because they made hydrocarbons inherent in the essential character of being American. These revolutions produced economic growth that shaped paths for political advances, particularly by providing the fiscal resources for significant progressive policies and programs. The result was the American Rise to economic and geopolitical preeminence. Unbeknownst to business and political leaders at the time, however, these four revolutions would also contribute the vast majority of greenhouse gas emissions, which ultimately created the climate crisis.

The second thread suggests that developments in the provision of energy security were closely linked with fundamental realignments in the American political system. Each of these energy and political transitions saw the national government provide a new type of security to the citizenry. Wood-fired steam and water power were associated with the First Political Age (1789–1860), which saw the federal government focus on national security. The introduction of coal was associated with the Second Political Age (1861–1932), which saw the addition of macroeconomic security to the list of government responsibilities. The arrival of oil was associated with the Third Political Age (1933–2008), which saw microeconomic security become part of the government portfolio. The gas revolution was associated with the Fourth Political Age (2009–present), which saw climate security become part of the government portfolio. In each political age, support for or opposition to an expanded governmental role in providing a new type of security became the key characteristic distinguishing the major political parties. Therefore, the period before the adoption of the US Constitution saw national security differentiate political actors; macroeconomic security filled this role in the First Political Age; microeconomic security came to the fore in the Second Political Age; and in the Third Political Age environmental security joined the fray. As hydrocarbons became more pervasive, they allowed the nation to provide national, macroeconomic, and microeconomic security simultaneously. As suggested above, this was something new in human history. The third thread contends that recurring generational patterns have also contributed to political realignments in the United States. These generational dynamics help explain when political ages begin and end. They also provide insight into why these political ages are separated into progressive and conservative cycles.

In chapter 4, the first chapter of part II, a final narrative thread explores whether America's sudden loss of unchallenged energy security and a cyclical conservative backlash produced severe adverse impacts (e.g., deep political dysfunction, troubling economic financialization, destabilizing income and wealth inequality, counterproductive educational inequity and infrastructural decay, deteriorating international prestige, and dangerous anthropogenic climate change) during the final cycle of the Third Political Age. The United States had just begun to address these problems at the outset of the Fourth Political Age, before a countercyclical election result destroyed its ability to make progress.

In the final two chapters, I suggest an agenda for maintaining American geopolitical centrality and guaranteeing a prosperous future. The United States must build upon two centuries of progressive change by reforming its political system, rethinking its approach to national security, and reconsidering how it fosters economic security. Most critically, to protect and expand upon the great benefits gained during the Hydrocarbon Age, the country must also introduce a sustainability revolution capable of providing both energy and climate security. Despite the obvious challenges, the country's natural advantages and the shape of its history provide reasons to believe it can succeed in accomplishing these objectives.

PART ONE

HYDROCARBONS AND THE AMERICAN RISE

Steam, National Security, and the First Political Age

Long before humans discovered how to leverage hydrocarbons, gaining access to ever-larger stores of energy had been the defining characteristic of our species' advance. For our progenitors, the quest for high-energy food was central to life itself. This was most obviously true when the hominid brain dramatically increased in size about 2.5 million years ago. This larger brain required far more calories, which meant that these early humans needed a high-quality, energy-rich diet. Fortunately, they were able to use the new capabilities associated with their evolved brains to catalog a far wider assortment of plants and hunt other animals more effectively. Because it was so much easier to obtain adequate food when humans worked in groups, early social organization was centered on hunting (and to a lesser degree gathering).[1]

Learning how to use and control fire was an important evolutionary step. Nineteenth-century scientists assumed that humans had figured out how to use fire effectively comparatively late in their development, perhaps only dating back to about forty thousand years ago. In 1999, however, Richard Wrangham proposed what he called the "cooking hypothesis," which suggests that this technological advancement occurred about 1.8 million years ago. In a more recent book, he bolstered the theory with evidence pulling from paleontology, archaeology, primate studies, ethnographies, and digestive physiology. In sum, Wrangham believes the control of fire allowed early hominids to cook their food, which led to numerous physical changes. Because cooked food is easier to eat and digest, it provided them with far more usable energy. As the guts of *Homo erectus* got smaller, there was more energy available to grow their brains. Cooking also killed bacteria and natural poisons, alleviating dangers from food-borne pathogens and leading to a more varied diet. Access to campfires also provided *Homo erectus* with the warmth that allowed for the shedding of body hair, which meant that they could run faster without overheating. Group hunting and cooking also had significant social benefits. Perhaps the biggest advantage was that humans with access to heated food reproduced more effectively, thus spreading across the land far more quickly.[2]

This first energy revolution, which provided early hominids with the ability to gradually diversify their omnivorous diet, provided the basis for the gradual human diaspora that led the species to inhabit nearly every corner of the planet. This led to the wholesale alteration of entire ecosystems, particularly as numerous mammal species were hunted to extinction—with few consequences for humankind. Richard Manning argues, "For a variety of reasons tied to humans' special basic abilities, we are unique among predators and foragers. We do not pay nearly as heavy a price for depleting our food sources. We have so many alternatives: we can simply switch food, or move to a new range, or both." So, from a relatively early stage in the development of the species, we haven't believed that there were ecological limits to our continued advance.[3]

About twelve thousand years ago the Neolithic Revolution, which represented the transition from hunter-gatherer groups to agricultural societies, began. There is a common misperception that this shift happened quickly, but in fact it occurred over several millennia. This transformation was far from entirely positive, which begs the question, why agriculture? As Tom Standage writes, "Quite why humans switched from hunting and gathering to farming is one of the oldest, most complex, and most important questions in human history. It is mysterious because the switch made people significantly worse off, from a nutritional perspective and in many other ways." The mythology suggesting that agriculture provided dramatic benefits, such as plentiful food and more free time, is not true. Farming is more efficient from the perspective of production per unit of land, but until very recently it was far more labor intensive, so that the amount of free time available for most people was dramatically decreased. It also generally led to the adoption of a less diverse diet because of the heavy reliance on grains. This, in turn, had numerous negative health impacts. Fewer key nutrients in the new diet meant that agriculturalists were far shorter and less robust compared with hunter-gatherers. Additionally, there were a variety of diseases associated with malnutrition that previously had been largely absent.[4]

So, once again, why did these human groups switch? The answer, it turns out, is that the change happened so slowly that humans didn't recognize that it was happening. Groups followed a gradual path from being pure hunter-gatherers to focusing on systematic cultivation. Perhaps the key reason why they traveled down this path was a significant change in the global climate. During the period from roughly 12,800 to 11,500 BP, the globe experienced the Younger Dryas, which was characterized by a cold, dry climate. This era ended with the arrival of the warmer and wetter Holocene, which paved the way for agriculture. This new period provided conditions that allowed some hunter-gatherer groups to become less mobile. The resulting sedentism meant that these groups often stayed in the same camp for most of the year, particularly near rich waterways that provided a profusion of fish or shellfish. During this proto-agricultural period, those who had previously gathered wild grains began to plant a few seeds to provide some protection against the impact of weather variations. Over time the energy-rich nature and easy storage of these grains made systematic planting more attractive. Thus, a food source that had previously been a small part of the hunter-gatherer diet slowly took on a more important role. Ultimately this led to population growth. While women in hunter-gatherer groups had to stagger their pregnancies because they had to carry young children until they could walk, sedentism made it easier to have more children. This meant that there were more mouths to feed, which required ever more planting of increasingly domesticated grains. So without really thinking about it, pockets of humanity adopted an entirely new way of life.[5]

In the coming millennia agriculture would spread to the rest of humanity, sometimes peacefully but often with farmers decimating the hunter-gatherers in their path (as with the Linearbandkeramik, who swept across Europe in just three hundred years). As crop cultivation was slowly marching across the planet, humans were also relying increasingly on domesticated animals to reduce the need to hunt for meat. Jared Diamond suggests that six factors distinguished the animals that would be domesticated: a flexible diet, fast growth, the ability to breed in captivity, a nonaggressive disposition, the lack of a nervous disposition, and a modifiable social order. These attributes led to the early domestication of sheep and goats, followed later by cows, pigs, and chickens. Combined with the three main domesticated cereals (wheat, rice, and maize), these sources would be the key pillars of early civilization. The intensification of farming led to greater population concentrations and the development of social stratification, largely because it became feasible to acquire and retain durable goods in a way that had been impracticable among mobile bands, who had had to carry everything from place to place. Food became the key barter currency in the acquisition of these goods. As a result, as Standage writes, "in the ancient world, food was wealth, and control of food was power." Ultimately this reality required centralized leadership capable of organizing farming communities (often structured around religious practices) to ensure their stability. As villages evolved into cities, better political organization was needed to manage increasingly complex agricultural practices, such as irrigation farming. This eventually led to higher yields, allowing individual farmers to produce more food than they needed, which provided the additional calories necessary to support government officials, religious leaders, and craft workers. Finally, food was paid to tax collectors and used for large state projects, such as irrigation systems, roads, city walls, or temples. Farming triumphed not because individual humans were better off as a result but, as Manning writes, because "biology and evolution don't care very much about quality of life. . . . What counts to biology is a species' success, defined as its members living long enough to reproduce robustly, to be fruitful and multiply. Clearly, farming abetted that process."[6]

The invention of the horse collar fifteen hundred years ago in China was the most important innovation since the beginning of the Neolithic Revolution because it tapped into a better energy source. The horse collar allowed farmers to largely discontinue using lumbering oxen to plow fields, instead relying on a far more agile creature. This meant that a single farmer could substantially increase the acreage he had under cultivation. As Manning writes, "This simple innovation probably had as profound a role in reshaping the world at that time as internal combustion and tractors have had in our own." Farmers were also able to take advantage of the more mobile horses to travel greater distances to markets, which led to a considerable expansion in the size of cities and overall populations. As Carolyn Steel argues, agricultural advancements and urbanization had long gone hand in hand, but this evolution picked up speed with the arrival of the horse collar. Another fundamental shift was that as humans more densely settled the planetary surface, different groups increasingly came into contact with one another. This eventually led to the creation of extensive trading networks.[7]

These trading networks were birthed by spices, which Arab traders initially carried overland to the Mediterranean world. Spices were exotic and lightweight, and they made food more enjoyable. Europeans were crazy for them. As Standage writes, "The pursuit of spices is [another] way in which food remade the world, both by helping to illuminate its full extent and geography, and by motivating European explorers to seek direct access to the Indies, in the course of which they established rival trading empires." By about two millennia ago, the trade routes extended throughout Eurasia, connecting the Romans, the Parthians, the Kushans, and the Chinese. Knowledge, culture, and goods flowed in both directions. Infectious diseases found new populations to attack. "The plague, which appears to have originated in central Asia, reached Caffa along the overland trade routes before being spread around Europe by Genoese spice ships." This Black Death ultimately killed off one-third of the Continent's population, but even this did not sour people's taste for spices. In fact, rather than reducing their consumption, Europeans anxiously tried to locate trade routes that would overcome Muslim dominance. To do so, they tapped into the naturally occurring energy flows available to sailing ships. During the Age of Discovery, European explorers using wind energy pioneered shipping routes around the Cape of Good Hope to India and across the Atlantic Ocean to the "undiscovered" Americas. As windmills and watermills were also becoming far more commonplace—used to mill grain and later as mainstays of the early manufacturing industries—these new energy developments underpinned the birth of modernity.[8]

About 1500, income disparities among nations were slight. Over the coming centuries, however, a combination of financial prowess and energy dominance led first the Dutch and then the British to gradually pull away from the rest of the world. These two nations also enjoyed important geographic advantages, as both had easy access to the Atlantic Ocean, which became the key byway of the first significant wave of globalization. In addition, they benefitted from nonabsolutist political institutions that were consistent with economic growth in an increasingly globalized world and allowed them to reap the largest rewards from modernity (while other Great Powers, such as the Portuguese and the Spanish, slowly faded after their initial dominance). The Dutch initially emerged as a trading power because of their transformative use of financial institutions. Amsterdam had led a monetary and fiscal revolution that included a system of public debt, the first truly modern central bank to manage monetary policy, a straightforward and effective excise tax, and a corporate governance model that fundamentally changed global economics. Just as importantly, the Dutch stood out because of their preeminence in the use of the most important energy sources of the time. They had the best full-rigged sailing vessels, thousands of windmills and watermills, and plentiful domestic peat supplies. These resources helped them build an expansive trade empire and a manufacturing base to turn imported goods into useful products. The Dutch East India Company controlled the key Asian trade routes for most of the seventeenth century, bringing goods ranging from spices to silks back to hungry European markets. As a result, the Netherlands became the planet's most powerful nation.[9]

When the British began to challenge Dutch trading supremacy, war was inescapable. From 1652 to 1674 the two nations fought three wars, with the Netherlands prevailing because of its superior fiscal and energy position. Late in the century, English aristocrats and London merchants helped William III, the Dutch Prince of Orange, overthrow James II. The most significant initial outcome was that the new sovereign brought modern finance to Britain, giving it an important advantage over other emerging powers, in particular the French. The other big result was an agreement that gave the Indonesian spice trade to the Dutch, while the British developed the Indian textile trade. Textiles proved to have a much bigger market, which aided British expansion. The island nation also benefitted from a superior geographic position. As Alfred Mahan wrote, "If a nation be so situated that it is neither forced to defend itself by land nor induced to seek extension of its territory by way of land, it has, by the very unity of its aim directed upon the sea, an advantage as compared with a people one of whose boundaries is continental." As energy and manufacturing technologies began to proliferate across the English Channel, the geopolitical situation shifted dramatically toward the British.[10]

INNATE Revolutions: Industrial
The British and the First Industrial Revolution

In the mid-eighteenth century, Britain was still a nation dominated by trade. Agriculture and manufacturing were larger sectors of the economy but clearly not as important for wealth creation. This began to change when industrialization began to transform the textile sector. Britain had both a domestic and a foreign trade network but lacked efficient methods of production to meet growing demand. As a result, its dominance in the textile trade had begun to falter. The First Industrial Revolution changed this trend: its "great achievement . . . was that it led to continuous growth, so that income compounded to the mass prosperity of today [i.e., the late sixties]." This was something historically new. Technological advancement and abundant energy supplies were at the heart of this revolution, which was characterized by the "substitution of mechanical devices for human skills; . . . inanimate power [taking] the place of human and animal strength; . . . [and the] marked improvement in the getting and working of raw materials." And, textile manufacturing was the key sector. As Eric Hobsbawn wrote, "No other industry could compare in importance with cotton in this first phase of British industrialization." Because of increasing competition in this area, there was a push to develop machines that could reduce the amount of skilled human labor needed. In quick succession, British inventors devised the spinning jenny and the water frame to increase output. When Samuel Crompton married the two, his machine increased output and transformed the textile industry. It soon became clear, however, that to fully take advantage of its newfound gains, England would need more power than traditional forms of energy could provide.[11]

The use of hydrocarbons is not modern. They weren't suddenly discovered in the past several hundred years. More than five millennia ago, the Mesopotamians tapped into bitumen that leached out of the ground. The Greeks and Romans used it for building and road construction. It was also highly useful in making war, smeared on the tips of burning arrows or thrown in simple grenades. In Roman Britain, blacksmiths burned coal in their forges. After the fall of the Western Roman Empire, this knowledge was largely lost, and firewood became central to meeting energy needs. Urbanization and population growth drove demand; the average city dweller might use the equivalent of a half ton of wood annually, either directly or through purchases of commercial goods. As Paul Roberts writes, "At some point the possibility of running out of fuel came to be feared as much as any other large scale disaster—war, drought, or plague." With its limited forests and relatively advanced industrial sector, wood supplies peaked in Britain in the late thirteenth century. At that time, coal briefly reemerged as blacksmiths and artisans began using it once again. Over the next couple of centuries, pressures on firewood supplies were eased when the Black Death ravaged the population. By the fifteenth century, however, the firewood crisis had returned.[12]

It was at this point that coal began to play its fundamental role in reshaping society. Nearly all forms of energy employed by humans are derived from solar radiation. As Thom Hartmann writes, "We are made of sunlight." The most important store of prehistoric sunlight is found in fossilized hydrocarbons. Plants and algae use photosynthesis to split water into hydrogen and oxygen, while also absorbing carbon dioxide. The oxygen is released back into the atmosphere, and the carbon and hydrogen are stored in the plants and algae in the form of glucose. Starting about three hundred million years ago, during the Carboniferous Period, on a much warmer and wetter planet, vast numbers of swamp plants died, sank, and were buried in low-oxygen environments. Over thousands of years, microorganisms turned this organic matter into peat. Then, over many epochs pressure and heat squeezed this peat until it matured into one of four types of coal. Lignite, also known as brown coal, is found nearest to the surface and is only about 60 percent carbon, having many impurities and a relatively low caloric output. Subbituminous and bituminous, the latter also known as black coal, are found progressively deeper and have fewer impurities and a higher carbon content. With carbon concentrations reaching upwards of 85 percent, subbituminous and bituminous coals represent more than 90 percent of what is mined today. Finally, relatively rare anthracite, or rock coal, has carbon concentrations of up to 98 percent and burns very cleanly.[13]

Britain enjoyed a large concentration of lignite near the surface, where it was relatively easy to extract. Although the adoption of this brown coal was slow, it eventually became apparent that it had key advantages compared with firewood: it was more abundant, it allowed for centralized production via mining, it was easier to transport, and most importantly, it had a higher energy density. It was an initial peaking of coal supplies, however, that finally led to a more fulsome embrace of coal. Between 1550 and 1700 the production of coal had increased tenfold in Britain. At the same time, it became essential to the entire national economy both as a domestic fuel and in emerging industries. The problem was that many of the brown-coal reserves near the surface had peaked and were producing less coal every year. Fortunately, miners had discovered new sources of bituminous coal and anthracite, which burned hotter and cleaner. However, these types of coal were located deeper, and deeper mines had a tendency to fill with water, which there was no effective way to remove.[14]

The solution to the dilemma not only provided access to vast coal reserves but also delivered the needed demand for all that coal. Steam power had been experimented with for hundreds of years, but it wasn't until the early eighteenth century that Thomas Newcomen constructed the first commercially viable steam engine to pump water to the surface. By March 1712 the Englishman "had improved the fortunes not simply of a few local capitalists, but of all humankind. The Newcomen engine may have been expensive, noisy, and comically inefficient. . . . The engine may have burned through more than a ton and a half of coal a day. But even at that, the new device was considerably cheaper than the alternatives." Its impacts were profound. Within forty years, coal production in Britain had doubled to more than six million tons. When the Scottish inventor James Watt perfected his significantly more efficient engine in 1775, output soon topped ten million tons annually. It was during this period that the steam engine made its fateful transition from a tool that made coal mining possible to the implement that would give birth to the First Industrial Revolution. The Age of Steam began when the textile industry rapidly adopted the coal-fired steam engine to power its looms. This in turn cemented the role hydrocarbons would play in furthering industrial growth. Soon the coal-steam combination "powered belt drives in countless factories."[15]

This chain of events combined to guarantee Britain's ascendancy in the First Industrial Revolution. While the nation faced some limited competition from continental Europe, even eventual manufacturing juggernauts like Germany lagged far behind. As Niall Ferguson writes, "In 1615, the British Isles had been an economically unremarkable, politically fractious and strategically second-class entity. Two hundred years later Great Britain had acquired the largest empire the world had ever seen, encompassing forty-three colonies on five continents; they had robbed the Spaniards, copied the Dutch, beaten the French and plundered the Indians. Now they ruled supreme."[16] But in the early part of the nineteenth century a new contender emerged in a most unlikely place: the other side of the Atlantic Ocean. During the early 1800s America began building its own industrial capacity. Within fifty years it was poised to challenge the British for dominance of the Second Industrial Revolution.

First Political Age
Progressive Cycle, 1789–1829

The American Republic was born over the course of a quarter century, between 1763 and 1789. Although it seems to have been largely forgotten today, the colonies that sought their independence during this period were not weak, poor, or immature. Quite the contrary. In the century and a half since the founding of Jamestown, the colonists had come to be one of the most prosperous peoples in the world. Many enjoyed a standard of living far superior to that of their brethren in Great Britain. The colonists enjoyed plentiful natural resources, ranging from rich soils to dense forests and essential metals. American port cities were slowly becoming key actors in Atlantic commerce, most importantly providing the critical foodstuffs and other supplies necessary to maintain slave plantations in the West Indies. Although they relied on foreign products to some degree, the ordinary farmer and his family were largely self-sufficient. The population of the colonies was doubling with every generation, with the "melting pot" slowly emerging as immigrants from Great Britain and northern Europe were joined by African slaves in the New World.[17]

Why did these prosperous people rebel against the world's most powerful nation? Contrary to what many Americans seem to believe today, it was emphatically not because of the large tax burden on the average colonist. It is true that in the dozen years before the Boston Tea Party the British had attempted to levy several taxes on their North American colonies to defray the costs of the French and Indian War. But these duties were so unpopular that all except the tax on tea were ultimately abandoned, and even that most famous one resulted in a considerable *reduction* in the price of tea. As Niall Ferguson writes, "On close inspection, then, the taxes that caused so much fuss were not just trifling; by 1773 they had all but gone." Thus the primary issue doesn't seem to have been taxes. Nor can we turn to an antigovernment rationale. The history of this period makes clear that most Americans were not rejecting the British Empire. On the contrary, most were emphatically asserting their firm desire to maintain longstanding connections with the mother country. The primary issue was the English Parliament's taxing Americans without their consent. In 1772 the revolutionary leader Sam Adams wrote, "The *absolute Rights* of Englishmen, and all freemen in or out of civil society, are principally, *personal security*, *personal liberty*, and *private property*." Americans were not suggesting that taxes were unjust, Adams argued, but that they were only just when the body politic joined together to levy them to promote their mutual interests. While they wanted to remain part of the empire, Americans were demanding that their colonial governments operate on the same level as the British Parliament.[18]

When it became clear that this demand would never be satisfied, the colonists began to move aggressively to sever their relationship with the Crown. Sam Adams was the most important actor in engineering this outcome. In December 1773 he masterminded the Boston Tea Party. A furious British Parliament reacted by closing the Port of Boston and stationing troops in the city. In April 1775 the American Revolution officially began when the Minutemen repelled a British expedition intended to capture militia stores in Concord, Massachusetts. The following February, Thomas Paine fanned the flames of independence when he published his wildly popular *Common Sense*. In this pamphlet, he forcefully argued that Americans were far less secure when they relied on the British to protect them and that there would be significant economic benefits if the colonies became sovereign states. Less than six months later the Second Continental Congress announced to the world that the colonists were no longer subjects of King George III but rather a self-governing people. "We hold these truths to be self-evident," they wrote, "that all men are created equal, that they are endowed by their Creator with certain unalienable Rights, that among these is Life, Liberty, and the pursuit of Happiness. That to secure these rights, Governments are instituted among Men, deriving their just powers from the consent of the governed, That whenever any Form of Government becomes destructive of these ends . . . it is their right, it is their duty, to throw off such Government, and to provide new Guards for their future security." As these words from Declaration of Independence show, national and economic security were at the very heart of the American experiment from the beginning.

Three months after the British defeat at Lexington and Concord, George Washington had taken command of the Continental army. In March 1776, after placing artillery on Dorchester Heights, he forced a British evacuation of Boston. Over the next few years the war moved from New England to the Mid-Atlantic states. During these years Washington succeeded, with a few notable exceptions, not so much by winning decisive battles but by keeping the Continental army together as a fighting force. Much to his chagrin, he became the master of the tactical retreat, which had the strategic benefit of keeping the British in the field long enough for the Americans to gain assistance from the French navy. When the French navy arrived in 1781, Washington was able to force British troops under General Charles Cornwallis to surrender at the Battle of Yorktown, essentially ending the American Revolution and promising national autonomy.[19]

It is perhaps not surprising that in a world that by almost every measure was far more dangerous then it is today, providing national security was considered the government's most important function. During the first dozen years of self-rule, however, the weakness of government institutions endangered national security, even as Washington was trying to hold together the Continental army and force the British to abandon the colonies. After the war concluded, it became ever more apparent that the Articles of Confederation, the country's first governing documents, were flawed. The government had no power to levy taxes, no executive authority, no independent judiciary, single-state veto power, no uniform currency, no power to regulate foreign trade or interstate commerce, and no way to repay state and national war debts. In 1787, when state delegates met in Philadelphia with the intention of revising the Articles, they quickly decided to instead frame an entirely new government more conducive to securing and growing the young nation. Through painstaking negotiation and compromise, the delegates agreed upon a plan that would create three branches of government: a bicameral legislature (with a House represented according to population and a Senate having equal representation for each state), a unitary executive, and an independent judiciary. The most radical portion of the entire document was, and is, the preamble, which reads: "We the People of the United States, in order to form a more perfect union, establish justice, insure domestic tranquility, provide for the common defense, promote the general welfare, and secure the blessings of liberty to ourselves and our posterity, do ordain and establish this Constitution for the United States of America." Those first three words, *We the People*, were revolutionary by the standards of the eighteenth century, in which everyone was ruled. While a strong central authority was created to back up these fine words and provide a more energetic government, this notion of self-rule came to define American politics.[20]

Not everyone liked the Constitution, or at least not everyone liked all parts of it. Thomas Jefferson, serving abroad as minister to France, wrote to his old friend James Madison, who is considered the Father of the Constitution, with two large concerns. First, he was concerned that the document contained no explicit bill of rights, "providing clearly . . . for freedom of religion, freedom of the press, protection against standing armies, restriction of monopolies, the eternal and unremitting force of the habeas corpus laws, and trials by jury in all matters of fact." He should not have been overly concerned, as one of the first orders of business for the new Congress was passing ten amendments guaranteeing such rights, nine authored by Madison himself. Jefferson was also unhappy that term limits had not been provided for within the document, fearing that the presidency would become a lifetime position. "I own, I am not a friend to a very energetic government. It is always oppressive." His fears would largely not come to pass. For more than two centuries the government has had an appropriate amount of vigor to protect national security and expand economic security. The real question, which I address later in this book, is whether it has enough dynamism to meet the challenges of the twenty-first century.[21]

* * *

The First Political Age began with the selection of George Washington as the nation's first president. Although Washington had been the de facto leader of the nation since 1775, this was an official act of the whole people to raise him to that lofty position. Given Washington's immense popularity, which made his elevation a foregone conclusion, it might be said that the election was more clearly a major realignment because of the selection from a crowded field of John Adams as vice president. Combined with Washington's elevation, this was a monumental repudiation of weak central government. There is little doubt that the policy direction of the nation fundamentally shifted when Washington took office. As Ron Chernow writes,

> Washington's catalog of accomplishments was simply breathtaking. He had restored American credit and assumed state debt; created a bank, a mint, a coast guard, a customs service, and a diplomatic corps; introduced the first accounting, tax, and budgetary procedures; maintained peace at home and abroad; inaugurated a navy, bolstered the army, and shored up coastal defenses and infrastructure; proved that the country could regulate commerce and negotiate binding treaties; protected frontier settlers, subdued Indian uprisings, and established law and order amid rebellion, scrupulously adhering all the while to the letter of the Constitution.[22]

While there was little controversy about the direction national security policy took, the same could not be said for macroeconomic security, which became the key issue separating the Federalist and Democratic-Republican political parties. Within the administration, Treasury Secretary Alexander Hamilton was charged with creating macroeconomic policies aimed at stabilizing an economy that had been struggling since the end of the revolution. The Hamiltonian program for economic modernization was threefold: (1) consolidation and repayment of state and federal war debts; (2) establishment of a national bank to facilitate financing this debt; and (3) encouragement of manufacturing and trade. Over the course of twenty-four months, Hamilton produced three state documents of historic import arguing on behalf of these policies. These would be critical to establishing an economic system that could eventually take advantage of the nation's vast supplies of hydrocarbons—although this clearly wasn't understood at the time.

On 9 January 1790 the *First Report on the Public Credit* was submitted to Congress. Hamilton patiently informed the legislators, most of whom had no experience with global finance, that restoring American credit was highly important for American security. "Loans in time of public danger," he wrote, "especially from foreign war, [will be] an indispensible resource." For the country to be deemed creditworthy by European bond markets, however, its war debts would have to be paid. This would mean full remuneration by the federal government of all outstanding bonds, including those owed by the various states. The assumption of state debt would clearly broaden the influence of the central government in the arena of macroeconomic policy. The total debt comprised about $50 million in national debts and another $25 million from the thirteen states. Hamilton believed that a portion of the revenues from the national tariff should be redirected to pay down this sum. Following extended negotiations with Democratic-Republicans, which included agreeing to seat the national government near Georgetown, Maryland, Federalists were able to gain adoption of this critical macroeconomic policy in Congress. Since that time the national government has never defaulted on its debt.

On 14 December 1790, Hamilton submitted *The Report on a National Bank* to Congress. In it, the treasury secretary promoted the creation of a central bank as the basis of his fiscal policy. It was intended to fund the national debt and to fashion a stable currency, which through lending would become an expedient medium of exchange. The bank was to be a public-private corporation, held privately but with the government holding 20 percent ownership. Although Madison and Jefferson questioned the legitimacy of such an institution, Hamilton was able to convince Washington that the creation of a bank was clearly constitutional. He suggested that the Necessary and Proper Clause (article 1, section 8, clause 18) provided the broad implied powers that were required to implement specifically enumerated powers. In a separate opinion on the constitutionality of the bank, Hamilton wrote: "The whole turn of the clause . . . indicates that it was the intent of the Convention, by that clause, to give a liberal latitude to the exercise of the specified powers. The expressions have peculiar comprehensiveness. They are, 'to make all *laws* necessary and proper for *carrying into execution* the *foregoing powers*, and *all other powers*, vested by the Constitution in the *government* of the United States, or in any *department* or *officer* thereof.'" Hamilton then traced the enumerated powers that would be more effectively exercised with assistance from a central bank, including setting and collecting taxes, borrowing money, coining money, regulating foreign currency, and regulating property. Washington sided with Hamilton in a decision that has reverberated down through the ages. As Chernow concludes, "Unlike his fellow planters, who tended to regard banks and stock exchanges as sinister devices, Washington grasped the need for these instruments of modern finance. . . . With this stroke, he endorsed an expansive view of the presidency and made the Constitution a living, open-ended document. The importance of his decision is hard to overstate, for had Washington rigidly adhered to the letter of the Constitution, the federal government might have been stillborn."[23]

Although the initial two reports were extraordinarily important for stabilizing the nation at the end of the nineteenth century, the third report, the *Report on Manufactures*, in 1791 would have the most significant long-term impacts. Congress had requested a study that would examine how to decrease dependence on foreign military arms, but Hamilton submitted a much broader report that outlined the establishment of an "American School" of economics. This system of economic nationalism came to represent the standard model for commercial growth. At its core was mass education, which had been widely adopted during the colonial period and would be expanded upon during the coming decades. Hamilton covered the other major policies in his masterwork, which challenged the belief that agriculture was the most important economic sector. While he agreed with Jefferson that farming would remain preeminent in the short term, he questioned the contention "that it has a title to any thing like an exclusive predilection." Manufacturing, he believed, would not only create new opportunities for capital but also provide critical assistance to agriculturalists, because industrial workers would create a new domestic market for surplus foodstuffs, which faced large import duties abroad. To create the industrial sector, however, required strong government action in three areas. First, the nation needed a central bank to stabilize the economic system and provide capital to promote growth in manufacturing enterprises. Second, an import tariff was necessary to protect infant industries. Hamilton also wanted direct government subsidies within specific sectors, but this was something that the government would not approve for more than a century. Third, government investment in internal improvements, most critically in transportation, was required to create a large national market. The Constitution had already put in place a key policy to fashion this marketplace by abolishing all internal tariffs and thus creating the legal basis for an internal system of trade. Supporting this American System would become the central platform of three successive political parties—the Federalists, the Whigs, and the Republicans. As the leaders of these parties worked to fulfill Hamilton's vision, they added key building blocks to the base of the Hydrocarbon Nation.

At the same time that Hamilton was establishing this economic strategy, Washington was setting the geopolitical course. Four years into his presidency, he proclaimed neutrality as war broke out once again between Britain and France. There was great pressure to offer support to the French, who had come to America's aid during the Revolution. Instead, Washington sent John Jay to England to sign a treaty normalizing trade relations, while demanding that the British withdraw from forts in the Northwest Territory and recompense losses from attacks on American shipping. This agreement ensured that the nation would not become involved in the European conflict. In his Farewell Address to the nation, Washington famously laid out his thinking on national security policy. He argued that the most important ingredient of safety was unity, imploring his fellow citizens to avoid sectionalism and destructive civil wars. He also warned against the danger of an overgrown military. The most important part of the address, however, built upon Washington's belief that "it is our true policy to steer clear of permanent alliances with any portion of the foreign world." He worried that habitual hatred or fondness for any nation could lead America into a war in which it had no real interest. "Against the insidious wiles of foreign influence," he wrote, "the jealousy of a free people ought to be constantly awake, since history and experience prove that foreign influence is one of the most baneful foes of republican government. . . . The great rule of conduct for us in regard to foreign nations is in extending our commercial relations, to have with them as little political connection as possible." Thus, he placed American economic development at the fore, believing it to be the true path to national greatness. He judged that the geographic position of the United States, protected from Europe's great powers by a wide ocean, allowed the nation to follow a different course. "Taking care always to keep ourselves by suitable 'stablishments on a respectable defensive posture," Washington concluded, "we ' safely trust to temporary alliances for extraordinary emergencies." This ' olicy was intended to make sure America was strong before it entered into ie international fray. It had worked brilliantly during his tenure; the question was whether future presidents would be able to keep the country on this path.[24]

Federalist policies were highly effective at stabilizing the young nation. Not only did the Hamiltonian economic plan improve American creditworthiness but while the Federalists were in power gross domestic product (GDP) increased by more than 80 percent, nearly twice the rate under their Democratic-Republican successors. Washington had kept the nation safe while charting a course for future presidents to follow. Despite these realities, however, the Federalist Party itself began to splinter within half a decade. This was the result of growing animosity between Hamilton and Washington's immediate successor, John Adams. Although Adams got the better of Hamilton in the policy arena when he negotiated a peaceful resolution to tensions between America and France (a brilliant move that allowed the nation to grow economically rather than becoming entangled in European conflicts, just as Washington would have desired), Hamilton ultimately scuttled Adams's reelection bid. In the process, Hamilton ensured the demise of the party he had founded.[25]

* * *

In 1800 the Federalists were booted from office. While historians have long argued that this election represented a significant realignment, the actual record calls this into question. It is true that the contest saw a different political party take the reins of government, with Thomas Jefferson receiving the support of a new coalition made up of southerners, farmers, and tradesman. Even though it was one of the most vitriolic elections in America's history, the transition from Federalists to Democratic-Republicans was largely peaceful, beginning the tradition that has only been transgressed once in the country's history. In his inaugural address, Jefferson made a plea for unity: "Let us . . . unite with one heart and one mind. Let us restore to social intercourse that harmony and affection without which liberty and even life itself are but dreary things. . . . Every difference of opinion is not a difference of principle. We have called by different names brethren of the same principle. We are all Republicans, we are all Federalists." Notwithstanding the shift in power, there wasn't a significant

policy redirection. Although there was a short-lived and relatively mild conservative retreat from economic modernization, the Democratic-Republicans
ultimately backtracked after the failure of their policies became apparent. And
Jefferson largely embraced Washington's national security approach, sharing
the belief that the nation was blessed to be "kindly separated by nature and
a wide ocean from the exterminating havoc of one quarter of the globe. . . .
possessing a chosen country, with room enough for our descendants to the
thousandth and thousandth generation." Although the economic and foreign
policy emphasis may have shifted somewhat, this hardly seems the stuff of a
major electoral realignment. Rather, I would argue it was simply the continuation of the same basic governing principles.[26]

The Democratic-Republicans opposed funding the national debt, the national bank, and protective tariffs, but Jefferson knew he could do little to
reverse the Hamiltonian economic program. Instead he focused mainly on
foreign policy, which he believed was the proper purview of the executive.
In this role, he made one excellent decision that forwarded American interests and one terrible decision that led to economic hardship and war. When
Jefferson took office, it was far from clear that the nation could protect the
frontier between the Appalachian Mountains and the Mississippi River. Only
four roads connected this region to the coast, and fewer than 10 percent of
the young republic's population lived there. As Stephen Ambrose writes, "It
seemed unlikely that one nation could govern an entire continent." Jefferson,
however, saw a future in which new technologies (in due course powered
by hydrocarbons) would aid western expansion and make governance feasible. As a result, he pursued a course during his first term that sought to secure
the strategic port of New Orleans for America. Louisiana at the beginning of
the nineteenth century was heavily contested, with the British, the Spanish, the
French, the Russians, and the Americans all vying to control parts of the territory and Native American tribes caught in the middle. The Americans had
an advantage, because they were slowly taking possession of pockets of land
beyond the Mississippi River. Without control of the river and New Orleans,
however, no nation could expect to really maintain political control over the
territory. This was apparent to Jefferson. As Ambrose maintains, "In an age of
imperialism, he was the greatest empire builder of all. His mind encompassed
the continent . . . he thought of the United States as a nation stretching from
sea to sea. . . . He rejected the thought of North America being divided up
into nation-states on the European model." Shortly after Jefferson took office,
when a secret treaty was agreed upon by the Spanish to return control of New

Orleans to the French, Jefferson determined to take a hard line. He let Napoleon know that any attempt to land French troops in New Orleans would lead to war and then instructed the American minister to France, Robert Livingston, to negotiate the purchase of the city. On Independence Day in 1803, the treaty between the two nations arrived in Washington, and to most everyone's surprise the purchase included not just New Orleans but the entirety of the Louisiana Territory. For a price of $15 million (3¢ per acre), the size of the nation had been doubled. Henry Adams later wrote that this act had no parallel in American diplomatic history. It was Jefferson's finest accomplishment in the White House, and it was critical to the American Rise.[27]

In the years that followed the Louisiana Purchase, relations with Britain and France deteriorated as the Napoleonic Wars raged in Europe. During this period, trade suffered as both nations interfered with shipping vessels from the United States, with the British going so far as to press American seaman into the Royal Navy. Rather than risk outright war, Jefferson recommended adoption of the Embargo Act of 1807, which called for the implementation of a nonimportation act passed by Congress the previous year and prohibited American ships from engaging in foreign trade. Secretary of the Treasury Albert Gallatin was opposed to the embargo, arguing that "governmental prohibitions do always more mischief than had been calculated. . . . As to the hope that it may . . . induce England to treat us better, I think it is entirely groundless." He proved correct on both counts. But Jefferson believed that politics outweighed economics, his desire to hurt the English clouding his judgment about what was actually in the best interest of the nation. In fact, he seemed willing to cause temporary harm to the American people if it meant he could strike out against the British monarchy.[28]

The embargo was an economic disaster for the United States, driving the nation into the worst depression since colonial times. Exporting made up a much larger proportion of the American economy than it did of the British economy, so the former suffered much more than the latter. While the agricultural South suffered somewhat, the economies of the commerce-oriented Mid-Atlantic and New England states were devastated. These two regions endured economic losses from reductions in both exporting and shipping. The nation's net export earnings declined by 80 percent, while its net shipping earnings fell by 45 percent. At the same time, the value of imports for domestic consumption fell by nearly 50 percent. The overall impact of the embargo between 1807 and 1809 was a 5 percent reduction in GDP. As Henry Adams wrote, "Financially, it emptied the Treasury, bankrupted the mercantile and agricultural class, and ground the poor beyond endurance. . . . Morally, it sapped the nation's vital force, lowering its courage, paralyzing its energy, corrupting its principles, and arraying all the active elements of society in factious opposition to government. . . . Politically, it . . . brought the Union to the edge of a precipice." Why Jefferson chose to follow this path has long been a mystery, largely because he never shared his thinking with the American public. Ultimately his Anglophobia seemingly came to the fore, and the embargo became a tool to break economic ties between the two nations. His desire to create an "agrarian paradise" ran counter to what was best for the nation financially. The embargo was a misstep that temporarily diverted the American Rise.[29]

In March 1809, Jefferson officially repealed the embargo. It was replaced, however, by a series of acts that maintained the earlier legislation's nonimportation features, while allowing exportation to Britain, which had the unintended impact of leading to an accumulation of American capital in English banks. Anglo-American relations continued to decline over the coming years, while the Federalist Party experienced a short-lived resurgence. The Democratic-Republicans under President James Madison chose to declare war in June 1812 not only to protect national independence but also because it helped unify them politically. The war was fought over the subsequent three years, before the Treaty of Ghent finally brought it to a close—although word of the peace did not reach America until after the Battle of New Orleans had elevated Andrew Jackson into the national pantheon.[30]

During the War of 1812, the ideals of Democratic-Republicans began to shift. Prompted by early military defeats, many younger members of the party began to favor a stronger army and navy. By the end of the war, Madison himself was returning to positions he had ostensibly abandoned more than twenty years earlier. In 1815, in his seventh State of the Union address, he laid before the American people his Madisonian Platform, which included four key elements: (1) maintenance of a strong defense posture; (2) creation of the Second Bank of the United States to market government securities, provide credit, and supply a uniform currency; (3) implementation of a moderate tariff structure to provide revenue and protect critical domestic industries; and (4) construction of internal improvements such as roads and canals. All but the last objective eventually became part of a postwar nationalist agenda, internal improvements failing to gain support only because Madison and his successor, James Monroe, believed a constitutional amendment was required before the federal government could fund such projects. Fortunately, state governments would fill this void.

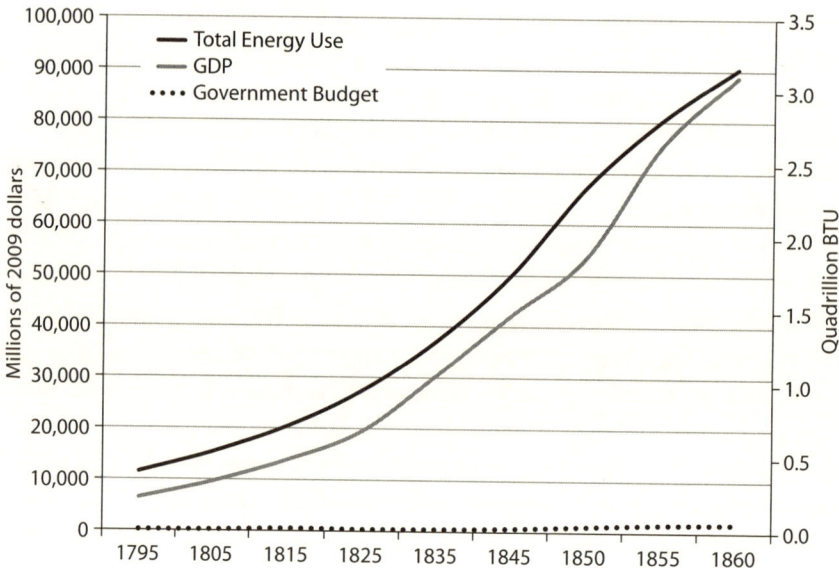

Figure 1.1. US Total Energy Use, GDP, and Government Budget, 1795–1860. US Energy Information Administration, *Annual Energy Review, 2009,* tinyurl.com/ycmr4qxk; *Budget of the US Government, Historical Tables, 2017,* tinyurl.com/ycvb35yg; Louis Johnston and Samuel H. Williamson, "What Was the U.S. GDP Then?," accessed 28 August 2017, tinyurl.com/yc2ev2nr.

Thus, with Hamilton vindicated and the path toward increased macroeconomic security restored, market confidence was reestablished and the American Rise got back on track. Total energy production and consumption would continue to grow, with water and steam power joining wood as the primary motive forces of the era. As energy security increased (fig. 1.1), it was closely mirrored by a significant economic expansion, with GDP growing approximately ninefold in the First Political Age. Separated from the great powers of the nineteenth century by a wide ocean and committed to a policy of isolationism, the federal government was able to remain extraordinarily small during this period. Although the Hamiltonian system would soon face challenges, prehydrocarbon energy sources joined macroeconomic security policies in building the foundation for the Hydrocarbon Age.[31]

INNATE Revolutions: Industrial
Water-Powered Mills and the American System of Manufactures

"The first American economy was built on water and wind," writes Michael Lind. The Atlantic coast had rivers ideal for the siting of waterworks, as well as sailing ships that could capture the winds to move goods to market. This was where the First Industrial Revolution arrived in the United States, via immigrants skilled in the industrial arts. One of the more important newcomers was an English factory worker named Samuel Slater, who had memorized engineering designs during his time in Britain working for Richard Arkwright, the creator of the water frame. Considered by some in the country of his birth to be an industrial spy, Slater played a vital role in advancing manufacturing in the United States, which earned him the moniker Father of the American Factory System. He first worked in Rhode Island, where he established the nation's first water-powered cotton-spinning mill, and then started increasingly sophisticated operations in Connecticut, Massachusetts, and New Hampshire. A member of an influential Brahmin family would improve upon Slater's work. Under the auspices of his Boston Manufacturing Company, Francis Cabot Lowell built the continent's first fully integrated factory in Waltham, Massachusetts. His company would become an influential innovator in both industrial and business techniques. Just four years after creating the company, Lowell died, and his partners began to look for a new mill site because there was insufficient water power available on the Charles River for their expanding operations. They founded a new town on the Merrimac River, which was named for Lowell and would become "the Manchester of America." Located on the Pawtucket Falls, the spinoff Merrimac Manufacturing Company widened a failed canal on the site to provide water power for fifty mills. By 1835, as Chaim Rosenberg writes, "with over one mile of factories filled with machinery, Lowell far surpassed every industrial town in America." The textile trade benefitted greatly from a protective tariff that had been levied on imported cloth back in 1816. While Lowell was the greatest of the factory towns, others soon dotted New England—in Lawrence, New Bedford, Fall River, Saco, Manchester, Chicopee, and Dover. Later, larger cities further south, such as Philadelphia, Baltimore, and New York, also entered the sector. By 1840 there were more than a thousand mills in the United States, three-quarters of them in New England. This industrial sector had succeeded in breaking America's dependence on foreign textile imports, creating a model for the establishment of a domestic marketplace that would be copied in other parts of the economy. Although steam would eventually replace water power in these mills, this would take many decades. As Lind writes, "As late as 1869, water provided 48.2 percent of the overall energy and steam power only 51.8 percent." More than any other region, New England had used water power to become the industrial heart of the nation. At the same time, the South was stagnating economically, as it remained wedded to plantations operating on slave power. This widened a sectional divide that would have greatly concerned George Washington.[32]

Although the First Industrial Revolution first reached America through textile mills, it was in another area that the government and the private sector began laying the groundwork for the American-dominated Second Industrial Revolution. This time, the key innovations came by way of France. By the early nineteenth century the French had been working to develop interchangeable parts within their armaments industry. At a time when firearms were all essentially handcrafted, the goal was to produce large numbers of weapons that were so exactly uniform that a part from one musket or pistol could be used in any other. In one of the great ironies of American history, it was the nation's greatest early Francophile—Thomas Jefferson—who brought the concept to this country. The irony lay in the fact that this was a vital step toward the mass industrialization and urbanization that would render impossible the creation of the agrarian paradise the Sage of Monticello struggled to fashion. In August 1785, Ambassador Jefferson wrote to Secretary of Foreign Affairs John Jay:

> An improvement is made here in the construction of the musket which it may be interesting to Congress to know, should they at any time propose to procure any. It consists in the making every part of them so exactly alike that what belongs to any one, may be used for every other musket in the magazine. The government here has examined and approved the method, and is establishing a large manufactory for the purpose. As yet the inventor has only completed the lock of the musket on this plan. He will proceed immediately to have the barrel, stock, and their parts executed in the same way. Supposing it might be useful to the U.S., I went to the workman. He presented me the parts of 50 locks taken to pieces, and arranged in compartments. I put several together myself taking pieces at hazard as they came to hand, and they fitted in the most perfect manner. The advantages of this, when arms need repair, are evident.

In the years to come, Jefferson would try to interest Secretary of War Henry Knox, but with little success. The idea, however, slowly worked its way into the War Department. It was initially championed by Major Louis de Tousard, a former French artillery officer, but it was truly integrated into strategic thinking by the first chief of ordnance, Colonel Decius Wadsworth, and his deputy, Lieutenant Colonel George Bomford. They were interested not only in interchangeable parts but in using machines to produce them. Their efforts were so successful that, as Joseph Wickham Roe wrote, the "system of interchangeable manufacture is considered to be of American origin. In fact, for

many years it was known in Europe as the 'American System' of manufacture."
In the early nineteenth century, government contractors like Eli Whitney and
Simeon North were tasked with investigating these new processes. Although
early historians trumpeted Whitney's role, more recently it has become clear
that the lesser-known North did far more to advance interchangeability and
mechanization. Within a short period of time, however, the government took
direct responsibility for advancing these new industrial techniques.[33]

In 1794, Congress had passed legislation creating the first national armory
at Springfield, Massachusetts; a second armory was created at Harpers Ferry,
Virginia, the same year. As David Hounshell writes, "The significance of events
at the Springfield Armory from 1794 to 1815 is that . . . armsmaking was transformed from a craft pursuit into an industrial discipline and the weapon from a
shop creation into a factory product." Under the leadership of Colonel Roswell
Lee, Springfield first introduced gauges to measure the uniformity of parts. This
was central to achieving interchangeability. Additionally, hand labor was replaced by special-purpose machines. Further south, at Harpers Ferry, John H.
Hall employed these techniques to fabricate the fully uniform breech-loading
rifle—the first weapon to achieve this distinction on American shores. Hall
tried unsuccessfully, however, to persuade army officials to mass-produce these
rifles. Hall's techniques eventually proliferated back to the Springfield Armory,
which by midcentury was producing fully interchangeable muskets. Although
they were more expensive to produce than handcrafted firearms, they were easily repairable in the field. During this period, Thomas Warner determined
that extraordinarily refined gauges were critically important to making truly
uniform components, so he focused much of Springfield's attention on this
problem. The combined efforts of the Springfield and Harpers Ferry armories
had transformed the production of firearms. Soon this new American System
of Manufactures began to make its way into the private sector.[34]

Not surprisingly, this transition began first in the arms industry. Samuel Colt
made uniformity and mechanization central to his production of revolvers,
hiring Thomas Warner away from the Springfield Armory. The company used
hundreds of special-purpose machines for each revolver, although workmen
still had to fit machined parts to ensure uniformity. One of the most important
aspects of Colt's operation was its subcontracting of workers to perform specialized tasks. When these workers ultimately left, the methods proliferated to
other industries. Starting in the 1830s, Eli Terry had adopted mechanization
to fabricate the ultracheap wooden clocks that fundamentally changed America's relationship with time; by the 1850s, 150,000 clocks were being produced

annually. Isaac Singer brought armory practice into the production of sewing machines. Singer successfully married these industrial methods with modern marketing and installment purchasing, becoming an early mass producer and altering the balance of power in the textile trade. With these and many other uses of the American System of Manufactures, the United States had become a major player in the First Industrial Revolution. More importantly it was poised to become the global industrial leader during the decades to come, when hydrocarbons also came to play a crucial part.[35]

INNATE Revolutions: Transportation
Steamboats and Canals

During the early Republic, the term *internal improvement* came to be applied mainly to transportation systems constructed by the government. As discussed above, the national government's role in funding these public works became the subject of one of the defining debates in the young nation. While the advocates of energetic central administration were largely unsuccessful in this arena, the country benefitted from aggressive state action. The first phase of the Transportation Revolution occurred on the waterways, where inland navigation was initially promoted primarily to link interior communities to Atlantic ports. Rivers were the nation's commercial arteries, but they were limited by geography and navigability. Communities grew up along rivers because they ensured a connection with the outside world. The reality, however, was that sailing ships could only travel a short distance upriver from the ocean. This meant that commerce was largely one-way, with smaller craft shuttling agricultural goods downriver.[36]

The introduction of the steamboat in the early nineteenth century represented a paradigm shift in two respects. First, the arrival of steam-powered watercraft marked a critical departure in terms of the energy source used for transport. Rather than continuing to depend on largely renewable energy sources (e.g., human, animal, wind, and water), as had been the case for thousands of years, transport began to rely increasingly on nonrenewable fuels. During the 1820s and 1830s, steamboats relied heavily on wood-generated steam, but because forests were rarely cropped, supplies began to plummet and prices soared. During the 1840s, therefore, almost all steamboats transitioned to coal. This was a critical turning point in American and world history, marking transportation as an ever-growing market for hydrocarbons. Second, the coming of the steamboat heralded the beginning of a complete rethinking of the nation's political and economic future (which railroads and the telegraph would radically accelerate). While providing regular transport services along the Eastern Seaboard was somewhat important, "it was in the trans-Appalachian West," wrote Louis Hunter, "that a new mode of transportation was most needed and it was the steamboat which supplied the need." By 1815, paddle steamers were able to ply the Mississippi and Ohio Rivers in both directions. The coming years represented the steamboat's golden age. It dominated trade between Atlantic ports, eventually overtook sailing vessels on the Great Lakes, and transformed the economy throughout the West. Towns like Pittsburgh and Cincinnati could ship products more than a thousand miles downstream to the Gulf Coast, while also receiving goods that were hard to transport by road. "By 1820, 69 steamships were navigating the western rivers, and the total peaked at 727 in 1855 . . . [with] 16,000 miles of steamboat routes." These new commercial connections helped speed the settlement of the Louisiana Territory, changing the entire economic and political makeup of the American Republic.[37]

One of the more important court cases in American history emerged from early competition over steamboat licenses. In 1807 Robert Fulton successfully navigated the Hudson River with a steamboat. The following year, he and his partner, Robert Livingstone, applied for and received a license from New York State for a steamboat monopoly. Among their numerous ventures, they hired Aaron Ogden to operate ferries between New York and New Jersey. Thomas Gibbons was also navigating the Hudson River between the two states, but he had a license provided by the national government under the auspices of a 1793 congressional act relating to the coasting trade and fisheries. Ogden sued in New York, arguing that because Gibbons had no state license, he was precluded from docking in the state. The state court found in favor of Ogden, but the decision ultimately reached the US Supreme Court. Writing for the majority, Chief Justice John Marshall argued that the US Constitution's Commerce Clause prohibited states from regulating trade where the national government had already done so. Article 1, section 8, clause 3, gave Congress the power to regulate commerce between foreign nations, among the states, and with Indian tribes. Because the Constitution provided no parallel power to the states, Marshall reasoned that it was solely a federal power. Given that the waterways were the primary arteries of early-nineteenth-century trade and that any definition of commerce would include navigation, Marshall wrote: "Steamboats . . . can no more be restrained from navigating waters, and entering ports which are free to such vessels, than if they were wafted on their voyage by the winds instead of being propelled by the agency of fire. . . . The act of a state inhibiting the use of either to any vessel having a license under the act of Congress, comes, we think, in direct collision with the [Act of 1793]." With this decision, Marshall established the Commerce Clause as the federal government's defining instrument for regulating the national economy. In the decades and centuries to come, this power would be used to dramatically increase the government's role in providing economic security.[38]

After the War of 1812, the pace of internal improvements increased significantly. This was largely because of a growing belief that government had a key role to play in these endeavors. The first step was to provide access to regions that were isolated because they weren't near navigable rivers. The immediate solution was canals. However, it was not the national government but visionary state leaders who were at the forefront. The quintessential example was the New York governor, DeWitt Clinton, who persuaded his state legislature to fund construction of a canal from Albany (on the Hudson River) to Buffalo (on Lake Erie) after President Madison reluctantly vetoed an effort to fund the project at the national level. The Erie Canal, which would stretch 363 miles, opened in 1825, just eight years after it was approved. It employed horses and mules to tow boats, because neither sailboats nor steamboats were deemed practical. Disproving critics like Thomas Jefferson and Martin van Buren, the canal was an immediate commercial success. It decreased the costs of transportation to the interior by 95 percent and significantly increased trade between the East Coast and the Midwest. Peter Bernstein writes, "By bringing the interior to the seas and the seas into the interior, the Erie Canal would shape a great nation, knit the sinews of the Industrial Revolution, propel globalization—extending America's networks outside our own borders—and revolutionize the production and supply of food for the entire world. . . . This skinny ditch in upstate New York would demonstrate that trade and commerce are the keys to the expansion of prosperity and freedom itself." The success of the canal spurred a flurry of canal projects in the Northeast and the building of the Wabash & Erie Canal, which linked the Ohio River to the Great Lakes. None of these other projects proved nearly as successful as New York's Erie Canal, however, largely because of the arrival of railroads.[39]

One of the most interesting aspects of transportation in early America was the near total lack of success enjoyed by roads. Given their later centrality, it is curious that they didn't take hold much earlier. The reason was largely a lack of government support. Unlike their British and European counterparts, America's national and state leaders would not approve significant funding for road building because of sectional jealousies (there were, of course, a few notable exceptions, such as the Cumberland Road). Roads were considered a local responsibility, but most farmers had no interest in spending time and effort constructing anything that was not necessary to reach local markets. Thus, it was left to the private sector to develop and operate a series of turnpikes that charged for access to relatively good roads. While a few turnpikes proved to be commercially successful, most "turnpike companies were not even able to collect sufficient tolls to provide for the maintenance and operation of the road, let alone make a contribution to the capital for the investors who had put up the money." With the arrival of railroads, even the most prosperous turnpikes found it difficult to compete, and the national road network continued to languish for more than a century.[40]

INNATE Revolutions: Transportation
A Nation of Railroads, Part 1

The national railroad network was one of the most important pillars of the American Rise. In fact, it was probably the most important political and economic development of the nineteenth century. Combined with the telegraph, railroads made continentwide governance practical. They promoted the settlement of the lightly populated West. While passenger travel and westward settlement were important, it was the transportation of freight over long distances that truly transformed the national economy. The dramatic increase in the flow of commercial products allowed American agricultural and manufactured products to reach world markets at a scale previously unheard of. This growth ultimately turned the nation into an economic and geopolitical powerhouse. During two political ages, railroads made all of this possible. And of course hydrocarbons would eventually underpin this entire system.

Christian Wolmar writes that "a railroad was a far more sophisticated concept than any previous invention, requiring several elements to come together: the technology, both for the traction and the track; the financing to pay for it; the permission of the state to build it; the creation of an appropriate legal framework; and, of course, the labor for construction." As had been the case with the First Industrial Revolution, the key component parts of this phase of the Transportation Revolution came together first in Great Britain. The development of this new system took off in the early eighteenth century with tramways, wooden tracks on which wagons were pulled by draft animals. As one might imagine, however, there were significant limitations to this approach that required innovative solutions. The first improvements were iron-protected rails for durability and a projecting rim on the wheels to avoid slippage. The next major advance was replacing draft animals with wood-powered steam engines. By the late 1820s the British had proven the technical and economic viability of railroads. It was at this point that the technology crossed the Atlantic Ocean, where it forever changed the course of history.[41]

Fear that Baltimore would be left behind in a fierce intercity competition to connect the Eastern Seaboard with the interior led merchants from that city to sponsor construction of the first American railroad. The advantages bestowed by the Erie Canal upon New York City, in particular, drove municipal leaders to begin construction in 1828 of the Baltimore & Ohio, which was initially intended to stretch to the Ohio River at Wheeling, West Virginia. As James Dilts writes, "It took courage . . . to reject the prevailing canal technology and choose a rudimentary form of mining transportation to fashion a long-distance internal improvement." Unfortunately for Baltimore, it would take twenty-four years to construct the B&O line, more than enough time for other cities to catch up and surpass it. The railroad would, however, continue in operation and ultimately range as far as Chicago, becoming one of the four main trunk lines on the Eastern Seaboard. Regardless of the B&O's lack of success at beating out rival cities, it helped prove that Americans were unabashed enthusiasts for railroads. By 1840 there were 3,328 miles of railroads in the country—two miles more than all the canals built to that point. Two decades later, on the eve of the Civil War, the mileage had increased nearly tenfold, to 30,636 miles.[42]

As Americans energetically undertook large-scale railroad construction, they also began to manufacture locomotives, which to that point had been imported from Europe. The early pioneer in this arena was an Irish immigrant named Isaac Dripps, who modified the British-designed John Bull engine for use in America. One of his most famous adjustments was the attention-grabbing cowcatcher, which was used to clear the tracks of wandering livestock. The Pennsylvanian Matthias Baldwin would eventually become the globe's preeminent engine maker, designing locomotives with other distinctly American features (like sparkcatchers for the wood-burning engines) and capable of conquering the nation's rugged terrain. This transition away from being solely an importer of cutting-edge transportation technology proved critical in the American Rise; it would be especially important as automobiles and airplanes took center stage in the upcoming century.[43]

The development of the railroads was a classic case of a public-private partnership. During its first two decades, the cost of building the network was approximately $310 million—equivalent to $9.5 billion in 2015. In the 1850s that amount more than tripled, to $1.15 billion—$39.5 billion in 2015. Private capital would never have been able to fund construction of the railroads at that scale without government assistance. During a period when the national government continued to shirk its responsibility to provide this help (the central reason the early network was so fragmented, with few state-to-state or regional lines), it was state governments that stepped to the fore. The most important aid came in the form of property rights delivered by eminent domain, but railroads also received tax exemptions, loans, and monopoly rights that were unheard of during this period. This made sense because state governments, particularly in the West, saw the railroads as central to the political and economic development of the nation. With its extractive institutions and reliance on waterways to connect plantations to cotton markets, the South was slow to adopt railroads. Not only did this result in its economic stagnation but it would contribute mightily to the region's downfall during the impending crisis that would inaugurate a new political age.[44]

First Political Age
Conservative Cycle, 1829–1861

The progressive cycle that had begun with the inauguration of George Washington came to an end with the minor realignment election of 1828. During that electoral contest, the New York senator Martin Van Buren fashioned the nation's first successful grassroots political organization. As suffrage was being extended to nonlandowning white men, southern planters joined western farmers and northern democrats to elect General Andrew Jackson to the presidency, at the same time establishing the Democratic Party. Jackson considered himself a reformer, although much of his agenda was focused on reversing key macroeconomic security decisions made by his predecessors. His views on the Second Bank of the United States were particularly dangerous. He believed the bank concentrated excessive financial strength in a single institution, served mainly to make the rich even richer, and favored northeastern states. When Henry Clay attempted to force an early recharter of the institution in 1832, Jackson rejected the legislation in a strongly worded veto message. Jackson was highly concerned that the institution concentrated control in the hands of a very few people who had shares in the existing bank, rather than providing for an open competition to purchase stock in the rechartered government corporation. "The bounty of our Government," Jackson wrote, "is proposed to be again bestowed on the few who have been fortunate enough to secure the stock and at this moment wield the power of the existing institution. I can not perceive the justice or policy of this course." His fear was that this type of concentrated economic power would create a plutocracy that could steer national policy in directions most beneficial to the wealthy—a far from unwarranted concern. In the most stirring passage of the message, the president argued against policies that created economic inequity:

> When the laws undertake to add to . . . natural and just advantages artificial distinctions, to grant titles, gratuities, and exclusive privileges, to make the rich richer and the potent more powerful, the humble members of society—the farmers, mechanics, and laborer—who have neither the time nor the means of securing like favors to themselves, have a right to complain of the injustice of their Government. There are no necessary evils in government. Its evils exist only in its abuses. If it would confine itself to equal protection, and, as Heaven does its rains, shower its favors alike on the high and the low, the rich and the poor, it would be an unqualified blessing. In the act before me there seems to be a wide and unnecessary departure from these just principles.[45]

What was unique about this veto was that it was based on purely political and ideological considerations rather than on constitutional grounds. This was the first time such a veto had been issued; it would certainly not be the last. Far more troubling, however, was that Jackson didn't work with Congress to craft a new bill that would provide the benefits of a central bank while also addressing some of his valid concerns. This signaled that although he didn't like the particular arrangements of the recharter, he was simply opposed to any national bank. Thus, at the end of his tenure four years later, he allowed the bank to expire on schedule. As had happened three decades earlier with the First Bank of the United States, this move away from national support for a central bank crippled the economic system, plunging the nation into a deep depression and weighing down the one-term presidency of the aforementioned Martin Van Buren. As Daniel Walker Howe writes, "The Panic of 1837 merged with that of 1839 into a prolonged period of hard times that, in severity and duration, was exceeded only by the Great Depression that began ninety years later."[46] While the nation struggled to regain its footing as laissez-faire economics dominated this conservative cycle, two issues came to characterize national politics: territorial expansion and slavery.

After four ineffectual years under the leadership of the Whig Party, another Jackson protégé came to power. Although largely forgotten by most Americans, James K. Polk is considered by historians to have been the country's best one-term president, and he is often listed as one of the ten best overall. In a move that signaled a renewed emphasis on the primary security issue of the

First Political Age, national defense, Polk focused his 1844 campaign on territorial expansion. After his narrow victory over Henry Clay, he set four goals for his term: (1) reestablish an independent treasury, (2) reduce tariffs, (3) acquire some or all of the Oregon Territory, and (4) acquire California and New Mexico. He accomplished all four goals, but the latter two "Manifest Destiny" policies became the hallmarks of his administration.[47]

After pursuing an aggressive line with the British, Polk determined that rather than risk war it was better to compromise by dividing the Oregon Territory along the 49th parallel, with the United States gaining control of the Columbia River and the British retaining Vancouver Island. From the time he took office, Polk also believed that war with Mexico could be avoided. Tensions between the two nations dated back to the Mexican War of Independence, when numerous Americans had suffered economic losses during the conflict but were never repaid. The reparations issue festered for years until Mexico finally agreed to pay $2 million (out of $6.5 million claimed) in 1839. Five years later, however, only three payments had been made, and American leaders were running out of patience. Polk wanted to obtain all the Mexican land north of the Rio Grande to El Paso, then west near the 32nd parallel to the Pacific Ocean. He offered to forgive the remaining amount owed and pay between $15 million and $40 million for the territory. Unfortunately, Mexico was not of a mind to negotiate. Despite heavy opposition from a Whig Party that was worried about the spread of slavery, Polk ultimately led the nation into war. American forces quickly took control of the territory at issue, but the Mexicans would not yield until General Winfield Scott successfully captured Mexico City. The Treaty of Guadalupe Hidalgo expanded the nation to encompass "all of present-day Texas, New Mexico, Arizona, Utah, Nevada, and California, as well as parts of Kansas, Oklahoma, Colorado, and Wyoming." Later the southern sections of New Mexico and Arizona were obtained peacefully through the Gadsden Purchase. All told, Polk had expanded the nation by one-third, adding more territory to the nation during his tenure than under any other president, including Jefferson. The result was a Rising America that was fully continental in scope and situated to exert its influence around the world at a time of its own choosing. From a geopolitical perspective, American access to both the Atlantic and the Pacific Ocean was one of the main reasons why it became the planet's dominant nation during the following century. But this wouldn't happen until it dealt with the most important internal political issue of the nineteenth century.[48]

Slavery was America's original sin, and it had long divided the nation. The

Constitutional Convention struggled mightily with the issue, until southern interests succeeded in protecting the slave trade for at least twenty years and were allowed to count slaves when tabulating the number of seats they would have in the US House of Representatives. Combined with the nonproportional nature of the Senate and the Electoral College, this provided southern states with a strong base of power that allowed them to dominate American politics for a half century. The result was that the southern planter class was able to stymie any attempts to eliminate the immoral institution of slavery. Regardless of the South's built-in power base, a nascent abolition movement had been evolving since the founding of the nation and had eliminated slavery in every northern state. In 1831, in the first edition of the *Liberator*, William Lloyd Garrison issued a call to arms to finally eliminate slavery everywhere. More than two decades later, on 4 July 1852, Frederick Douglass delivered a pointed address elucidating the principles at the heart of the abolition movement: equality and freedom for all. "I am not included within the pale of this glorious anniversary! The rich inheritance of justice, liberty, prosperity and independence, bequeathed by your fathers, is shared by you, not by me," he pronounced.

> This Fourth [of] July is *yours*, not *mine*. *You* may rejoice, I must mourn. . . . Fellow-citizens; above your national, tumultuous joy, I hear the mournful wail of millions. . . ! Is it not astonishing that, while we are ploughing, planting, and reaping, using all kinds of mechanical tools, erecting houses, constructing bridges, building ships, working in metals of brass, iron, copper, silver and gold; that, while we are reading, writing and ciphering, acting as clerks, merchants and secretaries, having among us lawyers, doctors, ministers, poets, authors, editors, orators and teachers; that, while we are engaged in all manner of enterprises common to other men, digging gold in California, capturing the whale in the Pacific, feeding sheep and cattle on the hill-side, living, moving, acting, thinking, planning, living in families as husbands, wives and children, and, above all, confessing and worshipping the Christian's God, and looking hopefully for life and immortality beyond the grave, we are called upon to prove that we are men!

He concluded that the country must be roused to denounce the institution of slavery, for the good of the nation.[49] During the subsequent years this sentiment began to be shared by an ever-growing number of people, particularly once it became apparent that the evil institution was also shackling economic progress.

But other voices were articulating a very different message. Senator John Calhoun, a former two-term vice president, under John Quincy Adams and then Andrew Jackson, became the leading proponent for the continuation of slavery. He contended that abolition and union couldn't coexist, because the South would never surrender its institutions. Calhoun argued, "There never has yet existed a wealthy and civilized society in which one portion of the community did not, in point of fact, live on the labor of the other." Regardless of the dubiousness of this historical assessment, it was the Frenchman Alexis de Tocqueville who noted during his journeys through the United States the true impact of slavery on economic development. "The colonies in which there were no slaves," he wrote, "became more populous and more rich than those in which slavery flourished. The more progress was made, the more was it shown that slavery, which is so cruel to the slave, is prejudicial to the master." In their recent book *Why Nations Fail*, Daron Acemoglu and James A. Robinson show that failing states share one central characteristic: they have extractive elites. These select few put in place institutions that inequitably extract economic resources, but in so doing they fail to keep pace with more equitable societies whose institutions promote innovation. According to these authors, the American South was the quintessential extractive society, falling further and further behind its northern counterparts. The southern-dominated Democratic Party, however, consistently opposed any change that would seek to create inclusive institutions that might close this gap. The Whig Party, meanwhile, was ultimately destroyed because it lacked any clear position on this most important issue. The nation's second major realigning election would shepherd into power a party that would finally work to destroy slavery and restore macroeconomic security as a focus of national policy. In so doing, this new party would shape the political institutions needed to birth the Hydrocarbon Age.[50]

CHAPTER TWO

Coal, Macroeconomic Security, and the Second Political Age

Although the United States had its fair share of growing pains during the First Political Age, there was no doubt that America was an up-and-coming nation. The physical size and the population of the country both quadrupled, and the size of the economy grew sevenfold. How did this happen without the national government consistently fostering economic modernization? First, the Mid-Atlantic and northern states made large investments in key infrastructure projects. Second, the private sector vigorously overcame the limitations of the national government. This combination contributed to a communications revolution, with the telegraph making the dramatically enlarged nation governable. It played an even bigger role in a multiphase transportation revolution, which saw turnpikes, canals, steamboats, and railroads create a truly national marketplace for American agricultural goods—while also promoting economic expansion in other sectors. These transformations, however, were not good for all Americans. Native Americans were steadily pushed off their land, and slaves continued to toil without political or economic liberty. Even among the white population, income inequality widened dramatically during this period. In the Second Political Age, these issues would tear the country apart before a painfully slow national healing process began.[1]

For decades, Henry Clay and others had labored to prevent a national rupture over slavery because they believed the Union would not survive it. In 1820, Clay had negotiated the Missouri Compromise, which prohibited the future entry into the union of slave states north of the 36°30′ latitude line. In 1850, Clay negotiated a second agreement that attempted to maintain a balance between slave and free states. Four years later, however, Senator Stephen Douglas championed and President Franklin Pierce signed the Kansas-Nebraska Act. This bill effectively repealed the Missouri Compromise and allowed the citizens of incoming states to determine with a popular vote whether they wanted slavery. This act of political cowardice supplied the spark that would lead directly to the outbreak of the American Civil War. US Supreme Court Chief Justice Roger Taney, writing for the majority in *Dred Scott v. Sandford* (1857), added dry wood to the flames when he argued that no one of African descent could be a citizen of the United States. These developments led to the rising influence of a new political party, the Republicans, and a new leader, Abraham Lincoln. At the outset of a new political age, Lincoln would help eliminate slavery and solidify macroeconomic security as a clear duty of the federal government. In so doing, he would create the conditions necessary for rapid economic growth fueled by hydrocarbon energy.[2]

Second Political Age
First Progressive Cycle, 1861–1897

The major realigning election of 1860, which put Abraham Lincoln in the White House, represented the sharpest departure between any two periods in America's political history. It signified the culmination of the Republican Party's rapid rise to power. The party was largely a creature of the Northeast, the Upper Midwest, and Far West, with a coalition of businessmen, skilled workers, professionals, and commercial farmers favoring its platform of economic modernization. William Strauss and Neil Howe argue that no Hero Generation emerged during this period, largely because of the timing of the Civil War crisis. While this argument is plausible, it is nevertheless clear that a partnership between the Transcendental (Prophet) Generation and the Gilded (Nomadic) Generation produced a *turning* that accomplished the expected objectives of a progressive cycle by creating institutions that established macroeconomic security as a core duty of the national government. The result was the beginning of a seven-decade period in which the *crisis* set the stage for an extended *high*, which saw the Republican coalition dominate national politics. The party finally began the long process of addressing the catastrophic racial divide within the nation as Lincoln's condemnation of slavery gave birth to a positive view of liberty "in which government had an affirmative obligation to ensure equality under the law." Before he had even taken office, seven American states seceded from the Union because they feared Lincoln would seek to eliminate slavery; eventually, four more states would join the new Confederate States of America.[3]

In his first inaugural address, the new president repeated his earlier position that he had no authority to remove the institution of slavery from states where it already existed. He directed much of his attention in this speech not to slavery but to union.

> I hold that, in contemplation of universal law and of the Constitution, the Union
> of these States is perpetual. . . . It follows from these views that no State upon its

own mere motion can lawfully get out of the Union; that resolves and ordinances to that effect are legally void, and that acts of violence within any State or States against the authority of the United States are insurrectionary or revolutionary, according to circumstances. I therefore consider that in view of the Constitution and the laws the Union is unbroken, and to the extent of my ability, I shall take care, as the Constitution itself expressly enjoins upon me, that the laws of the Union be faithfully executed in all the States.

Lincoln made clear that he didn't seek a violent restoration of the Union but that he had a constitutional duty to "preserve, protect, and defend." He concluded with one of the most beautiful passages in American rhetoric: "We are not enemies, but friends. We must not be enemies. Though passion may have strained it must not break our bonds of affection. The mystic chords of memory, stretching from every battlefield and patriot grave to every living heart and hearthstone all over this broad land, will yet swell the chorus of the Union, when again touched, as surely they will be, by the better angels of our nature."[4]

Unfortunately, not all Americans shared Lincoln's views and hopes. Little more than a month after his elevation, Confederate troops opened fire on Fort Sumter in Charleston Harbor. In the coming four years more than two hundred battles would be fought between Union and Confederate forces, with many more skirmishes engaging troops from both sides. The land war began badly for the Union forces, with several defeats in the Eastern Theater, but a naval blockade was established that would ultimately cripple the Confederate economy. The land war saw its turning point in the Western Theater, where a series of successful sieges and battles led by Generals Ulysses S. Grant and William T. Sherman crushed the Confederate forces. On 1 January 1863, Lincoln's Emancipation Proclamation made ending slavery a clear objective of the war. Later that year, the Confederate general Robert E. Lee's attempted invasion of the North ended at the Battle of Gettysburg. Lincoln visited the area several months later and delivered the best-known speech in the nation's history, telling the crowd at the battlefield dedication "that from these honored dead we take increased devotion to that cause for which they gave the last full measure of devotion—that we here highly resolve that these dead shall not have died in vain—that this nation, under God, shall have a new birth of freedom—and that government of the people, by the people, for the people, shall not perish from the earth." In 1864, Grant took control of the Eastern Theater and began the process of slowly grinding down the remaining Confederate forces. At the same time, Sherman marched south and lay waste to the heart of the Confederacy. In March 1865, with the end of the war in sight, Lincoln was sworn in to begin his second term in office. In his second inaugural, he articulated his vision for the country's future: "With malice toward none, with charity for all, with firmness in the right as God gives us to see the right, let us strive on to finish the work we are in, to bind up the nation's wounds, to care for him who shall have borne the battle and for his widow and his orphan, to do all which may achieve and cherish a just and lasting peace among ourselves and with all nations." The following month, on 9 April 1865, Lee surrendered to Grant at Appomattox Court House. Five days later, President Lincoln was assassinated —the last casualty in a bloody conflict that had claimed the lives of six hundred thousand Americans.[5]

Why did the South lose? At a basic level, it was because both the Confederate government and the individual states that joined the secessionists were opposed to policies aimed at providing macroeconomic security. As Michael Lind writes, the "provisions of the Confederate Constitution were carefully crafted to forestall the possibility that the new government would ever attempt anything like the programs of Alexander Hamilton and Henry Clay for national economic development." Not only did the South fail to create a banking system but its embargo on cotton exports was a self-inflicted wound that never convinced the British or French to come to their aid. Perhaps more importantly, in one of the first industrial wars the Confederacy was seriously outclassed. The North was able to fully leverage a growing manufacturing sector to mass-produce arms and ships, and its superior banking system allowed it to debt-finance purchases from Europe. The North also had far better telegraph and railroad infrastructures, the latter proving particularly important in a conflict in which the benefits of mobility were being explored. While hydrocarbons were yet to become a major factor, the Civil War demonstrated the potential of mechanized warfare.[6]

Although much of the history from this era focuses on the military conflict and ending slavery, the Republican dedication to providing macroeconomic security was just as important to the American Rise. President Lincoln was a committed economic modernizer: "My politics are short and sweet, like the old woman's dance. I am in favor of a national bank . . . in favor of the internal improvements system and a high protective tariff." Once in office, he was able to enact this vision. While most of the heavy lifting was done in Congress, President Lincoln happily approved the National Currency Acts of 1863 and 1864 (which established a system of national banks and a national currency), the Revenue Act of 1862 (which established a progressive income-tax structure), and tariff legislation that initiated a period of trade protection aimed at growing American industry. He also signed the Homestead Act, encouraging western settlement; the Morrill Land-Grant Colleges Act, which transformed higher education; the Pacific Railway Acts, which enabled construction of the transcontinental railroad; and a law creating the Department of Agriculture. With these dramatic legislative accomplishments, which represented one of the most significant policy punctuations in American history, Lincoln and his fellow Republicans shaped the early institutions that would promote the INNATE revolutions—which in turn would pave the way for significant expansions in national and economic security.[7]

Although Republican policies and programs had begun to establish the framework for an industrial economy, the party failed to create the institutions necessary to react to the difficulties endemic to the long transition away from an agrarian economy. While previous eras had seen too little federal government promotion of an industrial economy, future eras would be characterized by too little government regulation of increasingly powerful capitalist interests. This shift took hold almost immediately and led to the Panic of 1873, which was caused by railroad speculation and a string of bank failures. It was during the subsequent national crisis that President Ulysses Grant vetoed the Inflation Bill of 1874, an ill-conceived act that would have put more money into circulation ostensibly to help small western farmers. Grant rejected it because he feared damaging American creditworthiness and possibly unleashing economic forces the national government wasn't capable of battling. Although a prudent decision, it didn't pull the nation out of the Long Depression, the period of recurrent economic contractions that continued for more than twenty years. This Gilded Age saw a select few making huge fortunes and rapidly increasing real wages for skilled workers, but it was also characterized by great poverty among the millions of immigrants who were pouring into the country to seek economic opportunity. The South fell further behind economically in the postwar years as it failed to industrialize and became ever more dependent on agricultural exports like cotton and tobacco. And after federal troops were withdrawn during the 1870s, freed slaves were stripped of voting rights and lost any economic security that had been established during Reconstruction.

Instead of reacting to the growing crisis with more muscular national government providing microeconomic security, political leaders ushered in an ever more powerful private sector. This was the beginning of the corporatization of America, when two major developments resulted in an explosion in the number of these business associations and granted them unprecedented rights. The first of these developments started in the 1850s, when several states, most importantly New Jersey and Delaware, began passing laws that made it easy to create corporations; previously a state charter had been required. These corporations provided for the pooling of capital from a broad array of investors, while limiting shareholder liability to the amount of their investment in the company. Shareholders with common stock were empowered to elect a board of directors, who in turn appointed executives to manage the day-to-day operations of the company. Perhaps the most troubling aspect of these modern corporations was the executives' fiduciary duty to the shareholders. At a basic level, they were required to maximize shareholder profits even if that came at the expense of the public good. Much later, as the environmental impacts of hydrocarbon use became clear, this would prove disastrous.

The second development has had even more far reaching impacts. In 1886 the US Supreme Court redefined the nature of corporations in *Santa Clara County v. Southern Pacific Railroad*. There are several strange aspects to the case. First, it was actually a relatively benign tax case. Santa Clara County and Fresno County were suing the Southern Pacific Railroad and the Central Pacific Railroad for unpaid county taxes on fences that had been added to their landholdings. Second, this was one of a growing number of cases in which a corporate defendant used the Fourteenth Amendment's Equal Protection Clause, even though the amendment specifically refers to natural persons. Third, and most shocking, the key finding in the case appears, not in the majority opinion, but solely in the case headnote. None of the constitutionally appointed justices drafted this short summary; instead they relied on the court reporter Bancroft Davis, a procorporation former president of the Newburgh and New York Railway. Davis wrote in the headnote:

> One of the points made and discussed at length in the brief of counsel for defendants in error was that "corporations are persons within the meaning of the Fourteenth Amendment to the Constitution of the United States." Before argument, MR. CHIEF JUSTICE WAITE said: "The Court does not wish to hear argument on the question whether the provision in the Fourteenth Amendment to the Constitution which forbids a state to deny to any person within its jurisdiction the equal protection of the laws applies to these corporations. We are all of opinion that it does."

Thus, without any public debate, one of the least intellectually gifted Chief Justices and a clearly biased court reporter forever changed the shape of American history. From that time forward, corporations have used their constitutional rights to become ever more entangled with the political system. As "juristic persons," corporations have free-speech rights. These have been used to justify spending billions financing political campaigns and lobbying government officials, leading to disequilibrium that strongly favors the private sector. As Robert Engler argues, few sectors benefited more from the resulting systems of private government than the hydrocarbon industries, oil in particular.[8]

As the Long Depression dragged on and income inequality became more pronounced, there was a rising chorus of criticism regarding America's economic trajectory. One of the more influential commentators was Henry George, whose *Progress and Poverty* was widely read in the late nineteenth century. The book opened with the following:

> The present century has been marked by a prodigious increase in wealth-producing power. The utilization of steam and electricity, the introduction of improved processes and laborsaving machinery, the greater subdivision and grander scale of production, the wonderful facilitation of exchanges, have multiplied enormously the effectiveness of labor. At the beginning of this marvelous era it was natural to expect, and it was expected, that laborsaving inventions would lighten the toil and improve the condition of the laborer; that the enormous increase in the power of producing wealth would make real poverty a thing of the past. . . .
> Now, however, we are coming into collision with facts which there can be no mistaking. . . . The enormous increase in productive power . . . has no tendency to extirpate poverty.

While modern economists might argue about the merits of George's chosen solution, a single tax on the value of unimproved land, it is clear that he would have directed the revenues from this tax to provide both macroeconomic and microeconomic security. In identifying both the benefits and the challenges of the emerging hydrocarbon age, he suggested that government had a moral imperative to provide broad-based economic security. As the INNATE revolutions drove the economy forward during this progressive cycle, they drove economic growth and provided the fiscal resources necessary to make that objective politically achievable.[9]

INNATE Revolutions: Industrial
America and the Second Industrial Revolution

By the early 1800s coal had been an important energy source in Britain for centuries. For most of that time, however, Americans had little use for the stuff. Why? Because the colonists had access to more abundant traditional energy sources. In the eighteenth century, the most important of these were the continent's vast forests. During late colonial times, however, growing coastal cities began to import relatively small quantities of coal from England. Then independence came, and that trade faded. Following the war, the Richmond Basin, in Virginia, became the primary domestic source of coal. But the bituminous coal from this region was costly for two reasons. First, extraction relied on slave labor, which was often unreliable because plantation work always took precedence. Second, poor southern roads and canals provided insufficient options for getting the coal to market cheaply. As a result, steam-based manufacturing lagged in America. Instead, as we have already seen, the rapid industrialization that began in the 1830s was powered almost entirely by water. Except in the Pittsburgh region, 98 percent of large manufacturing firms used water power. In Pittsburgh, all of the large factories and most of the small ones were steam powered. And their operating costs were one-third those of any similar facilities in America. In fact, their plants were 25 percent more cost efficient than plants in Britain. What was the secret weapon? They had access to cheap and abundant coal.[10]

During the early nineteenth century, Pittsburgh was the only industrial city in America with enough bituminous coal to satisfy domestic needs and meet all manufacturing requirements; it was even able to ship the resource west along the Ohio River. Yet, the cost of moving the coal to major coastal cities remained prohibitive because of poor transportation connections to the other side of the Appalachian Mountains. By the time this situation changed, however, it was a different variety of coal that would be conveyed. Initially, miners largely overlooked anthracite. It was an unknown quantity for Americans and Europeans alike, with uncertain properties as a fuel source. It was also located only in mountains that were hard to reach. "Stone coal" was extremely hard and difficult to ignite, but it was remarkably pure (with carbon contents between 85 and 100 percent), provided tremendous heat with almost no smoke, and was long burning. During the War of 1812, when supplies of bituminous coal from Britain and Virginia were disrupted, there was some fleeting interest in using anthracite for manufacturing. After this short-lived attention began to diminish, a small cadre of true believers began building canals to transport the resource to eastern markets. They judged that its positive qualities would make it highly desirable if its prices were somewhat competitive with those of other resources. The most important of these visionaries were Josiah White and Erskine Hazard, as well as William and Maurice Wurts. Although both pairs ultimately succeeded in constructing their watercourses, they still faced great difficulty in actually selling the anthracite, because potential customers found it tricky to use. Marketing efforts to improve public awareness, as well as adaptations to grates and furnaces that made the coal easier to burn, saved the prospectors. With the technological advances in place, it became clear that consuming anthracite saved its users money because it demanded far less labor and burned more efficiently. It was half the cost of Richmond Basin bituminous coal. As a result, in the late 1820s anthracite consumption surpassed that of both imported and Virginia coal. By the early 1840s, more than 85 percent of all coal used in the United States was anthracite.[11]

Pennsylvania anthracite was shipped by canal and river through Philadelphia and New York to industrial cities and towns along the Atlantic Seaboard. With these transportation routes in place and new markets being opened all the time, the production of stone coal climbed sharply. While only two tons had been extracted in 1820, by 1849 more than 3,700 tons were being mined annually. During that period, prices continued to drop (from $7.42 per ton in 1829 to $3.46 in 1852) until anthracite was cheaper than all of the alternatives—wood, charcoal, and bituminous coal. As a result, it displaced those other fuel sources for household heating and an increasing array of industrial purposes. As Alfred Chandler Jr. wrote, "The opening of the anthracite fields thus provided the American Northeast with a constantly increasing supply of excellent coal at decreasing prices." A variety of industries benefitted from the availability of the resource, in particular the iron, glass, paper, and textile industries. With the arrival of this powerful new energy source, factories quickly emerged in each of these sectors, in many cases merging the American System of Manufactures and coal power. Not only did anthracite play a role in forges, it also increasingly drove the specialized machines that were critical to the success of these manufacturing works. Coal-fired steam power quickly became the key motive force in glassworks and paper plants before ultimately making its way into "baking, sugar refining, the brewing of malt liquors and the distilling of spirits, but also in the process of earthenware, plated ware, chemicals, and India rubber products."[12]

The rise of anthracite had an enormous influence on the national economy. Manufacturing grew from 17 percent to 30 percent of national output just during the 1840s, faster than during any other decade that century. It also saw significant increases in productivity, as well as the beginnings of tremendous growth in per capita incomes. Finally, it made the quickly urbanizing American Northeast an up-and-coming player in the First Industrial Revolution. Steam-powered cities like Boston, Worcester, Providence, Taunton, Fall

River, New Haven, Middletown, Buffalo, Utica, New York City, Jersey City, Baltimore, Philadelphia, and Pittsburgh were now driving the national economy. This would have a decisive impact on the outcome of the Civil War. As Barbara Freese writes, "The American Northeast's coal-fired economic surge before 1860 further deepened the political and economic divisions between the industrial North, dependent on coal-burning factories, and the agrarian South, dependent on slave-exploiting plantations." Given the North's 38-to-1 and 15-to-1 advantages in coal and iron production, respectively, a Union victory was virtually assured after the South failed to land an early knockout blow in the war. After the conflict, the largely intact northern economy was able to invest in western expansion and development. This meant that the South continued to fall behind economically as apartheid took hold, while the other regions leveraged hydrocarbons and innovation-friendly policies to drive national economic growth. In particular, the conditions were right for industrial cities in the Northeast to become the most important actors in the next phase of industrialization.[13]

* * *

The Hydrocarbon Age began in the mid-1860s, at the birth of the coal-fired Second Industrial Revolution. While coal had previously just been one energy source being consumed by Americans, as figure 2.1 shows, from that time on it would be essential to the American way of life. The country would never look back; instead it would determinedly extract other fossil fuels to push economic growth, which in turn would create new political opportunities. This commenced in parallel with the opening of the new frontier of industrialization. While iron, paper, glass, textiles, firearms, and sewing machines characterized the first phase of this revolution, the second phase was exemplified by capital goods industries, most importantly steel. In addition, as Joel Mokyr writes, while the "First Industrial Revolution, and most of the technological developments preceding it, had little or no scientific basis," the Second Industrial Revolution "accelerated [mutual feedback loops] between science and technology." This new industrialization also continued to build a bridge to the mass production that would take hold at the beginning of the twentieth century. According to Chandler, two preconditions were needed to set the stage for mass production: "First, the railroad and telegraphic network had to be built in order to permit large amounts of raw materials to move at a rapid but steady pace into the factories, and the finished goods to be shipped out as regularly and as speedily. Second, there had to be available large quantities of fossil fuel

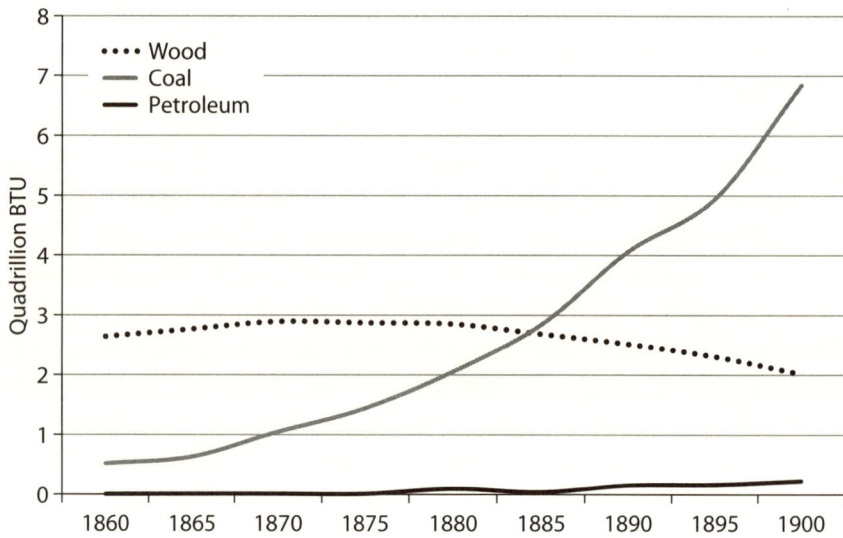

Figure 2.1. US Energy Consumption by Source, 1860–1900. US Energy Information Administration, *Annual Energy Review, 2009.*

(in this case coal) to provide essential heat and power. The new production processes, therefore, had their beginnings in the 1850s as the railroad and telegraph networks expanded and as coal supplies became abundant." Although this process began in the 1850s, it really accelerated during the decades after the Civil War. Although Britain, Europe, and Japan all grew during this period, it was America that became the dominant actor. The nation enjoyed vast natural resources, abundant labor, readily available capital, and an enormous domestic market for consumer goods. It exploited these advantages to build the largest economy in the world, a distinction it has not relinquished for well over a century.[14]

The single most important product of the Second Industrial Revolution was steel. Thomas Misa writes that steel was "at the center of U.S. social, economic, and institutional development" during this period. The Bessemer process was the technology that propelled steel into its preeminent position. Named for Sir Henry Bessemer, the method involved injecting air into molten pig iron to remove impurities. This was critical because it allowed for mass production of cheap steel. The most important early consumers of cheap steel were the railroads, which utilized steel rails during the 1870s and 1880s to build transcontinental railroads, which radically changed transportation in America. Steel also played an important role in the dramatic urbanization that occurred starting in the last two decades of the nineteenth century. It provided the strength needed for the infrastructure projects that made American-style urban density possible—skyscrapers, transit systems, and bridges. (Little did politicians or city planners know at the time that they were experimenting with some of the key tools twenty-first century leaders would need to battle climate change.) Additionally, steel reshaped modern warfare, first by providing case-hardened armor for battleships before ultimately making its way to land forces via the tank. Finally, innovations that produced steel for high-speed tools continued the ongoing revolution in factories by delivering more resilient machinery that could operate at higher temperatures and thus accelerate production. As Misa concludes, during the Second Industrial Revolution, "modern America took institutional and physical form as a nation of steel." Because a dense energy source was required to create the temperatures needed to produce high-quality steel, it was at this point that coal truly came into its own on the North American continent. Without coal there would have been no Second Industrial Revolution. And without a great deal of coal there would have been little chance that America would lead that epic economic transformation. Fortunately, the United States turned out to have some of the best and largest reserves on the planet.[15]

Although coal use had expanded greatly from the 1820s to the 1860s, at the end of the Civil War firewood still accounted for the lion's share of national energy consumption. In the coming half century, however, coal burning doubled every decade, and America became the largest producer in the world—a status it would retain for more than a century. "By 1900," Freese writes, "coal was the unrivaled foundation of U.S. power." Disturbingly, most of the expansion in coal use during this period was not in clean-burning anthracite. Instead, the more readily available and dirtier-burning bituminous came back to the fore in both the Appalachians and the Midwest. Because bituminous operations were less capital intensive, there was far greater competition for this market. This meant there were more actors operating smaller companies. That provided an opening for the labor movement, led by John Mitchell's United Mine Workers (UMW), to seek shorter hours and safer working conditions for miners laboring in this most dangerous job. By the twentieth century, the UMW was strong enough to also take on the powerful anthracite industry, in which owners had traditionally exercised more control over workers. In 1902, 150,000 striking anthracite miners created one of the first major energy crises in American history. With a Progressive Era public largely supporting the miners, President Roosevelt stepped in to force management to accept a compromise that was seen as a victory for organized labor. Just as critically, as Freese writes, "the 1902 strike was . . . a vivid lesson in how dependent the nation was on coal, and how deeply it could be hurt when supplies ran short." As the nation expanded to use other hydrocarbons, this reality became one of the guiding policy imperatives for political and business leaders.[16]

* * *

The transition from wood and water power to coal had innumerable consequences, but the most important was the dramatic growth in American social development. As discussed earlier, this is part of a story that is seldom told. While the technologies were critically important, without a foundation of cheap energy none of these gains would have been realized. The transformative role of steel, and manufacturing more generally, would not have been possible without coal to power the foundries. As Robert Engler put it, "Harnessing the power of nature . . . made possible the development and control of tools for carving an industrial civilization out of a frontier wilderness in a remarkably short order." To keep up with the growing demand from industry, mining operations continued to move steadily west, and coal fields were soon opened in states like Wyoming, Colorado, and New Mexico. "By deploying a modern, coal-based industrial system," Bruce Podobnik writes, "the United States became the first modern nation-state that was truly continental in scale." During this Second Industrial Revolution, America's transportation, energy, and industrial sectors would be transformed, signaling the true beginning of the American Century. As we will see, the availability of so much cheap energy solved many economic and political problems facing the nation and helped it become one of the world's great powers.[17]

INNATE Revolutions: Transportation
A Nation of Railroads, Part 2

At the outset of the Civil War, the fragmented American railroad infrastructure began to reveal core weaknesses, most importantly a relative lack of speed and track uniformity. The Union had an important advantage over the Confederacy, however, with twice as many railroad miles and far superior rolling stock. During the war, President Lincoln nationalized the most important railroads and appointed Daniel McCallum and Herman Haupt to overcome existing shortcomings. No similar effort was possible within the disjointed Confederate government, which would prove disastrous in the long run. As Christian Wolmar writes, "The Civil War was the world's first railroad war . . . [and] the iron horse turned out to be one of the most effective weapons of war invented by man, helping create a far more intense, deadly, and lengthy type of warfare." Railroads were involved in many aspects of the war. They rapidly transported troops and supplies to the battlefield. They were adapted to carry railroad-mounted guns. They were used as long-distance ambulances, helping to carry the wounded to urban hospitals. In the end, the railroads helped show how important mobility would be for modern warfare. At the same time, they helped jump-start a postwar economic boom.[18]

One of the key transitions during the war was in the motive power for trains. Until then, American locomotives had used wood-powered steam engines almost exclusively, but the Pennsylvania Railroad determined that coal was far more efficient and plentiful, and ultimately all other railroad companies would follow its lead. This meant that hydrocarbons were now dominant in the transportation sector. Vaclav Smil writes that "several attributes apply to all forms of fossil-fueled . . . transport. In contrast to traditional ways of moving people and goods they are not only much faster, often almost incredibly so. They are also incomparably more reliable, substantially cheaper, and capable of transferring at one time much larger numbers of passengers or much greater masses of goods." Thus, the ability to leverage massive coal resources for the postwar expansion of the rail network was crucial to the American Rise.[19]

As the war raged, the Republican Congress passed and President Lincoln enthusiastically signed the Pacific Railroad Act of 1862. After years of debate and southern intransigence, northern leaders were at last able to provide support for the first transcontinental railroad. The legislation provided the Union Pacific Railroad and the Central Pacific Railroad with land-grant concessions and financial incentives to begin construction (the former moving west, the latter working east). In Stephen Ambrose's estimation, "Next to winning the Civil War and abolishing slavery, building the first transcontinental railroad [from Nebraska to California] was the greatest achievement of the American people in the nineteenth century." Not only was the railroad one of the notable engineering feats in history but it put on display one of the central drivers of the American Rise, immigrant labor. The more than seventeen-hundred-mile railroad was largely built by the Union Pacific Railroad's Irish workforce and the Central Pacific Railroad's Chinese workforce. At the same time, it was newly arrived European immigrants who had crossed an ocean searching for economic opportunity who were driving demand for the lands opened by the railroad. From our contemporary perch, it is hard to truly appreciate the revolutionary nature of the transcontinental railroad. "When the Civil War ended in 1865, it took a person months and might cost more than $1,000 to go from

New York to San Francisco," writes Ambrose. "But less than a week after the pounding of the Golden Spike, a man or woman could go from New York to San Francisco in seven days. That included stops. . . . And the cost . . . by June 1870 was . . . $136 for first class [i.e., a Pullman sleeping car] and $65 for . . . emigrant class [i.e., a bench]." Similarly, freight costs fell dramatically, which helped create a truly continental economy. Not only was the nation now connected coast to coast but America had 40 percent of all the railroad mileage on the planet. This was an enormous advantage as the nation started leveraging its coal-fired industrial potential, providing an extraordinary domestic market for manufactured goods.[20]

In the half century that followed the Civil War, the American railroad network expanded at a nearly unimaginable speed. By the beginning of the First World War, US railroad mileage had increased from 35,000 miles to more than 250,000 miles. This was more mileage than in the entirety of western Europe, the most economically developed part of the world at the time. The railroads furthered the synergistic development of industrial capacity and also continued to fuel westward migration. Perhaps equally important in this period, was the role railroads played in the development of the corporation. The railroads grew into the largest companies in the nation, which had huge implications for their future. As discussed earlier, corporate leaders have a fiduciary duty to maximize shareholder profits even if this comes at the expense of the public good. During the Gilded Age, the potential negative consequences of this duty were vividly on display as the railroads worked feverishly to build ever-larger unified networks. As Wolmar writes, "During the final quarter of the nineteenth century, the railroads became, first, disliked and, then, widely resented. . . . They turned into the rapacious monopolist, reviled by almost everyone." This was partially the result of corruption, which was rampant. But more crucially, it was simply the natural result of unregulated large business enterprises more interested in turning a profit than in promoting the public good. This didn't win them many friends among the general public, particularly small western farmers, who felt they were essentially the economic hostages of the railroads. It was easy for them to demonize railroad barons like Daniel Drew, Cornelius Vanderbilt, and Jay Gould, who were mostly interested in getting rich.[21]

Given how reviled the railroads became, it is not surprising that critical voices demanding forceful government regulation compelled national leaders to take action. The problem was that they really didn't know how to proceed effectively. The Interstate Commerce Act of 1887, intended to challenge the railroad monopoly, instead dangerously weakened this key pillar of the American Rise. During its early years the Interstate Commerce Commission (ICC) was basically powerless, but that began to change as leaders of the Progressive Movement took control of the government. By the time this happened, however, the business case for railroads had changed dramatically. First, the massive undertaking of creating a coast-to-coast system had been mostly completed, and there were no new territories to conquer. Second, the need for speedier, more comfortable service was altering railroad operations. Third, tourism was becoming a potential profit center. Fourth, new technologies, such as all-steel cars and more powerful locomotives, were being incorporated. Each of these innovations presented new capital obligations, which, Wolmar writes, "exposed the fundamental reason that, across the world, [railroads] struggled to make a profit even in the good times."[22]

The ICC came to the fore just as these new financial imperatives were reshaping railroads. The commission's primary tool was the denial of rate increases, which made it harder for the railroads to react to the changing twentieth-century landscape. One of the key problems with these early regulators was that they believed railroads were operating in a vacuum, that they were an independent economic sector. But railroads were actually part of a larger transportation sector and would face increasing competition from a number of new actors, most importantly automobiles, trucks, buses, and airliners but also electric streetcars, subways, and interurban trains. According to Albro Martin, "The great tragedy of this failure of human beings to intelligently order their economic environment lies in the long-term effects on the railroad system as an enterprise. American railroads, quite literally, never got over the shock which archaic Progressivism's cruel repudiation of their leadership produced." Despite temporary resurgences during the two world wars, by the mid-twentieth century the railroads had entered into an inexorable slide. They would steadily lose passenger-miles to cars, buses, and airliners. They were slowly losing freight-miles to trucks, which accounted for a solid 10 percent of intercity cargo haulage by 1941. These trends weren't concerning to most leaders at the time. But given the far greater energy intensity of these new transportation platforms, many now lament government's failure to more forcefully support the railroads. That said, there is little doubt that the coal-fired iron horses had transformed the national economy. At the same time, coal-fired electricity was beginning to do its part to drive growth during the American Rise.[23]

INNATE Revolutions: Electrification
From Steam Engines to Electricity

In the late nineteenth century, Nikola Tesla prophesied: "The day when we shall know exactly what 'electricity' is, will chronicle an event probably greater, more important than any other recorded in the history of the human race. The time will come when the comfort, the very existence, perhaps, of man will depend upon that wonderful agent." This statement is striking not solely for its prescience; it is also striking for the revelation of how little humanity knew about electricity just over a century ago. Electricity had been known for millennia, revealing itself both as lightning and as electromagnetism. Europeans began investigating its true nature starting in the fifteenth century. Benjamin Franklin conducted independent experiments during the eighteenth century that ultimately led to his invention of the first electric battery. Yet electricity's exact character remained elusive.[24]

As the nineteenth century dawned, Alessandro Volta and Andre-Marie Ampere made important advances in knowledge about electricity. The latter provided a physical understanding of the electromagnetic relationship, positing the presence of an electrodynamic molecule—decades later this would lead to the discovery of the electron. But the British chemist and physicist Michael Faraday did more than anyone to unlock electricity's secrets. The almost entirely self-taught son of a blacksmith, he was consumed by a desire to convert magnetism into electricity. Utilizing earlier work conducted by Hans Christiaan Ørsted, he stumbled upon the principle of electromagnetic induction. During ten momentous days in 1831 Faraday conducted a series of experiments that demonstrated how to use commonplace magnets to produce an electric current. From these experiments, writes Maury Klein, "emerged ramifications that would soon change the world. . . . Faraday had laid the foundation for both the theory of electricity and its practical development." In the coming years, he would go on to invent the electric dynamo, which was the antecedent of modern power generators. The next step in the evolution of our understanding of electricity was to prove Faraday's theories mathematically. This fell to James Clerk Maxwell, a Scottish physicist who built upon his predecessor's work to complete the classical theory of electromagnetism.[25]

The first important application of electricity was for the telegraph, which had been so important in creating a truly continental economy and system of governance in America. For two millennia, humanity had been seeking ways to communicate more effectively over long distances. The Greeks developed a relatively sophisticated system of fire signals that could shrink geography, which remained the primary form of sharing information beyond the horizon for more than a thousand years. Flag-signaling systems developed in the mid-seventeenth century were very effective over shorter distances. These systems evolved into more complex semaphore telegraph networks, such as the one developed by Claude Chappe and his brothers in late-eighteenth-century France. Within fifty years their network would comprise more than five hundred stations, and messages could be sent across the entire country in about forty minutes (although the effectiveness was limited during poor weather or at night). The Chappe approach spread to other European nations during the early part of the nineteenth century, although it found limited use on this side of the Atlantic.[26]

Starting in the 1750s and continuing into the early nineteenth century, various investigators explored different types of electric telegraphs. These included electrostatic, battery-powered, and electrochemical telegraphs, none of which proved to have practical applications. Early experiments with an electromagnetic telegraph seemed to fare little better, since the distance that signals could travel was limited. It was at that point that an art professor and painter from New England came forward to solve the problem. His name was Samuel Morse. His inspiration struck while he was sailing home from Europe in 1832, and he spent several years developing his tabletop system. It included wires, batteries, a sender that relied on pulses of current (for which he developed Morse code), and a printing receiver. Perhaps the most important development, however, was the introduction of circuits at regular intervals along the wire to boost the signal and allow it to be transmitted over longer distances, an idea provided by Morse's New York University colleague Leonard Gale. In the early 1840s, Morse received research funding from the national government to develop a forty-mile experimental line from Washington to Baltimore. This first line was so successful that Morse and his partners were able to assemble the private capital necessary to build a larger network in the Northeast. The invention was so profitable that it soon spread across all of North America, and ultimately South America and western Europe as well. Using nothing more than little pulses of electricity, it altered the course of societal evolution.[27]

By the 1880s, inventors were transitioning from the telegraph to electric lighting, which signaled the beginning of the Electrification Revolution. While this revolution began with efforts to brighten the night sky with arc lights, it was the invention of incandescent light bulbs for indoor spaces that had the greatest impact, leading to the replacement of candles and lamps with a far more consistent form of illumination. A number of competitors sought a commercially practical version of this new technology, but it was Thomas Edison who would take most of the credit. Using profits from his quadruplex telegraph, Edison had built history's first industrial research lab at Menlo Park in the New Jersey countryside. In 1877, working there, he first became publicly known as the inventor of the phonograph. The following year the problem of the electric lamp came to his attention, and he quickly filed a patent for an incandescent light bulb using "a carbon filament or strip coiled and connected to platina contact wires." But his invention burned out quickly, and platinum wiring was expensive. His breakthrough came when his team tested a carbonized cotton thread inside a vacuum-sealed bulb. Although the cotton was ultimately replaced by bamboo, this new approach was far cheaper, and the bulb burned for more than a thousand hours. Edison quickly determined, however, that the key invention was not the bulb but the system to deliver electrons to the lights.[28]

To demonstrate this new technology to potential investors, Edison and his associates created the first central power station on the grounds of the research laboratory. "It included three Edison dynamos, each rated at 6 kilowatts and powered by an 80-horsepower steam engine. Overhead lines carried current to the twenty-five lamps that lit the laboratory." Commercializing this system with a profit-generating business model would prove to be the true challenge. On 17 December 1880 the Wizard of Menlo Park formed the Edison Electric Illuminating Company. Two years later the company established the first investor-owned utility with the Pearl Street Station in New York City, serving nearly a hundred customers with approximately four hundred lamps. By 1884 the number of customers had increased to more than five hundred (with ten thousand lamps). The business continued to grow rapidly thereafter. "The Pearl Street Station launched not only a new industry but a new era in American life," writes Klein, "one that would increasingly be dominated by electricity." And of course that electricity was all produced using coal-fired steam engines.[29]

* * *

A Hungarian immigrant to America was responsible for the next great leap forward in the Electrification Revolution. Nikola Tesla arrived in New York City in 1884, two years after the inauguration of the Pearl Street Station. With a recommendation in hand from a former employee of Thomas Edison, Tesla was offered a job by the great man. The two quickly fell out, however, both because their personalities clashed and because they had different ideas about the proper type of current to build a future electricity system. Edison was the leading proponent of direct current, which flows in only one direction and without substantial variation in magnitude. Tesla favored alternating current, which periodically reverses direction (dozens of times per second) and whose magnitude can be increased or decreased with a transformer, allowing it to travel much greater distances from the generator. While still living in Budapest, Tesla had envisioned an alternating-current motor based on a rotating magnetic field. He tried to convince Edison that alternating current would be much more efficient and could be conducted further, but the older man wouldn't budge. Feeling underappreciated, Tesla quit his job with Edison and obtained the financial backing necessary to found the Tesla Electric Company. In 1888 he delivered a paper before the American Institute of Electrical Engineers that would change the course of the electricity industry. He described his rotary motor and showed those in the audience how a working model, for which he had already obtained a number of patents, operated. The audience was bowled over. An immediate battle to gain control of the rights to Tesla's system was ultimately won by George Westinghouse.[30]

When he was still in his early twenties, Westinghouse entered the American consciousness with his invention of an air brake that used compressed air to rapidly stop trains. It was in the electricity field, however, where he would have his largest influence. "It is to be the everlasting glory of George Westinghouse," predicted Henry Prout, "that he saw the meaning of [the differences between direct and alternating current] before it was seen in a big way by any other man who combined in himself the qualities and the capacities to make use of them." In 1885, Westinghouse had read about the European development of the electric transformer, which made it possible to step up power magnitude for long-distance transmission and step down power magnitude for delivery to household light bulbs. Few understood the import of this new system at the time, but Westinghouse instantly grasped that this was the key to the efficient distribution of electricity. Generators wouldn't need to be located in urban centers but could be located further away and could take advantage of potential energy at remote waterfalls. He quickly obtained the patent rights for the use of transformers in America.[31]

Westinghouse quietly began the industrial insurgency that would see alternating current become the dominant form of electricity transmission. Edison was furious that anyone would dare challenge him. Regardless, Westinghouse slowly built up his system during the late 1880s and steadily gained ground because his approach was much more economical. Edison saw an opportunity to damage Westinghouse when he was approached to determine whether electricity could be used to humanely execute criminals in death-penalty cases. The Wizard suggested that only his competitor's machine was deadly enough to accomplish the goal, suggesting to the public that direct current was the only approach that would ensure public safety in electricity transmission. He followed this more subtle strike with direct attacks, personally impugning Westinghouse for outrageously invading a technical field that should rightfully belong to him.[32]

By the early 1890s, growing demand and national economic crises were shaping the rivalry between the two approaches. As the War of Currents was evolving, electricity gained yet another customer, electric street railways. The pioneer in this sector was Frank Sprague, a retired naval officer who constructed the first successful electric railway system in Richmond, Virginia. Sprague proved that electric streetcars could traverse hilly terrain, which meant that they could compete with cable cars. Within two years the nation had more than two hundred electric railways, and a decade thereafter they dominated urban transportation. Westinghouse was the outright leader because his No. 3 motor, designed by Benjamin Carver Lamme, provided the best railway traction. Very quickly, "in most major cities the [railway] companies became far and away the largest users of electric power."[33]

The continued Long Depression, which had worsened at the beginning of the new decade, further complicated the War of Currents by making it more difficult for Westinghouse and Edison to obtain much-needed capital. Westinghouse had overextended his business, which forced him to undertake an ultimately successful reorganization of the Westinghouse Electric and Manufacturing Company. Meanwhile the combination of financial difficulties and the growing popularity of alternating-current systems finally forced Edison to begin shifting away from direct current. In addition, Edison had decided to consolidate the Edison General Electric Company with Thomson-Houston, another major direct-current company. "Incorporated on April 15, 1892, the new General Electric Company boasted a capitalization of $50 million (five times that of Westinghouse)." The War of Currents was over, but the larger battle to dominate the Electrification Revolution was still undecided.[34]

* * *

As Westinghouse had foreseen, it was hydroelectricity that paved the way for the ascendancy of central power. Edison's direct-current system had been based on having many centrally located generators, which would limit the distance electricity had to travel to consumers. Tesla and Westinghouse believed that there was no need for this distributed system, particularly if everyone adopted alternating current. They decided to test their theory first on the Niagara River in upstate New York. The famous falls had long been considered a prime location for water power. The water flowing out of Lake Erie and into Lake Ontario cascades down more than 330 feet, with more than 90 percent of this drop occurring in a stretch of just one mile above and below Niagara Falls, which itself has a 165-foot drop. It was estimated at the time that the power potential of the cataract was enough to run the entire industrial sector of the nation. One benefit of the river's location was that it was close to already thriving manufacturing cities, so it had a ready market nearby. In the mid-1880s, Thomas Evershed conducted initial work on a design to take advantage of the river's immense power. Rather than a central power station, he wanted to construct a 2.5-mile tunnel with more than two hundred waterwheels fixed to the walls to produce an estimated 150 megawatts of power for associated mills. With a projected cost of $10 million, "the project was not only huge but unprecedented." In 1889 the Niagara Falls Power Company and Cataract Construction were created to implement the plan. The key actor in this endeavor was a New York banker named Edward Dean Adams, who became involved in the project at the behest of the influential financier J. P. Morgan.[35]

Adams was a descendent of the family that had produced two American presidents. He had a solid academic background and had overseen several very successful railroad reorganizations. Adams discarded Evershed's plan to use waterwheels and build mills in favor of a central power station. After a bitter contest between General Electric and Westinghouse to build the generators for the facility, Adams finally accepted the latter's bid because it utilized alternating current. Completed in 1895, the station was a huge success. As Klein writes, "To the astonishment of almost everyone, the demand for power from the facility grew so rapidly that the company had to add an average of two new . . . turbines every year until the station housed ten units by 1900. . . . By 1904 the long-cherished dream of harnessing the falls for power had become a spectacular reality as the last of these units went online." The availability of cheap electricity resulted in rapid industrial growth in the region, while also delivering commercial power to Buffalo. Far more importantly, the Niagara hydroelectric project proved the viability of large central power plants. Another monumental development occurred the year the Niagara Power Station No. 1 opened, when General Electric and Westinghouse adopted a landmark cross-licensing agreement, which meant that the two companies would stop fighting over technology and focus on market competition. Combined with the opening of Niagara, the deal started the ball rolling for the next phase in the Electrification Revolution, the arrival of central power. But this would happen during another stage of the Second Political Age.[36]

Second Political Age
Second Progressive Cycle, 1897–1921

The minor realignment election of 1896 led directly to the establishment of an American Empire. Three decades into the Hydrocarbon Age, rapid economic growth had provided ample resources for the country to pursue a more aggressive national security policy (fig. 2.2). This was a significant change in the geopolitical orientation of the nation. After following Washington's policy of isolationism for more than a century, the nation began to flex its muscles. This change had been in the wind for more than a decade. In 1885 the Congregationalist minister Josiah Strong wrote in his book titled *Our Country* that the United States had a duty to spread the American values of civil liberty and spiritual Christianity throughout the world. This popular book endorsed shifting from isolationism to imperialism to accomplish this objective. Fortunately, the ideas Strong preached—territorial expansion, colonization, and cultural assimilation—were never fully adopted by a country that was correctly more interested in enlarging its economic interests. A half decade later a far more sensible and ultimately influential voice joined the debate, making a very different case for American expansion. Rear Admiral Alfred Thayer Mahan was arguably the most important geopolitical theorist of modern times, particularly because he grasped the military and economic importance sea power would have during the twentieth century. In 1890 he wrote in a persuasive article titled "The United States Looking Outward" about America's failure to compete for emerging global markets: "Our self-imposed isolation in the matter of markets, and the decline of our shipping interest in the last thirty years, have coincided singularly with an actual remoteness of this continent from the life of the rest of the world. . . . Whether they will or no, Americans must now begin to look outward. The growing production of the country demands it. An increasing volume of public sentiment demands it." This essay helped spark the emergence of the United States as a major actor during a critical phase of globalization, a move that was so successful that by the outbreak of the First World War the American economy was twice the size of Great Britain's.[37]

William McKinley's ascendance to the presidency in 1897 marked the transition from the American Republic to the American Empire. This was the re-

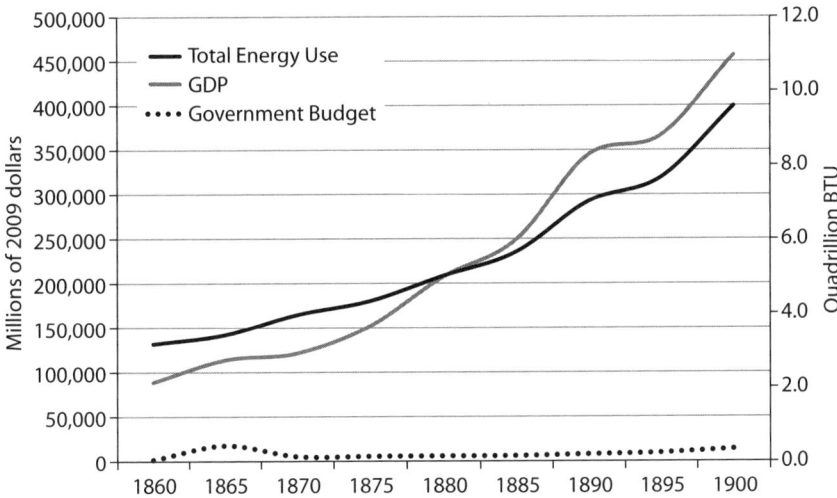

Figure 2.2. US Total Energy Use, GDP, and Government Budget, 1860–1900. US Energy Information Administration, *Annual Energy Review, 2009*; *Budget of the US Government, Historical Tables, 2017*; Louis Johnston and Samuel H. Williamson, "What Was the U.S. GDP Then?," 2017.

sult not solely of the nation's economic growth but also of the political realignment that paved the way for a marked amplification in the nation's geopolitical engagement in world affairs. As Paul Kennedy writes, it was during this period that America became one of the world's foremost nations—along with Britain, Germany, and Russia. Although it was a major actor working largely outside the great-power system, it had tremendous advantages that established it as a sleeping giant:

- Vast raw materials (including coal and oil)
- Lack of geographical constraints
- Absence of foreign dangers and no fixed alliances
- Rich agricultural lands
- Rapid industrialization
- Inward flow of foreign direct investment
- Sizable domestic markets, reducing dependence on trade
- Large population with a high per capita income level
- Modern infrastructure (including railroads and telegraph)

The nation's only significant weaknesses were its relatively immature financial and political institutions. Shortly after McKinley took office, the American people began to react passionately to Spanish atrocities in Cuba. At the urging of Congress, he took the country to war just weeks after the mysterious sinking of the USS *Maine* in Havana Harbor. The war lasted only three months, but in that time America removed the Spanish from power in both Cuba and the Philippines. The US Navy proved decisive. Commodore George Dewey crushed a Spanish squadron in Manila Bay, the crucial step in removing the Spanish from power after more than three hundred years of colonial rule. Another American naval force, under Captain Henry Glass, captured the island of Guam without any resistance. Finally, squadrons under Rear Admiral William T. Sampson and Commodore Winfield Scott Schley destroyed the Spanish navy's Caribbean Squadron. In the meantime, land forces were able to force the evacuation of Santiago, which ultimately led the Spanish to sue for peace.[38]

Although the United States would follow a somewhat different path from those of past empires, the Spanish-American War still signaled a fundamental change in its approach to national security. According to Arthur Schlesinger, "The Spanish-American War [projected] the United States irrevocably into the world of great powers." With new territorial interests in the Caribbean and the Far East, the nation was moving rapidly in the direction that Mahan had been pointing. Two years after the war, McKinley sent five thousand American troops to China to intervene in the Boxer Rebellion and instituted an "Open Door" trade policy in Asia. As this change in America's geopolitical posture was taking hold, a momentous shift was also occurring in domestic politics as power began to swing toward the executive branch. As Schlesinger wrote, when "Congress . . . forced a cautious president into war with Spain [it] inadvertently conspired against its own authority." McKinley's direct supervision of the war marked a departure in presidential control of foreign policy that would soon become institutionalized. The combination of a new geopolitical orientation and concentration of power in the executive signaled a transition to empire.[39]

*   *   *

During the Gilded Age, hydrocarbon production and consumption in the United States grew enormously. Total energy consumption increased by approximately 300 percent, with nearly all of that growth accounted for by a massive escalation in coal use. As had been the case during the First Political Age, this growth in energy utilization was accompanied by tremendous economic growth. Between 1860 and 1896, as can be seen in figure 2.2, America's GDP grew approximately fourfold, with per capita GDP growing by nearly two-thirds. As a result, political tensions were relatively muted during this period despite rising income and wealth inequality and the evolution of corporate power structures. This all began to change in the new progressive cycle. Increased energy security had undergirded major economic development, but over time it had also shaped pathways for a new political movement to demand changes. While the electorate had been worried about these disturbing trends for quite some time, in 1896 they had finally elected representatives willing to take on the embedded wealthy class.

In 1895, Lester Ward had written that the key problem the country faced was not too much government but too little—an idea that flew in the face of the reigning free-market ideology. Writing about the rise of an American plutocracy, Ward suggested that it was human nature to strive for wealth acquisition and that when restricted to useful purposes this appetite would continue to provide significant benefits to society. In this he differed from his contemporary Thorstein Veblen, who argued that "conspicuous consumption" undercut productive activity. Since in Ward's view it was impossible to suppress acquisitive passions, he believed it was better to focus on controlling its effects. What concerned him was that because wealth had become the key measure of individual worth, it had become easier to forsake the poor, who were seen as lacking merit. He argued that this was where government should step forward. "Government was instituted to protect the weak from the strong in this universal struggle to possess," he wrote, "or, what is the same thing, to protect society at large. . . . From this point of view, then, modern society is suffering from . . . undergovernment, from the failure of government to keep pace with the [changes] which civilization has wrought [and to] perform its primary and original function of protecting society." Ward pointed in particular to the lack of government oversight of companies enjoying monopoly power regarding the production of fossil fuels. He worried that because hydrocarbons had become so fundamental to economic growth, these corporate interests could hold society hostage. Interestingly, he didn't mention the fact that these same energy sources were not only the foundation of America's economic advance but also provided the fiscal resources government needed to protect society. Regardless, his popular and influential call for government to deliver microeconomic security to the weak roared onto the national agenda during the second half of the Second Political Age.[40]

This reformist program didn't fully begin to take shape until Theodore Roosevelt was thrust into the Oval Office upon President McKinley's assassination in September 1901. At forty-two, Roosevelt was the youngest president in American history—although he had already served as civil service commissioner, New York Police commissioner, assistant secretary of the Navy, governor of New York, and vice president of the United States. The new president was extremely popular with the public, who saw him as a bona fide war hero for leading his volunteer unit of Rough Riders during the Battle of San Juan Hill in Cuba. Roosevelt had long been a committed reformer, but now he had the power to turn his vision into reality. On 3 December 1901 he made his progressivism clear in an impassioned speech denouncing the trusts. He opened by contending that there was a widespread feeling that the big corporations were damaging the general welfare. While he didn't believe monopolies should be arbitrarily eliminated, he did believe that government had a responsibility to supervise and regulate their operations. "Great corporations exist only because they are created and safeguarded by our institutions," he said, "and it is therefore our right and our duty to see that they work in harmony with these institutions." He argued that the key challenges for contemporary leaders were to improve social conditions and address labor problems arising in the new industrial economy. "The chief factor in the success of each man—wage-worker, farmer, and capitalist alike—must ever be the sum total of his own individual qualities and abilities. Second only to this comes the power of acting in combination or association with others. . . . To be permanently effective, aid must always take the form of helping a man to help himself; and we can all best help ourselves by joining together in the work that is of common interest to all."[41]

The following year, these sentiments formed the basis of Roosevelt's Square Deal, a collection of policies aimed at controlling corporations, protecting consumers, and conserving the natural environment. As Doris Kearns Goodwin writes, "Under Roosevelt's Square Deal, the country awakened to the need for government action to allay problems caused by industrialization." In working to implement this new vision, Roosevelt worked with progressive congressional leaders to expand the government's role in providing macroeconomic security, while adding microeconomic security to its portfolio. And the results were considerable. He broke up corporate trusts. He signed legislation to regulate railroads, food producers, and drug manufacturers. He recognized the rights of labor. He adopted a comprehensive conservation program. And he supplanted the spoils system with an expansion of merit-based civil service reforms. Roosevelt's hand-picked successor, William Howard Taft, would build upon these actions by breaking up more trusts, regulating railroad rate increases (which had unintended negative consequences), conserving additional natural lands, and making it possible for the poor to safely deposit money via the postal service.[42]

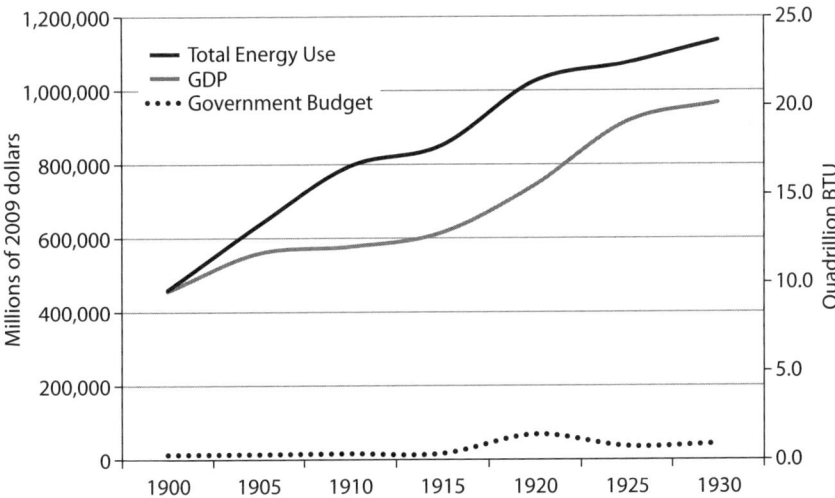

Figure 2.3. US Total Energy Use, GDP, and Government Budget, 1900–1930. US Energy Information Administration, *Annual Energy Review, 2009; Budget of the US Government, Historical Tables, 2017;* Louis Johnston and Samuel H. Williamson, "What Was the U.S. GDP Then?," 2017.

After he left office, Roosevelt delivered a speech in Osawatomie, Kansas, calling for a new nationalism that would see more muscular government stepping forward to solve the country's problems. Roosevelt built upon ideas developed by Herbert Croly, who had made the highly persuasive argument that the only way to achieve the Jeffersonian ideal of democracy and equality was to leverage Hamiltonian big government. "Really sincere followers of Jefferson are obliged to admit the superior political wisdom of Hamilton's principle of national responsibility. . . . The national public interest has to be affirmed by positive and aggressive action." In particular, Croly advocated for a national minimum wage (which Monsignor John Ryan had called a "living wage") that would increase the standard of living for every American. Theodore Roosevelt opened his address with the suggestion that the key aim of Progressivism was to further human betterment through enlightened government regulations and programs. "We work in a spirit of broad and far-reaching nationalism when we work for what concerns our people as a whole. We are all Americans. Our common interests are as broad as the continent. . . . The national government belongs to the whole American people, and where the whole American people are interested, that interest can be guarded effectively only by the national government. The betterment which we seek must be accomplished, I believe, mainly through the national government."[43]

In 1912, after a falling out with Taft, Roosevelt would run for the presidency again on this New Nationalist platform. His decision to run under the banner of the new Progressive Party effectively split the Republican vote and helped elect another progressive, New Jersey's Democratic governor, Woodrow Wilson. While he shied away from the Rooseveltian belief in big government in favor of the Jeffersonian idea that the best government consisted in as little governing as possible, Wilson argued, "Freedom today is something more than being let alone. The program of government of freedom must in these days be positive, not negative merely. . . . And the day is at hand when it shall be realized on this consecrated soil—a New Freedom—a Liberty widened and deepened to match the broadened life of man in modern America." Wilson drew upon this philosophy to kick off another period of rapid legislative progress that augmented what Roosevelt had accomplished. As figure 2.3 shows, with total energy use and national income continuing to rise quickly, Wilson had the fiscal resources to follow this path. Although the national government budget remained relatively small as a percentage of GDP, in real terms it had grown considerably during the Progressive Era. During his first term, Wilson introduced a prime-ministerial model of legislative leadership. That model, which he believed was necessary to successfully govern the ever-expanding nation, helped him compile one of the most impressive legislative records in American history. His New Freedom agenda focused on policies that would advance macroeconomic security. He gained passage of the Underwood Tariff Act (reinstating a federal income tax and lowering the basic tariff rate), the Federal Reserve Act (re-creating a national banking and currency system), and the Clayton Anti-Trust Act and the Federal Trade Commission Act (both of which expanded government powers to prevent unfair business competition). Wilson also worked in the microeconomic security area to increase labor's role in corporate decision making, which reinforced his ideas about strengthening democratic forces in American life. While the outbreak of the First World War curtailed steps forward in this area, the stage was set for future government action.[44]

* * *

Although significant progress was made in domestic politics under the Roosevelt and Wilson administrations, both presidents were also working to build upon McKinley's outward-looking foreign policy. During his tenure, Roosevelt engineered three major advances in American power. First, he worked behind the scenes to gain access to the isthmus at the center of Panama to build a canal that would link the Atlantic and Pacific Oceans, which he argued was the most significant foreign policy accomplishment of his presidency. This was certainly correct, as the new canal transformed international commerce and greatly benefitted the American economy. Second, after a series of debt crises in Latin America, Roosevelt modified the Monroe Doctrine. His Roosevelt Corollary stated that the United States would reserve the right to intervene in the affairs of any nation in the Western Hemisphere that failed to maintain economic and political order. Roosevelt did not hesitate to apply this new policy on a number of occasions. Finally, he negotiated a peaceful settlement to the Russo-Japanese War in 1905. This unprecedented presidential diplomacy marked a new role for America among the great powers and earned Roosevelt the Nobel Peace Prize.[45]

Woodrow Wilson's foreign policy began less auspiciously, as he fumbled an attempt to intervene in Mexico after the murder of a newly elected reformist president. He found his footing after the outbreak of the First World War in Europe. He successfully kept the Americans out of the conflict for nearly three years, but after the Germans resorted to unrestricted sinking of American merchant and passenger vessels, he led the nation to war in 1917. Although he had wanted to avoid war, once it began, he embraced the necessity for total war. Congress delegated the president a great deal of authority to wage the war, and what it didn't provide, Wilson simply took. "The president became responsible for organizing and controlling the industrial economy and for coordinating the transportation and communication industries so they could meet the requirements of the military commitment." American forces ultimately helped turn the tide, leading to a cease-fire the following year. While Wilson was less involved in day-to-day military strategy than McKinley (because he delegated more to competent civilian and military leaders), he was far more engaged in mobilizing the nation at home. After the cease-fire, Wilson spent more than half a year in Europe (more time abroad than any president before or since) trying to negotiate a lasting peace by creating the League of Nations. While he succeeded diplomatically abroad, his deteriorating health contributed to the League's failure to gain acceptance within the US Senate. Nevertheless, he still won a Nobel Peace Prize.[46]

By the end of the progressive cycle, the nation's leaders had taken critical steps to redefine the government's national security role while cementing its responsibility in the macroeconomic security arena. America took on new roles in geopolitics, although it remained hesitant to broadly engage the other great powers. In the macroeconomic sphere, key institutions had either been reinstated or newly created, most importantly the Federal Reserve System, finally reestablishing a central bank for the country. Together with the Lincolnian platform, it resulted in a country that was temporarily far more economically stable. Finally, the United States had begun to take small steps toward providing microeconomic security, particularly with regard to recognition of and protection for labor unions. While relatively limited advances were made, the debate had been joined, and microeconomic security began to provide a key distinction between the major political parties. Thus, by 1921 the nation had adopted measures aimed at maintaining its founding principles while addressing the challenges of the Second Industrial Revolution. Underpinning all of these political developments were new advances in the ongoing INNATE revolutions. The extraordinary intensification in hydrocarbon use during these years continually changed the meaning of modern life in America.

INNATE Revolutions: Electrification
Arrival of Central Power and Electrical Distribution

As America entered the second half of the Second Political Age, coal was providing an ever-increasing share of the nation's energy (fig. 2.4). Although coal-burning stoves were reluctantly adopted at first, because of their heating efficiency they had made their way into homes across the country for both cooking and heating. These cast-iron behemoths had transformed daily life, eliminating the need to constantly chop wood and allowing users to cook standing up rather than crouching over an open flame. Still, they were not without their challenges. It took an hour a day to properly maintain a coal stove. Cooking without being able to easily control the heat production was difficult. Plus, there were concerns about interior air quality. Coal had also come to be used for illuminating the interior of houses. This was accomplished by baking coal at a large facility to produce coal gas, which could then be piped to homes for use in incandescent mantles. Coal gas competed with Standard Oil's kerosene, the former largely triumphing in the growing cities as the latter was relegated to rural areas. As Barbara Freese writes, one of the most important outcomes of this victory was that "people got used to the idea of obtaining their fuel 'from the outside' . . . instead of from their own privately stored fuel stocks, and local governments began to see energy distribution as a valid municipal function."[47] This set the stage for central power's second act, building upon the success of Niagara Station. But this time coal would provide the key energy input. It would launch the Electrification Revolution, which would change the very nature of American progress.

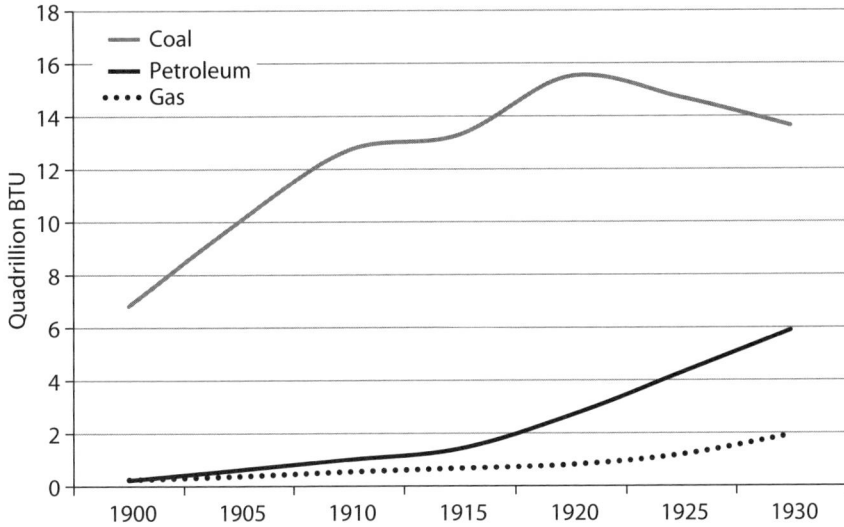

Figure 2.4. US Energy Consumption by Source, 1900–1930. US Energy Information Administration, *Annual Energy Review, 2009.*

The key technology needed to create massive centralized power stations burning coal was the steam turbine, which differed in essential ways from the existing steam engine. The steam engine relied on pressure; the steam turbine depended on kinetic energy. It conducted the heat energy in steam through nozzles to drive a rotating turbine to produce electric power. While the concept dated back to Hero of Alexandria, the great experimenter of antiquity who lived and worked during the first century CE, it wasn't until 1884 that Sir Charles Parsons filed a patent for a practical device. Although his initial machine was relatively weak, he spent the next quarter century working to produce a high-speed motor that would generate large quantities of electricity. His efforts paid off handsomely, as his turbo generators increased in capacity from 4k W in 1885 to 5,000 kW in 1910.[48]

The steam turbine found its first application on large ships, ultimately being designed into famous vessels like the *Mauretania*, the *Lusitania*, and HMS *Hood*. It was in electricity production, however, that the steam turbine really made its mark. In the early decades of the twentieth century it would slowly replace existing reciprocating steam engines because it was 80 percent smaller, ten to twenty times more powerful (initially achieving speeds of 2,000 rpm compared with the rather pedestrian 100 rpm of the older technology), and less expensive to operate, and it needed limited repairs. The primary reason it was so inexpensive to run was that its efficiency was continually increasing. According to Klein, "In 1902 it consumed 6.7 pounds of coal to produce one kilowatt-hour of electricity. By 1928 the new turbogenerators used only 1.76 pounds. . . . The larger the unit, the more cost-effective it became, and it seemed to have no upper limit as to size. Burning much less coal not only saved money but cut the amount of smoke pollution in the city."[49]

As the steam turbine was changing the reality of electricity production, population growth and urbanization were expanding the market for this commodity. Between 1880 and 1920, approximately twenty million Europeans crossed the Atlantic Ocean to seek new economic opportunities. America's cities were rapidly expanding to take in these new arrivals, many of whom were taking jobs in urban factories. These metropolises found themselves competing to attract workers and consumers primarily through policies that attracted or retained major employers, but there was also an evolving effort to improve living conditions—moving beyond the merely utilitarian focus that had prevailed in previous decades. These endeavors included promoting better housing and creating needed civic spaces like parks and libraries and government buildings. More importantly, political leaders sought to improve urban infrastructure in an attempt to remove the stigma that had been associated with cities since the early years of the republic, when, for example, Thomas Jefferson wrote: "I view great cities as pestilential to the morals, the health and the liberties of man. True, they nourish some of the elegant arts, but the useful ones can thrive elsewhere, and less perfection in the others, with more health, virtue and freedom, would be my choice." With government-funded sewers, garbage collection, water distribution, concrete roads, sidewalks, and public transit, cities became much more pleasant places to live. At the same time, they began to go vertical, with skyscrapers beginning to rise up and financial and service businesses rushing to fill these new spaces.[50]

As large coal-fired generating stations rapidly spread across the urban landscape, using electricity became a way of life in the growing cities. By the early twentieth century, more and more metropolitan areas were constructing local electric grids, which served both commercial and residential customers. As more houses gained access to electricity, there was a growing market for new consumer products. "By revolutionizing production and manufacturing," writes Klein, "electricity made possible the rise of the consumer economy that was to dominate the twentieth century and transform every corner of American life." For the first time, electric machines began to enter American offices and homes. Refrigerators, irons, vacuum cleaners, coffeemakers, washing machines, toasters, and many more appliances were available. Better lighting boosted productivity in high-rise office buildings, and in the coming decades new technologies would revolutionize the business world.[51]

* * *

As demand for electricity skyrocketed in the late nineteenth and early twenti-eth centuries, the challenge became how to best provide efficient distribution. This was when one of the largely forgotten founders of our modern world stepped to the fore—and if he is remembered, it isn't fondly. His name was Samuel Insull. In 1881 the twenty-one-year-old Englishman arrived in Amer-ica to become private secretary to Thomas Edison. Within a few decades he became the "premier financier-entrepreneur of his time, making Edison's basic idea of electricity generated by central stations work in thousands of cities, thus making the modern metropolis possible." A famously hard worker who could keep pace with his benefactor, Insull quickly became secretary of most of the great man's companies and essentially took over the financial reins.[52]

After the Pearl Street demonstration, Insull was charged with managing the production of power-station equipment. He consolidated industrial efforts in Schenectady, where he "created a small revolution in the mass production of parts (long before Henry Ford was to be credited with the concept) that al-lowed the volume of shipped products to go way up and the price per part to go way down." It was during this period that he came into contact with the journalist and railroad magnate Henry Villard, from whom he learned a great deal about the world of finance as the older man worked to merge the Edison Lamp Company and the Edison Machine Works. Three years later, in 1892, J. P. Morgan stage-managed the creation of General Electric. Rather than stay-ing and becoming the third ranking executive in the new corporation, Insull took a 50 percent pay cut to become the chief executive at the utility company Chicago Edison. What seemed like a backward step would put him in a posi-tion to revolutionize the entire electricity industry.[53]

When Insull arrived in the Windy City, only a few thousand of its popula-tion of one million had electricity. Prices were so high that most Chicagoans could not afford this extravagance. Although his first step to solve this prob-lem had been to build the largest central power station in the world, he soon realized that it was time to switch generating technologies. The utility was still using the reciprocating steam engine, but Insull decided it was time to shift to steam turbines. In so doing, he pioneered the use of this new technology for large-scale power production. He cajoled General Electric into building him a five-megawatt generator, which produced two times more electricity than any steam engine ever had. This amount of power was staggering, equal to the labor of one hundred thousand workers. And it was cheaper than what his competitors were providing.[54]

Insull realized, however, that steam turbines alone were not enough. Electricity was still too expensive. The main reason was the low load factor at power stations, which meant that they only operated at peak capacity for a few hours a day. Thus the huge capital investment in these generators wasn't being used efficiently. Insull set about trying to solve this problem, which was crucial to lowering prices and ensuring that electricity wouldn't remain a luxury good with a relatively fixed market. As Peter Fox-Penner writes, Insull developed a four-pillar approach that changed the entire economics of electricity distribution. First, he determined that power would be cheaper if its use were aggregated into the largest possible web of interconnections—what he called massing consumption and we now call the grid. Because apartment dwellers only used electricity at night, there was a need to attract new customers in off-peak times. The key was variety. Insull started with electric trolleys, which provided a significant surge in demand during the morning and evening rush hours. Factories were using larger amounts of electric-powered equipment, which added a key customer in the middle of the day. Large office buildings would soon obtain lighting. The best restaurants and hotels were electrified. In the coming decades the customer base would continue to grow.[55]

Second, Insull became one of the most important proponents for utility monopoly power. He believed the infant industry had to seek benefits from economies of scale in production. This meant avoiding laying multiple sets of expensive power lines or allowing power stations to sit idle. This wasn't possible if there was competition. While many see things differently today, utilizing monopoly power was crucial at the time. Third, understanding the political difficulties associated with running a monopoly much better than most utility operators, he argued that the sector needed the protection of government regulation if it was to obtain stability. Finally, "Insull and the industry's finest marketing force sang 'the gospel of consumption,' urging customers to buy ever more power and giving them discounts when they did." This was made possible by a new technology, the demand meter, which allowed the utility to track how much energy a customer was using. Edison had simply sold electricity as a service, but Insull realized that it was critical to charge for the amount used. The goal was to incentivize more consumption, thus smoothing out demand during nonpeak times and allowing for the most efficient use of large power stations. As Phillip Shewe writes, by the second decade of the twentieth century this strategy had made Chicago the most electrified city in the world, with more than two hundred thousand customers. As a result, it also had the cheapest rates. It was a model for the rest of the country.[56]

Fatefully, Insull soon set out to conquer the rapidly growing suburbs. This meant wiring thousands upon thousands of single-family homes. These residences had twice as many power outlets, and their generally wealthier residents used far more electricity. Then Commonwealth Edison, as the company had been renamed, began to expand beyond Chicago. By 1929 Insull operated utilities and suburban and transurban distribution grids in thirty-two states, controlling one-eighth of the nation's power. After the Great Crash, however, he became a scapegoat for the entire economic collapse. Franklin Roosevelt repeatedly attacked him as a greedy monopolist who wasn't interested in the public good. After his empire collapsed under the weight of the economic downturn, Insull was tried for mail fraud and embezzlement. His indictment and trial engrossed the nation, but his rags-to-riches story won over the jury, and he was acquitted of all charges. As Forrest McDonald writes, "For his fifty-three years of labor to make electric power universally cheap and abundant Insull had his reward from a grateful people: he was allowed to die outside prison." This was a tragic end for a person who had such an outsized role in expanding modernity in America. Fox-Penner summarizes his impact as follows: "Insull, perhaps more than any other single person, changed American life. Over the span of the next four decades nearly every urban home and shop got electric power and lights." This coal-fired Electrification Revolution increased productivity across the economy; more importantly, it transformed day-to-day existence. While these changes were happening, developments in the manufacturing sector were also transforming the nation.[57]

INNATE Revolutions: Industrial
Rise of Mass Production

The triumph of assembly lines and mass production is synonymous with one name, Henry Ford. During the early twentieth century, as the automobile was reshaping America, he was just one of many inventor-entrepreneurs who were attempting to capture this new consumer market. Ford believed that it was vital not only to produce relatively inexpensive automobiles but also to pay his workers enough that they would be able to purchase one of his cars, thus helping to create a market for the product. Fordism became a weighty industrial ideology during this era. To realize the first part of his vision, Ford adopted many of the advances associated with contemporary New England armory practice, such as mechanization, division of labor, and interchangeability. More importantly, however, he and his senior engineers introduced significant innovations that profoundly changed the industrial world.[58]

During the second decade of the twentieth century, the Fordist system was developed at the company's Model T factory in Highland Park, Michigan. The first step was laying out the plant machinery in sequential order to improve efficiency. Because speed and accuracy were central principles, the engineers who actually used these specialized machines designed them on-site. Allowing experienced managers to rearrange the machines to find additional efficiencies was a main contributor to the success of the Ford approach. As David Hounshell writes, "The principle of speed was apparent . . . everywhere . . . in the Ford plant." By 1912 these advances had helped the company quintuple production, from 13,840 to 82,388 units annually, in just four years. A new car was being produced every forty seconds.[59]

In 1913, Ford's second major advance—the assembly line— began working its way into the factory. Initially, gangs of workers had moved around the plant conducting specialized tasks. While this had clearly been highly effective compared with earlier efforts, there were still problems associated with workers' completing their tasks at different speeds. According to Hounshell, "In 'moving the work to the men,' the fundamental tenet of the assembly line, the Ford engineers found a method to speed up the slow men and slow down the fast men." At first the assembly line was used in only one part of the factory, but it was so successful that the company began rapid experimentation to integrate the system across the plant. Within a year there were three full lines in operation for chassis assembly, being fed by existing subassembly lines. The company added conveyor systems and gravity slides to speed production even more. The overall influences were even more dramatic than the systematic organization of machines, with production increasing sevenfold, to 585,388 units annually, in four years. Even more astounding, the price per unit fell from $950 to $360 in that time frame. In Hounshell's words, "Ford had given the world the first system, in the fullest sense of the expression, of mass production: single-purpose manufacture combined with the smooth flow of materials; the assembly line; large-volume production; high wages initiated by the five-dollar day; and low prices."[60]

One of the aspects of mass production that is often glossed over is the huge power requirements needed for these factories. For example, after Ford opened a second factory the company also inaugurated a thirty-megawatt power plant and the world's largest foundry. The electrification of manufacturing in America represented an important shift, as the Electrification Revolution provided the Second Industrial Revolution with a superlative form of energy. Electricity's "easily adjustable flow allows for previously unsustainable precision, speed, and process control." By eliminating the bulky shafts that transferred steam power to hundreds of machines and replacing them with less obtrusive wiring, it also allowed for the changes in factory configuration that defined mass production. The adoption of this new technology was speedy, growing from 5 percent of all installed mechanical power in 1899 to 80 percent in 1929. During the twentieth century, industrial electrification allowed for the production of an ever-widening web of consumer products. It also resulted in a major increase in the energy expenditures of the manufacturing sector.[61]

The nation's evolving dominance of the Industrial Revolution starting in the early nineteenth century had been crucial to the American Rise. While the American System of Manufactures made the country an important actor in the First Industrial Revolution, it was mastery in the production of steel (combined with access to immense stores of coal) that led to its preeminent position during the Second Industrial Revolution, which saw it grow the largest economy in the world. The nation maintained this advantage by showing the way in the adoption of mass production, which, together with two world wars, made America the leading geopolitical power on the planet. Corporations certainly followed this path to make money; the federal government supported these developments because they raised standards of living across the board. Throughout the nineteenth and twentieth centuries, national leaders firmly believed that using hydrocarbons to expand the manufacturing sector was the path to prosperity. And they were right. This combination of fossil fuels and industry helped make the country great. The continuation of the Transportation Revolution would also play a role on the nation's path to prominence, first through the developments of roads and later through the creation of a commercial airline industry.

INNATE Revolutions: Transportation
The Rise of Highways, Part 1

There is a recognized mythology in America that the country can thank (or demonize) President Dwight Eisenhower for its enormous interstate highway system. The story goes that his experiences first as a young officer traveling across the nation by road and later as a five-star general traveling across Germany by autobahn led him to forcefully support construction of this network when he became president. Those experiences may have influenced his decisions in office, but the foundations for the highway system had been laid decades earlier. In fact, the debate had truly begun during the late nineteenth century, before Eisenhower was born or automobiles had widely entered the national consciousness. This is not to suggest that Eisenhower had no role, but other actors played more significant parts. Their combined efforts helped establish the automobile as the dominant Transportation Revolution modality of the twentieth century. As a result, they also ensured that America would dramatically expand its use of petroleum. But the rise of the US national highway network began with the development of a far more sustainable approach to personal transportation.

One of the most important early manufactured products to emerge from the Second Industrial Revolution was the bicycle. It brought together many of the processes advanced by the American System of Manufactures with the power density and advanced materials of the late nineteenth century. As Hounshell writes, "Clearly the bicycle industry as a staging ground for the diffusion of armory practice cannot be over emphasized." One reason the bicycle was so critical was that it placed unique demands on New England mechanics, leading to advances in overall industrial skills. Another reason was that it required innovations like sheet-metal stamping and electric resistance welding. Finally, the national obsession with these two-wheeled wonders provided an extremely large customer base. The Western Wheel bicycle was one of the key consumer goods that began to call into question the assumption that American manufacturers could only compete in low-end markets. Perhaps more importantly, it began a national passion for speed that would eventually see foot-propelled vehicles surpassed by mechanically propelled vehicles.[62]

It was bicycle, not automobile, enthusiasts who started the clamor for improved roads in 1880. As discussed in the previous chapter, national and state politicians largely ignored American roads during the nation's first century. The League of American Wheelman hoped to change that when they began the Good Roads Movement, lobbying politicians for better roads to promote safety and personal convenience. Within a decade, several states had passed legislation to help fund construction of these roads, stipulating, however, that the money could only be used to build roads connecting towns and villages and thus facilitating transport between places. In response to mounting pressures, the Department of Agriculture also established an office (which would ultimately morph into the Bureau of Public Roads) to begin investigating road construction and management and to provide technical assistance to states.

General Roy Stone, who had close ties with the Good Roads Movement, was named the division's first director. He believed it was an established fact that the nation had the "worst roads in the civilized world, [which was] a crushing tax on the whole people" because it limited economic productivity. Since there was so little knowledge about the current state of the nation's roads, he focused on creating a clearinghouse for road information. The organization also constructed "object-lesson roads," which were short stretches built with a high degree of competence and with good materials, to demonstrate the benefits of good road construction. While these steps had a limited initial impact, they ultimately provided the office with immense influence over national highway policy.[63]

As the federal government was beginning to provide the organization and programs necessary to create world-class roads, major advances in the internal-combustion engine resulted in widespread adoption of the automobile. Work on these contraptions dated from the 1860s, and in the following decade a German named Nikolaus Otto created the four-stroke engine. Initially powered by coal gas, in 1885 it was modified to use gasoline and integrated into Karl Benz's first automobile. The gasoline internal-combustion engine would eventually triumph over other approaches for powering cars. In 1892, Rudolf Diesel conceived a new engine that used high pressure to more efficiently ignite its fuel. Because the diesel engine was heavier than the internal-combustion engine, it was initially used primarily in trucks, buses, trains, and ships; in Europe it would ultimately also be used for smaller vehicles. By the turn of the century, inventors like Émile Lavassor and J. B. Dunlop added new innovations to the automobile design that made it widely marketable. The middle class developed an early appetite for cars and had the money to reject streetcar and trolley systems in favor of these individual conveyances. In 1900, annual sales numbered eight thousand; a dozen years later they numbered more than nine hundred thousand. As Henry Ford entered the market with his much cheaper Model T, the growing car culture slowly became quintessentially American and a manifestation of twentieth-century democratic ideals.[64]

After years of halting developments, major advances were made in national road policy during the Progressive Era. This was largely because of Logan Page's ascendance to leadership of the roads office. It made sense that an engineer like Page would come to power during this era, as "progressive reforms were characterized by an attempt to replace political corruption with honest, efficient administration, a rational desire to base action on information, a reliance on voluntary cooperation, and a sense of social justice." The agency became outwardly much less political during this period, instead making decisions based on science. Page ensured that engineers would make most of the decisions involving highways. This was achieved in part through internal staffing and in part by helping to found the civil-engineer-dominated American Association of State Highway Officials (AASHO). This organization was "part lobby, part professional association, part quasi-political agency." Because Congress regularly looked to it for advice on highway matters, "no effective national highway policy could be enacted without its agreement." This transition had three critical outcomes. First, roads receiving federal funding would be much more professionally designed and constructed. Second, the engineering staff tended to ignore questions of funding—which would prove to be politically costly at times. Finally, the engineering focus made it much less likely that the key question of whether a road should actually be built would be asked before construction commenced. Engineers are often more interested in determining *how* to do something than in ascertaining *whether* to do it. This would lead to serious problems in the coming decades.[65]

In 1916, Page and AASHO secured passage of the Federal Aid Road Act, the first national legislation for highway funding. The bill promoted rural roads, assuming that cities could afford to build their own roads, and required that the national and state governments share building costs—the ratio was one to one. Approximately a million miles of roads would be eligible for funding. As Earl Swift writes, "Perhaps most important, the bill reserved most big decisions for the states. They'd initiate the projects, choose routes, and do the actual building." Additionally, the bill created the Bureau of Public Roads within the Department of Agriculture to administer national-highway funds. This bureau would have the authority to mandate minimum requirements for road construction. Finally, the legislation rejected funding mechanisms such as gasoline taxes or registration charges; thus, federal contributions would come out of general revenue.[66]

After a number of years of delay owing to the First World War and its immediate aftermath, the nation began to make real progress on rural roads during the 1920s. It was in this period that Thomas MacDonald, arguably the most influential person in the history of highway politics, began his more than three-decade tenure leading the Bureau of Public Roads. "MacDonald was above all a stereotypical engineer," writes Seely. "It was this engineering image, the picture of the expert, that let MacDonald implement the [bureau's] ideas." The "Chief" successfully worked to forge a durable national consensus among the highway community, particularly with regard to the focus on rural roads.

According to Swift, "He was positive that the federal government could not afford to build a system of long-distance highways and, from the same tight treasury, provide for state roads that would serve local populations. The latter were more important [economically]." A national economic calamity would soon shift this perspective. But first another transportation modality would appear.[67]

INNATE Revolutions: Transportation
Emergence of Commercial Airlines, Part 1

America's involvement with airplanes started in the late nineteenth century, but its clear superiority wouldn't materialize for more than a half century. A passion for flight helped propel the national drive to world leadership, both militarily and economically. As trains and automobiles had done before, this player in the Transportation Revolution would also substantially transform the lived human experience. In 1886, the American inventor Samuel Pierpont Langley, building on the work of Sir George Cayley and Otto Lilienthal, resolved to build the first powered airplane to be piloted by a human. Langley received fifty thousand dollars from the War Department to finance his experiments—the first national grant for aviation research. A decade later he was the first person to build a heavier-than-air powered machine capable of sustained flight; his two uncrewed airplanes were powered by compact steam engines. In 1903 Langley and Charles Manly built a vehicle designed for launch from a riverboat-mounted catapult. Two test flights saw the aircraft, piloted by Manly, plunge into the Potomac River.[68]

At the same time that Langley and Manly were experimenting with powered flight, a parallel effort was being undertaken by two bicycle mechanics from Dayton, Ohio—Wilbur and Orville Wright. From the time their father, a well-known bishop, brought home a rubber-band-powered toy helicopter, the brothers had been fascinated by flight. In 1890, after joining a bicycle club, the two opened a bicycle repair shop. Bolstered by their own innovative line of bicycles, the shop thrived and provided the Wrights with extra money to fund their early experiments with gliders and airplanes. News of the death of Lilienthal, the famed glider pilot, "renewed an interest in human flight that had lain dormant in Wilbur and Orville since they first read of the German's accomplishments." They began studying existing publications that dealt with aeronautics, concluding that the current literature was inadequate and did not properly address the key problem of aircraft controls.[69]

In 1899 Wilbur Wright had a breakthrough in his thinking about aerial control techniques. He constructed a biplane kite with control lines that allowed him to twist or warp the wings, so that he could steer the kite left or right. Over the next year, he and Orville designed and built a full-size glider based on this design. In September 1900 the brothers loaded the glider on a train and headed to Kitty Hawk, North Carolina, to test it in the steady winds on the Atlantic coast. Despite less lift than expected, they were able to fly the glider up to four hundred feet (surpassing earlier distances) and balance it while in flight. The next month, they returned to Ohio with plans for an improved glider. In July 1901 they traveled to Kitty Hawk with their new glider, a craft with a twenty-two-foot wingspan, making it the largest glider ever flown. The summer's experimentation proved disappointing, however, and the Wrights returned to Dayton in August greatly discouraged. That fall they constructed a wind tunnel and utilized specially designed lift and drag balances to test the characteristics of approximately two hundred airfoils. They soon discovered that earlier airfoil configurations had been highly inefficient, while a wing shaped like a parabola produced a great deal more lift. This discovery proved to be a turning point in their research.[70]

In August 1902, Wilbur and Orville returned once again to North Carolina, this time with a newly constructed glider based on the data they had acquired from their meticulous wind-tunnel testing. They quickly were able to fly the new craft more than five hundred feet, besting all previous attempts. During their experiments that summer they also invented the movable rudder, which was employed to prevent slippage during turns. The final piece of the control puzzle, the movable rudder cleared the way for the Wrights' effort to develop a powered aircraft. Upon returning to Dayton, they began intensive research aimed at designing a lightweight gasoline engine capable of producing eight horsepower and fabricating a viable propeller. By May 1903, with the help of the machinist Charlie Taylor, they had an operating engine and two propellers. In September, they boarded the train yet again for the trip to Kitty Hawk with a powered airplane that had a wingspan of forty feet. On 17 December, after a multitude of problems assembling the new aircraft, including cracks in the wood propellers that required designing metal replacements, Orville Wright made the first controlled and powered airplane flight, traveling 120 feet in twelve seconds. By the end of the day, Wilbur had flown the *Wright Flyer* 852 feet in fifty-nine seconds. The world would never be the same. However, as Roger Bilstein writes, "this incredible and historic feat was, even more incredibly, virtually ignored for nearly five years." This lack of interest prompted the Wrights to continue the development of a more advanced aircraft while conceiving a strategy to gain acceptance of the new technology. Their stepwise approach envisioned the use of aircraft for military reconnaissance, exploration, transportation of passengers and freight, and finally sport. In 1908 the Wright-A (a thirty-horsepower, two-person aircraft) began flight testing, and within a year it was accepted by the War Department for military applications. Two years later the Wright-B took flight, the modified design capable of attaining flight speeds of up to forty-five miles per hour.[71]

Unfortunately, the Wright brothers soon found themselves in a patent fight with the Herring-Curtiss Company. Glenn Curtiss had arrived on the aviation scene in 1904, when he won a contract to build a dirigible engine for the US Army. He formed an airplane manufactory three years later with Augustus Herring and Alexander Graham Bell that would compete against the Wright Company in the burgeoning new air business. In 1909, an epic patent battle began when the Wrights filed a complaint seeking to enjoin the Herring-Curtiss Company from selling or using flying machines for aerial exhibitions. The complaint alleged that Curtiss's *Golden Flier* violated the basic patent the Wrights had taken out based on their wing-warping technology design. The legal fight that ensued lasted nearly a decade, putting America far behind several European nations that were investing heavily in aircraft design.[72]

After the outbreak of the First World War, America fought to catch up with its European counterparts. In 1915 the National Advisory Committee on Aeronautics (NACA) was created "to supervise and direct the scientific study of the problems of flight with a view to their practical solutions." The organization would ultimately use advanced wind tunnels and an aggressive flight-testing program to broaden the understanding of aeronautics. During the war years, however, its most significant accomplishment was mediating the ongoing patent dispute between Curtiss Aeroplane and the Wright-Martin Company. The committee worked out a cross-licensing agreement that finally ended the long battle and opened the door for mass production of aircraft in support of the war effort. With trains and automobiles, as we have seen, the national government adopted a measured approach to fostering their development. In the half century that followed the First World War, a different path was taken to aid the expansion of commercial aviation. This included acting as an early client, investing in technological research, building critical infrastructure, and finally providing vital regulatory oversight.[73]

Moving beyond investing in research through the NACA, the national government nurtured early airlines through the US Airmail Service. This organization, housed within the Post Office Department, provided expedited mail delivery and oversaw the development of a navigational system employing guidance beacons that allowed pilots to fly at night. In 1925, Congress passed the Contract Air Mail Act to promote the expansion of commercial airlines by contracting with them to carry airmail. The following year Congress passed the Air Commerce Act, which instructed the secretary of commerce to foster air commerce; designate and establish airways; establish, operate, and maintain aids to air navigation (but not airports); arrange for research and development to improve navigation aids; license pilots; issue airworthiness certificates for aircraft and major aircraft components; and investigate aircraft accidents. The act resulted in the creation of the Aeronautics Branch (which became the Bureau of Air Commerce in 1934). Under the auspices of this organization, the Air Commerce Regulations were promulgated to codify the safety rules decreed by the statute. This government-ordered standardization was designed to improve the safety of commercial transport aviation, thereby enhancing the viability of new airline passenger services. At least initially, it was far from successful, because the technology was still seen as inherently unsafe.[74]

Significant safety advances in both technology and infrastructure were made during this period, most importantly in instrumentation and radio. In the early 1930s, rapid improvements were made to cockpit instruments such as altimeters, airspeed indicators, rate-of-climb indicators, compasses, and artificial horizon systems. Although radio had been used for more than a decade, perhaps the most noteworthy development made at the time was the utilization of radio for navigation. In 1932 more than eighty radio beacons became operational across the country, which helped pilots locate airports in poor weather conditions. About this same time, the first modern airliners were unveiled, many of which had been initially developed for the military—an evolutionary path that would be maintained for several decades. The Boeing 247 was the first to introduce key new technologies (e.g., all-metal construction, a fully cantilevered wing, and retractable landing gear) in an airliner that could carry ten passengers and cruise at 155 miles per hour. The plane that changed the world and enabled airlines to make money carrying passengers, however, was the Douglas Aircraft Company's DC-3. T. A. Heppenheimer writes that "the DC-3 had a lot in common with Ford's Model-T. Both were simple, rugged, and inexpensive; both introduced ordinary Americans to their respective modes of transportation; and, both made their way in a world where supporting infrastructure stood at a minimum." The DC-3 improved upon earlier innovations with an airframe made of a strong aluminum alloy, which resulted in more room inside the passenger cabin. It could accommodate twenty-one passengers and cruise at 180 miles per hour, and it had a range of nearly fifteen hundred miles (twice that of the Boeing 247). With operating expenses half those of its competitors, the DC-3 was a commercial success on a scale not previously seen. At its peak it had 80 percent of the domestic market share and was operated by more than twenty foreign airlines. It was arguably the most successful passenger airliner in history.[75]

Despite all these advances in technology and policy, however, there continued to be a relatively large number of deaths resulting from airline crashes. It would take a major political realignment to change this dynamic and provide regulations capable of ensuring safety. This would allow airlines to compete with railroads and ocean liners in transporting passengers over long distances. Although early airlines were minor consumers of hydrocarbons because they had such a small market share, this proportion of fossil fuel usage would continue to grow in coming decades, particularly as the nation became the geopolitical and economic leader in the forthcoming political age. Before this came about, however, the nation experienced the joys and devastation of the final cycle of the Second Political Age.

Second Political Age
Conservative Cycle, 1921–1933

In 1916 Brooks Adams delivered a compelling speech to the American Academy of Arts and Letters entitled "The American Democratic Ideal." Adams was a member of America's original political dynasty: his great-grandfather and grandfather had both served as president, his father was the ambassador to Great Britain, his brother Charles was a Union general and railroad president, and his brother Henry was a prominent political commentator and Pulitzer Prize–winning author. The Harvard-trained historian had himself written several well-regarded books, including *The Law of Civilization and Decay* and *America's Economic Supremacy*. In this speech before the Academy of Arts and Letters, he criticized America's focus on individualism rather than the national good: "I have doubts [about] the American 'democratic ideal.' . . . No organized social system, such as we commonly call a national civilization, can cohere against those enemies which must certainly beset it, if it fail to recognize as its primary standard of duty the obligation of the individual man and woman to sacrifice themselves for the whole community in time of need." He used as an example Germany's policy of universal military service, which had provided it with the might to fight a world war on multiple fronts. While he disdained German aggression, he bemoaned America's inability to sustain this type of collective sacrifice unless faced by a nearly cataclysmic national crisis like the Civil War. The result was a political system in which individual interests and local concerns usually won out, meaning that science-based legislation focused on the national interest had little chance of being adopted. Although recent administrations had had more success, this was the exception that proved the rule. "In our national legislature," Adams wrote "the instinct of unity, continuity, and order seldom prevails over individualism or disorder, with the result that our collective administration of public affairs may not unreasonably be termed chaotic." He concluded by telling the audience that the United States must strive to create a political culture capable of self-sacrifice if it hoped to preserve itself. Rather than heeding this advice, during the conservative backlash to come the nation engaged in such an orgy of individualism that it very nearly led to the civilizational collapse that Adams feared.[76]

This new conservative cycle saw the Republican Party turn to a more assertive laissez-faire approach that abandoned many of the macroeconomic security policies championed by Lincoln, Roosevelt, and Wilson. This new Republican brand successfully controlled the White House and Congress from 1921 to 1933. The Roaring Twenties were marked by tremendous economic growth and increased affluence for many Americans, who enjoyed themselves enormously despite a constitutional prohibition on alcohol consumption. One problem with this expansion, however, was that it was fueled to a large degree by stock and real estate speculation. This financialization of the economic system was possible because Republicans had eliminated many of the Progressive Era's business regulations. A second problem was that the benefits of this expansion were not shared by all Americans, with many families falling behind as income inequality grew. The result was growing social unrest. At the same time that these dangerous trends were taking hold on the domestic front, the nation was backing away from its role as a geopolitical leader. This isolationism would persist even as fascist nations increasingly destabilized the global order.

While campaigning for the presidency in 1920, Senator Warren Harding told the nation that after years of domestic progressivism and global war it was time to return to the perceived stability of the past. "America's present need is not heroics, but healing," he said, "not nostrums but normalcy; not revolution but restoration . . . not surgery but serenity." When the dark-horse candidate emerged from the Republican Party's nomination fight and then won a stunning victory in the general election, he brought back isolationism and laissez-faire economics. He is little remembered for those policies, however, but is recalled largely because of his administration's involvement in a hydrocarbon-related scandal—although it wasn't fully uncovered until after Harding died of an embolism two years into his term. Upon taking office, Harding had attempted to reverse the conservation policies advocated for by his predecessors. The key actor in this turnaround was Interior Secretary Albert B. Fall, a strong advocate of developing western lands. Montana senator Albert Walsh opened an investigation of the cabinet official, uncovering that Fall had accepted bribes to lease US Navy petroleum reserves at Teapot Dome in Wyoming and Elk Hill in California; he was later convicted and sentenced to a year in prison. This attempt to revert to an economic Wild West, with government and business colluding to pilfer natural resources, was indeed a return to normalcy.[77]

When Vice President Calvin Coolidge ascended to the presidency, he largely continued Harding's laissez-faire and isolationist policies. Perhaps more significantly, he expanded upon his predecessor's governing style, taking a hands-off approach to both legislation and regulation. Although he maintained a close relationship with the press and pioneered the use of radio, there was little substance to either his domestic policy or his foreign policy agenda. According to Sidney Milkis and Michael Nelson, "Coolidge demonstrated that the increasingly powerful and prominent executive office could be used as effectively by a president who wanted only to reign as by one who, like TR and Wilson, wanted to rule." Coolidge's successor, Herbert Hoover, didn't exactly fit into the new Republican mold. He came into office seeking reform of tariffs, taxes, conservation policies, and the organization of the federal government. But his lack of personal warmth and a philosophy that held that solutions to societal problems lay not in government but in encouraging private actors to develop more equitable and rational economic arrangements swayed few on either side of the aisle and certainly had almost no impact in the business community. When a stock-market crash in October 1929 shook the nation and started the Great Depression, however, Hoover uncompromisingly maintained the laissez-faire ideology even in the face of mass human suffering. This great technocrat, who had previously served as food administrator during the First World War and later as commerce secretary, refused to introduce or support any legislation that involved government intervention. Even as the economy failed to self-correct and one-quarter of the working population was unemployed, he declined to take action. The resulting economic, political, and social instability set the stage for the opening of a new political age and the most significant progressive cycle to date. At the same time, oil rapidly became the preeminent hydrocarbon—which would have weighty economic and geopolitical consequences for the American Empire.[78]

CHAPTER THREE

Oil, Microeconomic Security, and the Third Political Age

Approximately 150 million years ago a process similar to the one that created coal began in the seas, in this case with single-cell plants and planktonic animals that sank to the ocean bottom upon dying and were buried beneath sand and mud. As peat is a precursor to coal, kerogen is a precursor to oil. When subjected to pressure and heat over long periods of time, kerogen is changed into petroleum within the oil window—depths from 2,500 to 16,000 feet. While "rock oil" had been used for various purposes dating back five thousand years, in the late nineteenth century its key markets were illumination and lubrication. The American Century began shortly after the nation discovered and began exploiting its massive domestic oil resources. Although coal would remain a critical fuel source for the INNATE revolutions, it was just a matter of time before oil would become central to manufacturing, transportation, and agriculture.[1]

The most important person in the establishment of an American petroleum industry was George Bissell. He came up with the idea for oil drilling. In 1859 his Seneca Oil Company struck oil in Titusville, Pennsylvania. Neither America nor the larger world would ever be the same. "The enthusiasm for oil seemed to know no limits," writes Daniel Yergin, "and it became not only a source of illumination and lubrication, but also part of the popular culture." In the coming decade, the Pennsylvania Oil Region experienced a progressively intensifying oil fever. It was the arrival of John D. Rockefeller and his Standard Oil, however, that marked the true turning point. Rockefeller made over "an agrarian republic, so recently torn by a bloody civil war, into the world's greatest industrial power." Standard Oil developed relationships with the burgeoning railroads to cheaply distribute oil, and because it was so big it could negotiate discounts on transportation that none of its competitors could match. After gaining this critical beachfront, Rockefeller began to consolidate the oil industry and create the greatest monopoly in human history. As Yergin writes, "By 1879 . . . Standard Oil was triumphant. It controlled 90 percent of America's refining capacity. It also controlled the pipelines and gathering system of the

Oil Regions and dominated transportation." As it was obtaining control within the domestic sphere, Standard was also establishing itself as a key exporter.[2]

Although America would become the largest producer of oil, there would be a worldwide competition to develop petroleum resources and sell kerosene for the illumination market. Three major actors emerged to challenge Standard Oil's position. In the Caucuses, Ludwig Nobel, whose brother Alfred had invented dynamite and created the prize that still bears the family name, became the Russian Rockefeller. He helped the Russians develop and organize the oil industry in the Baku region. He also conceived of and operated the first bulk oil tanker, which allowed Russian oil to more easily reach global markets. It was an Englishmen, Marcus Samuel, who first sought to create a market for petroleum-based kerosene in the Far East. His Shell Oil Company, backed by the Rothschild family, sought to develop a network of storage tanks to distribute the precious fuel in Asia. More importantly, Samuel designed a markedly improved tanker and gained near monopoly control of kerosene shipment via the Suez Canal. In the Dutch East Indies, Royal Dutch Petroleum was created by Aeilko Jans Zijlker and expanded by Jean Baptiste August Kessler to develop a source of kerosene produced much closer to the large Asian markets. In the final two decades of the nineteenth century, these corporations engaged in an "oil war" to corner the market for kerosene. However, petroleum was still a small player in an age led by coal. At the turn of the century, the oil sector faced a dire challenge from an emerging electricity industry, which was undercutting its control of the illumination market. What was needed was a new user for petroleum. As fate would have it, a far larger customer appeared about this time: the internal-combustion engine.[3]

The gasoline-powered internal-combustion engine came along at the perfect time for an oil industry looking for a new customer base. As this new market took off, however, there was suddenly a decline in supplies as Pennsylvania fields reached peak production—a sign of things to come. Before the situation became too critical, however, numerous new discoveries put the United States on course to become the leading global petroleum producer. At the end of the nineteenth century, new fields tapped near Los Angeles became the most important in the nation. "By 1910," writes Yergin, "[California's] output would reach 73 million barrels, more than that of any foreign nation, and 22 percent of total world production." The next virgin fields were found in Texas, where the Spindletop salt-dome formation near Beaumont became one of the great gushers in history, producing seventy-five thousand barrels per day. "The boom at

Spindletop was to be repeated many times over in the Southwest . . . beginning with other salt domes along the Gulf Coast of Texas and Louisiana. But the Gulf Coast was about [to meet] its match in Oklahoma." For the next quarter century, Oklahoma fields like Glenn Pool near Tulsa would produce half of the region's petroleum before Texas regained its throne in the late 1920s. Meanwhile, the Mellon family, which had risen to prominence as the key bankers to industrial concerns in the American Midwest, was taking control of the Gulf Oil Corporation and turning it into a fully integrated organization that could compete with Standard Oil. Soon other entrants, like the Texas Fuel Company (Texaco) and the Sun Oil Company, would also be created, and each would begin to carve into Standard's market dominance.[4]

It was during this period that Ida Tarbell, working for the highly influential *McClure's* magazine, began her groundbreaking investigation of Standard Oil's controversial business practices. Starting in November 1902, her stories appeared for twenty-four successive months and became rhetorical fodder for the entire Progressive Movement. When turned into a book, the series became one of the most important treatments of American business ever published. It was perfectly timed with Teddy Roosevelt's emergence as a trustbuster. "Immediately after Roosevelt's election in 1904, his administration launched an investigation of Standard Oil and the petroleum industry. The result was a searing critique of the trust's control of transportation." In 1906, Roosevelt's administration formally brought suit under the Sherman Antitrust Act in *U.S. v. Standard Oil*. After the five years it took for the case to make its way through the federal courts, the Supreme Court found that the company had conspired to restrain petroleum commerce and ordered its dissolution. Standard Oil was quickly split up into Exxon, Mobil, Chevron, Sohio (eventually the American branch of British Petroleum), Amoco, and Conoco. The net impact on the oil business was positive, as the new competition in this sector allowed the new companies to respond more flexibly to changes in the industry. The most significant early example was Amoco's adoption of thermal cracking in the petroleum refining process. Thermal cracking is a process by which crude oil is simultaneously pressurized and heated to approximately 650 degrees, which breaks the hydrocarbon's molecular bonds to allow the removal of impurities, doubling the amount of gasoline produced from a single barrel of oil. Dr. William Burton, the director of manufacturing at Standard Oil, had discovered the process two years before the dissolution, but the parent company had denied funding to pursue it on a larger scale. The new Amoco, however, was no longer shackled and gave the go-ahead for a technology that inaugurated the Gasoline Age.[5]

* * *

The American government began thinking seriously about the link between hydrocarbons and national power during the mid-nineteenth century, when its naval steamships required a far-flung network of coaling stations. The country's initial foreign relationships aimed at securing coal for the US Navy later would shape national strategy as it sought to capture an ever-growing share of global oil production. This quest began after the First World War, the first true oil war. In the years leading up to the war, the British Royal Navy had transitioned to oil-powered ships at the behest of Admiral John Fisher and First Lord of the Admiralty Winston Churchill. This gave the British a strategic advantage in the surface war, as their ships could be refueled more quickly, were faster, and had superior range. During the transition, the Royal Navy had become a key supporter of opening new fields in the Middle East under the auspices of the new Anglo-Persian Oil Company. Although reserves in this region were still somewhat unproven, Churchill believed they could be critical to providing a diverse regional supply of petroleum. In a speech before Parliament, he proclaimed the growing importance of this strategic resource: "If we cannot get oil, we cannot get corn, we cannot get cotton and we cannot get a thousand and one commodities necessary for the preservation of the economic energies of Great Britain. . . . On no one quality, on no one process, on no one country, on no one route and on no one field must we be dependent. Safety and certainty in oil lie in variety and variety alone." As his vision for the future began to shift British policy, petroleum began to reshape global geopolitics; the true repercussions would not be clear until the fall of the empire.[6]

After the outbreak of the war, oil also became important in the land conflict. It was most important in transportation, with automobiles, trucks, and motorcycles moving troops and supplies. The British also introduced the first armored tanks during the war, and although they were rather crude, they did lead to a decisive victory against the Germans at the Battle of Amiens. Finally, the Wright Brothers' invention truly arrived with the war. Airplanes were used for reconnaissance and bombing, while specialized fighters helped create national heroes like Manfred von Richthofen (a.k.a. the Red Baron) and Eddie Rickenbacker. As the war progressed, however, high demand for petroleum resulted in a supply crisis, which, in turn, led the British to adopt a short-term energy strategy focused on securing adequate oil supplies from multiple sources. While the Allies suffered from stretched supplies, the Germans were faced with the complete collapse of their reserves. This was one factor that led to their surrender in November 1918. As Lord George Nathanial Curzon said at the time, "The allied cause had floated to victory upon a wave of oil." It was now clear that oil would be the single most important natural and strategic resource of the twentieth century. As Yergin writes, "The Great War had made abundantly clear that petroleum had become an essential element in the strategy of nations; and the politicians and bureaucrats, though they had hardly been absent before, would now rush headlong into the center of the struggle, drawn into the competition by a common perception—that the postwar world would require ever-greater quantities of oil for economic prosperity and national power."[7]

While the sleeping giant on the other side of the Atlantic already had the world's largest reserves, in the years after the war a struggle emerged to gain control of new fields in the Middle East and Mexico. Even though the United States produced two-thirds of the world's oil, American companies entered the fray. They were outraged by Old World imperialism, embodied in the San Remo Agreement of 1920, which aimed to divide Middle Eastern reserves among the European powers. Standard Oil of New Jersey (the future Exxon) became the nation's first standard-bearer in the region, with the explicit backing of the Wilson administration, which sought an "open door" in the area. Under President Harding, Secretary of Commerce Herbert Hoover manipulated the creation of a syndicate of major oil companies to push American claims. This Near East Development Company was presided over by Standard Oil of New Jersey president Walter Teagle, whose goal was to expand the American oil industry's interests globally. The first big fight was over the newly created Iraq, where the European-owned Turkish Petroleum Company had already secured a concession from the British-established monarch, King Faisal. After six years of negotiations, Teagle successfully gained a contract that gave the Americans a 23.75 percent share of Iraqi oil, with the same portion going to Royal Dutch Shell, Anglo-Persian, and the French. The balance went to the founder of the Turkish Petroleum Company, Calouste Gulbenkian. These parties also agreed to cooperate in future oil operations in the area that had encompassed the former Turkish Empire, which would come to account for all major Middle Eastern oil fields except those in Iran and Kuwait. This momentous Red Line Agreement (named for a red line drawn on a conference map around the relevant geography) seemed to establish a pathway to the orderly development of the region's petroleum reserves. In reality, however, it was but the opening shot in a decades-long clash between the parties and several interlopers.[8]

As auto ownership climbed sharply in the United States (by 1929 three of every four cars in the world were owned by Americans) and gas stations began to change the landscape, there was a rush to meet the demand with new domestic supplies. The latter years of the second decade of the twentieth century had seen few new discoveries, leading to concerns about future shortages. In the subsequent decade, however, innovative technologies changed this outlook, as torsion balances, magnetometers, aerial reconnaissance, micropaleontology, and seismographs were all introduced into the oil business. At the same time, drilling was advancing, with wells reaching depths of ten thousand feet. This led to the unearthing of large fields like Signal Hill in California and Greater Seminole in Oklahoma, which, combined with the introduction of thermal cracking in the refinery business, alleviated most scarcity fears. One other major development was Harry Doherty's effort to educate businessmen about the dynamics of petroleum reservoirs, which ultimately convinced operators to reduce the number of wells punched into a field to maintain pressure and ensure that more oil would be recovered.[9]

While the American oil business was expanding, additional supplies were being developed in Mexico and Venezuela. In 1910 the Englishman Sir Weetman Peason's Mexican Eagle Petroleum Company had opened the biggest field on the planet at Potrero del Llano 4, along with many other finds throughout the "Golden Lane." While this made Mexico the world's second largest producer, the Mexican Revolution and an apparent peaking of key fields created great uncertainty about the nation's future. Into this void drove Venezuela. Royal Dutch Shell had been operating in that country's Lake Maracaibo region since 1913, but after the First World War a full-fledged conflict erupted to gain a concession to drill in the region. Three companies emerged victorious: Royal Dutch Shell, Gulf Oil, and Pan American. In just eight years, from 1921 to 1929, Venezuela's annual oil production increased from 1 million barrels to 137 million barrels. This allowed it to replace Mexico as the second leading producer in the world.[10]

In 1931 the campaign to expand production in the Western Hemisphere returned home when a small-time actor in the American oil industry named Columbus "Dad" Joiner found the largest oil field anyone had ever seen to that point. "Ultimately," writes Yergin, "the East Texas reservoir proved to be forty-five miles long, and five to ten miles wide, 140,000 acres altogether. The field became known as the Black Giant. Nothing to compare with it had ever been discovered in America. And the boom that followed made all the others

—in Pennsylvania, at Spindletop, elsewhere in Texas, at Cushing, at Greater Seminole and Oklahoma City, at Signal Hill in California—look like dress rehearsals." The oil glut produced by this find, however, led to state government intervention to bring supply back in line with demand and to provide price stability. With massive available supply and prices fluctuating wildly, the new reformer in the White House decided to take action at the national level. President Franklin Roosevelt tasked Interior Secretary Harold Ickes with restoring order to the oil business. In so doing, he would often do battle with state actors like the powerful Texas Railroad Commission. Under the aegis of the US Constitution's commerce power, an Oil Code provided Ickes with the authority to set quotas for each state's monthly production; this arrangement was later reinforced by the Interstate Oil Compact. These production regulations were augmented with a new national tariff on imported oil to prevent a flood of cheap foreign oil from destabilizing the market. At the same time, Ickes also led a federal effort to wrest control of offshore oil leases away from the states, which would make possible real strategic planning with regard to this critical resource. When he became president, Harry Truman signed an executive order completing this takeover. For five years a number of states, led by Texas and Louisiana, challenged the new approach. But in 1950 the US Supreme Court found in favor of the federal government. Petroleum executives, who had long fought government regulation, were largely supporters of these efforts by the Roosevelt and Truman administrations. The result was stability within this crucial economic sector. The long-term consequences for the New Deal Coalition, however, would be dire. As Tyler Priest writes, "The politics of oil and race . . . weakened the Gulf of Mexico region's allegiance to the national Democratic Party and brought oil interests into the Republican Party's emerging national coalition. The oil bloc in this coalition would be led by Texas and California, the home to five future presidents."[11]

In the years leading up to the outbreak of the Second World War, developments on the international stage continued to shape the petroleum industry. These included a skirmish between the shah of Iran and the British over the concession to drill in Iran, which was ultimately settled with a new agreement that provided guaranteed revenues to the Persian leader. On the other side of the world, the Mexican government under President Lázaro Cárdenas finally nationalized oil production and became an important supplier of the Axis powers, which proved to be a major strategic problem for the British Empire, which relied upon Mexican and Venezuelan oil. Meanwhile, a New Zealander named Frank Holmes began exploring an entirely new region, the Arabian

Peninsula. He would eventually receive the financial backing of Gulf Oil and Standard of California (SoCal), which allowed him to guide the two companies to concessions in Kuwait and Bahrain, respectively. Then a SoCal-Texaco joint venture, which wasn't a party to the previous Red Line Agreement, signed the Blue Line Agreement with Saudi Arabia's Ibn Saud to explore and develop oil reserves in his nation. This proved to be the ultimate crown jewel in the oil business. "The gaining of the concession by an American company," writes Yergin, "would inevitably begin to change the web of political interest in the region." The United States would slowly be pulled down the rabbit hole of Middle Eastern geopolitics—from which it has yet to emerge. In 1938 the Damman Well Number 7 revealed that there was an enormous amount of petroleum in the region and sealed America's fate. Before this new treasure trove could be fully exploited, however, the greatest war in world history intervened.[12]

Third Political Age
First Progressive Cycle, 1933–1969

In the years leading up to the Third Political Age, a rancorous battle over the role of individualism was being waged among members of the nation's political and intellectual elite. At its heart, this was really a discussion about whether the American people supported positive or negative liberty, or some hybrid combining both. This discussion helped inform the growing debate about whether the government had a duty to provide its people with microeconomic security. In 1922 Herbert Hoover published *American Individualism*, in which he argued that the surest path to progress was through the nation's own distinctive form of individualism. "Our individualism differs from all others," he wrote, "because it embraces these great ideals: that while we build our society upon the attainment of the individual, we shall safeguard to every individual an equality of opportunity to take that position in the community to which his intelligence, character, ability, and ambition entitle him . . . that through an enlarging sense of responsibility and understanding we shall assist him to this attainment." Given the momentous transitions in American life that were shaking society to its core, however, this position seemed quaint. How would these bygone ideas bear up as the economy was rapidly shifting from agrarianism to industrialism? Would they maintain relevance as millions moved from small towns to big cities? What impact would they have on the nation's environment as the pace of life increased? In sum, would an American individualism rooted in the idea that equal opportunity can be safeguarded by a strong social and political culture be sustained in the modern world? Although Hoover's view was increasingly capturing the Republican Party, others believed the entire idea was bankrupt.[13]

In 1931, two years after the Great Crash, Charles Beard published an essay titled *The Myth of Rugged American Individualism*. He argued that Hoover's idea that the country's history was absent government interference was erroneous. In fact, business leaders believed that the economic stability provided by government regulations led to their own enrichment. "For forty years or more there has not been a President, Republican or Democratic, who has not talked against Government interference and then supported measures adding more interference to the huge collection." Yet, they had long demonized this interference to maintain the fiction of negative liberty. What the wealthy feared was that without this philosophical framework they might be taxed to provide support to those who were left behind in a rapidly changing economic environment. Beard concluded: "Whatever merits [American individualism] may have had in days of primitive agriculture and industry, it is not applicable in an age of technology, science, and rationalized economy. Once useful, it has become a danger to society." What the country needed to pull itself out of depression and then grow into the future was a new belief in the power of government to be an agent of constructive change.[14]

In 1932, when he hit the campaign trail to seek the presidency, Franklin Roosevelt, then governor of New York, became the key advocate of this new positive liberty. He championed providing all Americans with microeconomic security. On 22 May, in a commencement address at Oglethorpe University, he made his case to the generation that would lead the charge to radically change the role of government. Roosevelt explicitly raised the specter of continued economic insecurity, which had already eroded the financial reserves of average Americans. He told the students: "With these saving has gone, among millions of our fellow citizens, that sense of security to which they have rightly felt they are entitled in a land abundantly endowed with natural resources and with productive facilities to convert them into the necessities of life for all of our population. More calamitous still, there has vanished with the expectation of future security the certainty of today's bread and clothing and shelter." FDR hammered away at the financial opportunists who had brought the nation to its knees, accusing them of selfish behavior that shunned any public duty. He also disparaged the dogmatic belief that the economic system would eventually self-correct, reasoning that it not only ignored massive suffering but placed far too much faith in a discredited laissez-faire approach that put little trust in government's ability to provide needed direction. The candidate argued for direct social and economic planning, contending that the nation would unite in favor of such an approach to achieve broad civilizational goals. Roosevelt concluded with a powerful call to arms, implying that the alternative to action might be revolution: "The country needs and, unless I mistake its temper, the country demands bold, persistent experimentation. . . . The millions who are in want will not stand by silently forever while the things to satisfy their needs are within easy reach. We need enthusiasm, imagination and the ability to face facts, even unpleasant ones, bravely. We need to correct, by drastic means if necessary, the faults in our economic system from which we now suffer. We need the courage of the young."[15]

On 23 September, Roosevelt expanded upon these thoughts in a speech before the Commonwealth Club, a venerable public-affairs forum of which his opponent had been a longtime member. He argued that the rise of individualism and the propertied class dated from the election of 1800 despite the fact that "even Jefferson realized that the exercise of . . . property rights might so interfere with the rights of the individual that the Government . . . must intervene, not to destroy individualism, but to protect it." In a brilliant passage, Roosevelt expounded on the contemporary need for government to provide microeconomic security. America had differed from the rest of the world in that the productiveness of its agriculture and the expansiveness of its territory ensured that even when financial panics destabilized the economic system, people did not starve, because they could feed themselves. But with the closing of the frontier and the transition to industrialization, this earlier protection against economic insecurity had been largely eliminated. Roosevelt suggested that to protect individual liberty and democracy, enlightened government administration was needed to act as a counterweight to oligarchic business interests. "Whenever in the pursuit of [personal gain] the lone wolf, the unethical competitor, the reckless promoter . . . declines to join in achieving an end recognized as being for the public welfare . . . the Government may properly be asked to apply restraint." Thus, FDR had embraced a platform to complete the macroeconomic security agenda of the previous political age, while providing a microeconomic security agenda in a new political age. He completed his argument with an appeal for collective action to make this vision a reality: "Faith in America, faith in our tradition of personal responsibility, faith in our institutions, faith in ourselves demand that we recognize the new terms of the old social contract. We shall fulfill them . . . lest a rising tide of misery, engendered by our common failure, engulf us all. But failure is not an American habit; and in the strength of hope we must all shoulder our common load."[16]

In November, Roosevelt was carried into office by an electoral coalition that supported his economic plans. This alliance of labor unions, urbanites, minorities (African Americans, Jews, Catholics, and new immigrants), intellectuals, and populist farmers would hold sway for decades. Roosevelt won throughout the South because he made the calculated decision to ignore the continuing tragedy of southern apartheid, which allowed him to carry every southern state. He also swept every other region except the Republican stronghold in the Northeast, where the incumbent carried six of nine states. Important generational shifts also coalesced in Roosevelt's favor. Although William Strauss and Neil Howe suggest that FDR was a member of the prophetic Missionary Generation, his personality traits placed him more firmly with the nomadic Lost Generation. This accounts for his pragmatic decision-making style, which brought the nation through two earth-shattering crises. His true generation came of age as wild young adults during the Roaring Twenties, but most emerged after the Great Crash as "clear-eyed managers [and] selfless protectors."[17] Roosevelt was elected just as the largest generation so far in American history was entering the electorate in substantial numbers. The heroic GI Generation would quite literally be the foot soldiers in both an economic recovery and a global war. Throughout, they would largely remain loyal to Roosevelt and his Democratic successors, although they would also help elect a war hero from the opposing party as president. With this demographic, sectional, and generational coalition in place, the Democratic Party not only transformed how the country provided national and macroeconomic security but also added microeconomic security to the list of government responsibilities.

In the period after Roosevelt's election, but before he officially took office, the United States was gripped by a banking crisis that severely deepened the ongoing depression. The new president opened his inaugural address by telling all Americans that it was "the time to speak the truth, the whole truth, frankly and boldly. Nor need we shrink from honestly facing conditions in our country today. This great Nation will endure as it has endured, will revive and will prosper. So, first of all, let me assert my firm belief that the only thing we have to fear is fear itself—nameless, unreasoning, unjustified terror which paralyzes needed efforts to convert retreat into advance." Roosevelt had reason for such a hopeful preamble, in view of the underlying strengths of the hydrocarbon-fueled American economy. As figure 2.3 shows, growth in total energy consumption had continued to drive the economy forward through the final thirty years of the Second Political Age. The problem was that not only did government spending continue to significantly lag behind the economic expansion but basic regulatory protections had been eroded. Thus, Roosevelt argued, the most important early goals of the Third Political Age were to restore stability in the banking system and put the nation back to work, the latter to be achieved mostly by funding a wide array of public-works projects. Was this constitutional? He responded with a resounding yes. "Our Constitution is so simple and practical," he told his fellow Americans, "that it is possible always to meet extraordinary needs by changes in emphasis and arrangement without loss of essential form. That is why our constitutional system has proved itself the most superbly enduring political mechanism the modern world has produced. It has met every stress of vast expansion of territory, of foreign wars, of bitter internal strife, of world relations." Nevertheless, he suggested that if Congress didn't take action, he was willing to declare a state of emergency and take on extraconstitutional powers. This would not prove necessary, as members of the legislative branch had largely heard the message sent by the citizenry through their resounding defeat of the Republican approach. They seemed to share Roosevelt's conclusion: "We do not distrust the future of essential democracy. The people of the United States have not failed. In their need they have registered a mandate that they want direct, vigorous action. They have asked for discipline and direction under leadership. They have made me the present instrument of their wishes. In the spirit of the gift, I take it."[18]

Unlike Hoover, Roosevelt was blessed with a first-class temperament and a willingness to experiment that made all the difference during his twelve years in office. Beginning with his now-famous Hundred Days, Roosevelt built upon Wilson's earlier policies to fundamentally redefine the purpose of the national government. Like Wilson, Roosevelt became legislator-in-chief. His policies were aimed at laying the foundation for steady growth and to avoid the boom-and-bust cycle that had been the hallmark of America's economic history. At the heart of the New Deal was an overhaul of the government-business relationship. The Glass-Steagall Act of 1933 created a firewall between investment and commercial banking; it was augmented two years later by a banking bill that allowed the Federal Reserve Board to set reserve requirements for commercial banks. The Securities and Exchange Commission and the Federal Deposit Insurance Corporation provided stability within the financial sector. The National Labor Relations Act empowered unions. Roosevelt also adopted Keynesian economics, pumping large amounts of money into job-creation programs. The "alphabet soup" agencies tasked with providing unemployment relief included the Civil Works Administration, the Civilian Conservation Corps, the Federal Emergency Relief Administration, the Public Works Administration, the Tennessee Valley Authority, and the Works Progress Administration. Not only did these agencies employ millions of people who badly needed work, they also constructed the infrastructure that would support the nation's economic growth during the coming decades. Finally, Roosevelt began the process of offering Americans microeconomic security, with unemployment insurance, Social Security, and Aid to Families with Dependent Children. At long last, these programs tapped into the vast fiscal resources made available by the hydrocarbon-powered INNATE revolutions. As can be seen in figure 3.1, the government budget began a steep climb as this fundamental shift in policy took hold during the Third Political Age. The results were tremendous. Unemployment insurance proved to be a key automatic stabilizer during economic downturns. Without Social Security, nearly half of elderly Americans had lived in poverty, but the program ultimately reduced their proportion to about 10 percent. Until the program was gutted in the 1990s, Aid to Families with Dependent Children provided critical income to single mothers, most simply trying to raise healthy children.[19]

*   *   *

Roosevelt's New Deal set the nation on the path to recovery, but the Second World War actually ended the Great Depression. During the period before the United States entered the war, Roosevelt worked behind the scenes to position the nation to take advantage of any conflict. He was forced into this posture by the isolationist turn the American people had taken after the close of the First World War. Despite the high costs of German aggression and Japanese expansionism, the nation remained committed to a policy of nonintervention. Even after the fall of France, a huge majority of Americans were still committed to staying out. Considering that within a half decade the United States would be the world's most powerful nation, it is shocking how ill-prepared it was for that role. Although his people balked, Roosevelt started to react more assertively as the war worsened in Europe. He charted a course that would make American leadership a reality. He began by exchanging fifty American destroyers for British bases in the Atlantic Ocean, which still contained the world's most important trade routes. After winning reelection to an unprecedented third term, Roosevelt boldly proposed a program that would be implemented under the auspices of the Lend-Lease Act. In a famous press conference on 17 December 1940, he promoted the program in simple terms anyone could understand: "Suppose my neighbor's home catches fire, and I have a length of garden hose four or five hundred feet away. If he can take my garden hose and connect it up with his hydrant, I may help him to put out his fire. Now what do I do? I don't say to him before that operation, 'Neighbor, my garden hose cost me $15; you have to pay me $15 for it.' What is the transaction that goes on? I don't want $15—I want my garden hose back after the fire is over." Using this logic, under the Lend-Lease Act, America would provide the British with military arms that would be either returned or paid for after the war.[20]

In his annual address to Congress the following month, Roosevelt continued to build his case for a more interventionist policy with regard to a war being waged on four continents. He suggested that American security had never been so threatened by external forces, with the democratic way of life being challenged by fascist dictators. For this reason, he believed the country needed to unite behind all-out assistance to the Allies. "The happiness of future generations of Americans," he said, "may well depend upon how effective and how immediate we can make our aid felt." Moving forward, he argued that the objective should be to establish a world order based on four essential human freedoms: freedom of speech, freedom of worship, freedom from want, and freedom from fear. He declared that it was the duty of the United States to lead by example, in so doing making the case that the watchword of American national development had been progressivism: "Since the beginning of our American history we have been engaged in change, in a perpetual, peaceful revolution, a revolution which goes on steadily, quietly, adjusting itself to changing conditions without the concentration camp or the quick-lime in the ditch. . . . This nation has placed its destiny in the hands, heads and hearts of its millions of free men and women. . . . Freedom means the supremacy of human rights everywhere. Our support goes to those who struggle to gain those rights and keep them. Our strength is in our unity of purpose. To that high concept there can be no end save victory."[21]

In the following weeks, Roosevelt and his legislative associates worked to gain passage of the Lend-Lease Act. Signed into law on 11 March 1941, it was a genuinely inspired act, perhaps the most important policy to emerge from the Roosevelt administration, because it stabilized the Allies, mobilized American industry, and began pulling the nation out of the Great Depression. By the time the United States was an active participant in the conflict, its war industries were already producing the superior second- and third-generation weaponry that made a huge difference in the effectiveness of the armed forces. The Japanese attack on Pearl Harbor in December finally brought America into the war. It was Adolf Hitler's decision to declare war four days later, however, that truly changed the course of history. Hitler's declaration, resulting from his general ignorance about the strengths America would bring to the Allied cause, was without doubt one of the worst political and military calculations ever made. The decision would ultimately be catastrophic for fascism. Four months before the attack, Roosevelt had met with Britain's prime minister, Winston Churchill, to begin drafting a war plan that would make Germany the primary target of Allied forces and make the defeat of Japan a secondary goal. It was a testament to American power that it was able to successfully engage in two theaters, on opposite sides of the globe, at the same time.

To an even greater degree than Wilson two decades earlier, Roosevelt effectively took over management of national economic and social affairs during the war years. He mobilized both industry and labor to prosecute a total war. Natural resources were extracted at a furious pace, most importantly the nation's vast oil reserves. Domestic prices on key commodities were frozen, and rationing programs were instituted. As the sleeping giant finally emerged from its long slumber, it began to turn the tide of war in both Europe and the Pacific. After initial engagements with the Germans in North Africa, Sicily, and Italy, the Allies launched the D-Day invasion of France in June 1944. Within a year, Hitler was dead and the Soviets and Americans had forced a German surrender. A few months after V-E Day, President Harry Truman (who had succeeded to the presidency after Roosevelt's death) made his fateful decision to end the Pacific war by dropping two atomic bombs on Japan. The Japanese officially surrendered on 14 August 1945, ending the bloodiest conflict in human history. America emerged from the war in a dominant position. The US military was the strongest in the world. The mainland had been untouched by the conflagration, while the infrastructure of America's major competitors (particularly Germany and Japan) had been almost entirely destroyed. The nation suffered far fewer battle deaths than other major powers, with an even larger gap in civilian deaths. Finally, the economy was far stronger than before the fighting commenced because the massive spending on armaments had pulled the country out of depression and turned it into an industrial behemoth.

* * *

The Second World War was the first large-scale oil war in history. Hydrocarbons had played a role in several previous conflicts, but they hadn't been one of the primary causes and they hadn't been essential for such a broad range of the weapons being employed by the combatants. While petroleum was clearly a crucial strategic resource in the European theater, it was equally true that the conflict was not initially driven by the need to secure reserves. This wasn't the case in the Pacific theater, where the Japanese need for oil was one of the main reasons for its attack on Pearl Harbor. Japan had no hydrocarbon reserves, so it was totally dependent on imports to power its industrial rise. The United States was the most important supplier, but Americans became progressively more disenchanted with this arrangement after the Japanese invasion of China, with nearly three-quarters of Americans favoring an embargo on the export of war materials. As Japan turned to Southeast Asia in 1940, it also launched itself on a collision course with the United States. Surprisingly, this was an outcome that was embraced by a growing number of Japanese government officials, who believed confronting the Western powers was an inevitable outcome of its effort to build its Asian empire. As America began to build a two-ocean navy to prepare for a potential conflict, the Japanese admiral Isoroku Yamamoto began the planning process for a surprise attack on Pearl Harbor. When the United States finally instituted an embargo, the die was cast: on 7 December 1941 the infamous surprise attack took place, and the American Empire was thrust into the conflict. It emerged victorious four years later in large part because of its access to immense oil reserves and its ability to build the huge numbers of petroleum-powered tanks and planes and ships necessary to defeat the Axis powers on multiple fronts.[22]

Although the first two years of the war in Europe were not aimed at capturing oil fields, when the Germans invaded the Soviet Union in June 1941 one of their primary objectives was to capture the oil fields at Baku, which was not surprising given that Hitler believed oil was the single most important commodity of modern times. Fortunately, the push bogged down when the Battle of Stalingrad diverted attention from capturing the oil fields, with the Soviets ultimately capturing the entire German Sixth Army. This was arguably the single most important turning point of the war, because it denied the Nazi war machine the petroleum it needed to survive. Hitler was forced to rely upon the manufacture of synthetic oil, an energy-intense process developed during the previous decade whereby coal was processed to produce a liquid fuel usable in internal-combustion engines. Throughout the final two years of the war, a campaign was waged between Allied strategic bombers tasked with destroying synthetic-fuel plants and the Germans tasked with keeping them in operation. Over time the success of the Allied effort and the loss of the Rumanian oil fields at Ploesti crippled Hitler's war-making capability.[23]

Much credit for the American victory in the Second World War must go to a person who is little remembered today, Interior Secretary Harold Ickes, who in the 1930s had restored order in an oversupplied oil industry. In May 1941, the day after announcing an "unlimited national emergency," Roosevelt made Ickes the petroleum coordinator for national defense. His task was to transform an industry long concerned about overproduction into one that could cope with a lack of supply, which meant that he needed to find innovative ways to discover new resources and accelerate production in existing fields. Initially the big demand was in Britain, where supplies were dangerously low because of the successful German U-boat campaign against American shipping. These supply problems became even worse after the United States entered the war, when the submarines began targeting tankers in American coastal waters with devastating accuracy. Ickes worked with the oil industry to increase supply to counter these problems, with great success. As Yergin writes, "America's overall production record was very good: from 3.7 million barrels a day in 1940 to 4.7 million barrels per day in 1945—a 30 percent increase. . . . Altogether, between December 1941 and August 1945, the United States and its allies consumed 7 billion barrels of oil, of which 6 billion came from the United States." Another policy that helped the war effort was rationing. Although Americans had come to see oil consumption as a right, by early 1942 Ickes had begun issuing bans on gasoline use for certain activities. Before the end of the year a nationwide rationing system was put in place. Because of these various activities, the United States had sufficient petroleum supplies to wage war in two sprawling theaters with unprecedented success. The Germans had tried to fight a far more contained two-front conflict in two world wars, but without the necessary fuel reserves they had failed in both attempts. Although other advantages made it possible for the United States to achieve this seemingly impossible task, there is no doubt that without hydrocarbons the American Empire wouldn't have emerged from the war in such a dominant position.[24]

* * *

In his final State of the Union address, President Roosevelt provided a roadmap for American economic development in the coming decades. Although some of his proposal for a second bill of rights based on the entitlement to economic security would not be adopted, it has continued to serve an aspirational role for progressives. The president contended that the nation's enlarged size and industrial growth had rendered the first bill of rights, which was focused on protecting political privileges, inadequate to ensure that all citizens would have an equal opportunity to pursue happiness. Because he believed individual liberty required economic security, he put forward eight economic rights that society should strive to protect:

- Employment, with a living wage
- Food, clothing, and recreation
- A decent income for every farmer
- A business environment free of unfair competition and monopoly
- Housing
- Medical care
- Protection from fears of old age, sickness, accident, and unemployment
- A good education

As his successors worked to augment earlier legislation aimed at accomplishing these lofty objectives, the country enjoyed the longest period of widespread prosperity in its history. This period has been called the Great Compression, a term coined by the economic historians Claudia Goldin and Robert Margo. Americans during this era enjoyed economic affluence, relative economic equality, and a comparative lack of political polarization. The Nobel Laureate Paul Krugman writes that three key policies made the Great Compression possible. First, the richest individuals and corporations were heavily taxed. In 1944, for example, the top marginal rate for individuals was 94 percent on income over twenty-five thousand dollars (although this fell steadily to 70 percent by the end of the period). Second, wages set by the government during the war years remained high after the conflict ended. Third, labor unions were able to negotiate better wages at a time when many American workers were members. Krugman writes, "None of the bad consequences one might have expected from a drastic equalization of incomes actually materialized after World War II. . . . And the era of equality was also a time of unprecedented prosperity, which we have never been able to recapture." What Krugman misses in his analysis is the overarching benefits provided by rapidly accelerating total energy consumption (see fig. 3.1), which led to dramatic increases in economic growth and in the size of the federal budget. What was fascinating about the two progressive cycles of the Third Political Age was that the renewal and expansion of macroeconomic policies and the inclusion of broad-based microeconomic policies finally resulted in significant and sustained increases in the government budget. While the national economy grew by more than 600 percent between 1930 and 1980, annual federal expenditures increased from 4 percent to 24 percent of GDP. Major economic growth had been seen in past eras, but this type of expansion in government spending was entirely new in America. Hydrocarbons were supplying the necessary fiscal resources to provide national, macroeconomic, and microeconomic security to the citizenry all at the same

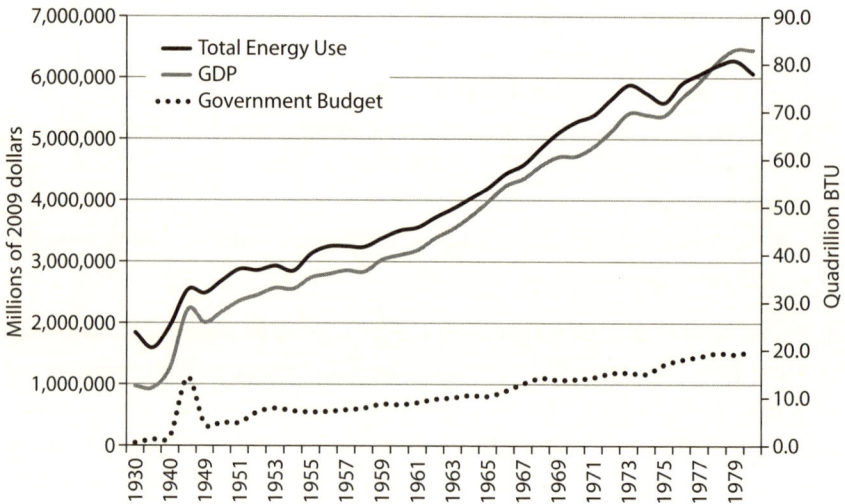

Figure 3.1. US Total Energy Use, GDP, and Government Budget, 1930–1980. US Energy Information Administration, *Annual Energy Review, 2009; Budget of the US Government, Historical Tables, 2017;* Louis Johnston and Samuel H. Williamson, "What Was the U.S. GDP Then?," 2017.[25]

time, which was something new in human history. Just as importantly, the political philosophy of the American people had finally evolved to encompass positive notions of liberty as fundamental to the national character. During the Great Compression, progressives actually stood conservative antigovernment rhetoric on its head. Ironically, most contemporary conservatives point to the 1950s as a golden age but are seemingly ignorant of the policies that led to it.[25]

Although the government clearly had the resources to provide broad-based social insurance programs, during the postwar years an intellectual battle emerged between progressives, who wanted to expand upon the New Deal, and conservatives, who wanted to reverse course. One of the more eloquent champions of the former group was John Kenneth Galbraith, who wrote at length about the need for social balance. He suggested that twentieth-century America was characterized by the overproduction of private goods and the underproduction of public services. He was particularly concerned about the failure to adequately develop both the social and the physical infrastructure necessary to maintain national cohesiveness and future growth in the Industrial Age. He believed America had become "a community where public services have failed to keep abreast of private consumption. . . . Here, in an atmosphere of private opulence and public squalor, the private goods have full sway." The negative impacts for societal well-being, he contended, were exceptionally acute. Because production of consumer goods had become so central to the development of the American economy, the nation had failed to make expenditures on the public goods necessary to maintain future societal growth. One particularly troubling outcome was continued inequality of opportunity. The children of wealthy and middle-class families had a distinct advantage over the children of poor families, particularly because the education system was so inequitable. Interestingly, by today's standards income inequality wasn't a large problem. Galbraith's arguments that there was still so much to do would help drive a second wave of institutional developments in the 1960s.[26]

As this evolution in progressive governing was happening, however, a conservative movement was gestating. One of the most important voices of this nascent crusade was the Nobel Laureate Milton Friedman, who argued that there was too much government regulation of business and that it was stifling growth. In this he was advocating for the return to the individualism that had reigned supreme prior to the Great Depression. The "preservation of freedom," he wrote, "is the protective reason for limiting and decentralizing governmental power. But there is also a constructive reason. The great advances of civilization, whether in architecture or painting, in science or literature, in industry or agriculture, have never come from centralized government." Regardless of the dubious historical support for that statement, it found an important audience among the American electorate and, more critically, among financial elites. They cheered Friedman's notion that government should essentially provide only three basic services: preservation of the rule of law, enforcement of private contracts, and promotion of competitive markets. This philosophical approach held that it was most important to protect economic freedom, and if that meant income inequality increased, that was simply because those falling behind lacked a work ethic. While these ideas would fail to shift elections or policy in the near term, within two decades they would provide the ideological underpinning for a conservative backlash against government provision of either macroeconomic or microeconomic security—and most certainly environmental security. In the meantime, however, a series of American presidents would lead the way in advocating for major advances in the latter two policy areas.[27]

When John Kennedy entered the White House, he had significant plans for building upon previous microeconomic security programs. In his speech accepting the Democratic Party's nomination for the presidency in 1960, he framed these plans by saying, "We stand today on the edge of a New Frontier —the frontier of the 1960's—a frontier of unknown opportunities and perils —a frontier of unfulfilled hopes and threats. . . . Beyond that frontier are the uncharted areas of science and space, unsolved problems of peace and war, unconquered pockets of ignorance and prejudice, unanswered questions of poverty and surplus." He described the election as a choice "between the public interest and private comfort—between national greatness and national decline—between the fresh air of progress and the stale, dank atmosphere of 'normalcy'—between determined dedication and creeping mediocrity." A mythology has arisen that because he was so focused on battling the Soviets in the Cold War, Kennedy's domestic agenda largely failed. The record seems to tell a different story. As Theodore White pointed out, during Kennedy's thousand days in office more legislation was passed than at any other time since the 1930s, much of it aimed at extending microeconomic security. This included an expansion of unemployment benefits, Social Security coverage, food distribution to the poor, school lunches, rural electrification, soil conservation, crop insurance, farm credit, and the minimum wage; funding for urban housing, highway construction, hospital construction, and other public-works projects; and conservation measures such as the Clean Air Act of 1963 (which established funding for the study and cleanup of air pollution), water pollution control measures, and the growth of the national park system. With these latter policies, the nation saw environmental security begin to emerge on the national stage.[28]

After Kennedy was assassinated, his successor, Lyndon Johnson, introduced his own blueprint for growing microeconomic security. In May 1964 he delivered the commencement address at the University of Michigan, in which he argued that it was time to dedicate the nation's great wealth and power to the task of elevating national life. "The Great Society," he said,

> demands an end to poverty and racial injustice. The Great Society is a place where every child can find knowledge . . . a place where leisure is a welcome chance to build and reflect . . . a place where the city of man serves not only the needs of the body and the demands of commerce, but the desire for beauty and the hunger for community . . . a place where man can renew contact with nature . . . a place where men are more concerned with the quality of their goals than the quantity of their goods. But most of all the Great Society is not a safe harbor, a resting place, a final objective, a finished work. It is a challenge con-

stantly renewed, beckoning us toward a destiny where the meaning of our lives matches the marvelous products of our labor.

While his predecessor had successfully gained approval for a large number of incremental advances in microeconomic security, Johnson leveraged his congressional mastery to gain a stunning collection of legislative achievements.[29]

In 1954 the Supreme Court had dealt the first real blow to southern apartheid in *Brown v. Board of Education*, declaring state laws establishing separate public schools for black and white students to be unconstitutional. In the subsequent years, the civil rights movement had continued to push for the total elimination of racial segregation and discrimination. This had led to new waves of violence in southern states opposed to integration of the races, with nonviolent protests aimed at swaying wider public opinion finally forcing the nation to take the issue seriously. The defining moment for the movement was the March on Washington in August 1963, where Reverend Martin Luther King delivered his "I Have a Dream" speech. Standing before the Lincoln Memorial, he opened by emphasizing that one hundred years after the signing of the Emancipation Proclamation, African Americans still were not free. The time was ripe, he told the 250,000 people assembled on the National Mall and the millions watching live on television, for all Americans to be guaranteed the rights to life, liberty, and the pursuit of happiness. "So I say to you," he proclaimed, "even though we face the difficulties of today and tomorrow, I still have a dream. It is a dream deeply rooted in the American dream. I have a dream that one day this nation will rise up and live out the true meaning of its creed: 'We hold these truths to be self-evident, that all men are created equal. . . .' I have a dream that my four little children will one day live in a nation where they will not be judged by the color of their skin but by the content of their character." Although President Kennedy had failed to act on campaign promises regarding civil rights, this speech helped push his administration to support civil rights legislation in the year before his assassination. It was left to his successor, himself a southerner, to take on this politically contentious issue and make King's dream a reality.[30]

After becoming president, Lyndon Johnson masterminded passage of both the Civil Rights Act of 1964 and the Voting Rights Act of 1965. These momentous bills officially ended southern apartheid, although they would take decades to fully implement. While it is more than fair to say that even today key objectives of the legislation remain only partially fulfilled, the benefits for African Americans and the nation as a whole have been tremendous. From a political perspective, however, passage of these civil rights acts proved devastating for Democrats, because it eliminated support for their party throughout the South. In the next ten presidential elections, only Republicans and Democratic governors from southern states were able to capture the White House. It would take more than two decades for the civil rights legislation to fully impact the Democratic hold on Congress, but when it did it, power was decisively handed over to the Republicans. Interestingly, this southern backlash came at the same time that the benefits enjoyed from the destruction of extractive institutions were transforming the regional economy. While this process is also far from complete, the South is emerging as a growth region despite state governments' continuing to stymie efforts to provide adequate economic security at a local level.

In 1965, Johnson gained approval for eighty legislative proposals. This second wave of progressive lawmaking resulted in significant expansions of the government's role in supplying macroeconomic and microeconomic security. Medicare provided federal funding for health insurance for older Americans. Medicaid provided funding to provide medical care for low-income Americans. The Elementary and Secondary Education Act for the first time in history provided federal aid for K–12 education, including the newly created Head Start program. The Economic Opportunity Act, part of Johnson's War on Poverty, provided money for a wide array of local antipoverty programs. The Department of Transportation was created to consolidate various agencies while providing new funding for public transit and highway safety. The Department of Housing and Urban Development was created to assist low-income Americans in finding adequate housing and to provide funding for neighborhood-renewal projects. Finally, Johnson significantly upped the ante by championing a slew of bills to provide environmental security. "We have always prided ourselves on being not only America the strong and America the free," said Johnson, "but America the beautiful."[31] During his presidency he began to turn this rhetoric into reality with a series of new environmental laws, including the Wilderness Act of 1964, the Land and Water Conservation Fund Act of 1965, the Solid Waste Disposal Act of 1965, the Motor Vehicle Air Pollution Control Act of 1965, the National Historic Preservation Act of 1966, the Endangered Species Preservation Act of 1966, the National Trails System Act of 1968, the Wild and Scenic Rivers Act of 1968, the Aircraft Noise Abatement Act of 1968, and the National Environmental Policy Act of 1969. While many environmentalists believed these bills were half measures, they would help form the foundation for subsequent legislation that still remains vital today. Environmental security had clearly arrived as a key component of the national debate over the role of government in a modern society. In the coming decade, under three different presidents, the nation would bolster these early efforts, but a vitriolic conservative backlash would place the very health of the planet at risk.

* * *

At the same time that the nation was enjoying unparalleled economic security at home, American leaders were wrestling with the new challenges of world leadership. Having won the war, the nation sought to win the peace by containing communism and promoting accelerated globalization. Petroleum would be central to achieving these objectives. As David Painter writes, "Controlling oil helped the United States contain the Soviet Union, end destructive political, economic, and military competition among the core capitalist states, mitigate class conflict within the capitalist core by promoting economic growth, and retain access to the raw materials, markets, and labor of periphery nations in an era of decolonization and national liberation." This process began in March 1947, as communism threatened to spread to non–Eastern Bloc countries, when President Truman announced a new national security doctrine that would come to bear his name. In this initial phase of the Cold War, he suggested that every nation must be permitted to choose whether to be governed by free institutions or a totalitarian regime. "I believe that it must be the policy of the United States," he said, "to support free peoples who are resisting attempted subjugation by armed minorities or by outside pressures. I believe that we must assist free peoples to work out their own destinies in their own way." While he favored financial aid aimed at providing economic stability, he held the door wide open for military support and engagement. The previous year, George Kennan, the US deputy chief of mission to the Soviet Union, had sent his famous "Long Telegram" arguing that the Soviets were incapable of foreseeing "permanent peaceful coexistence" with the Americans because they believed security could only be achieved through a "patient but deadly struggle for total destruction of rival power." The only solution was to show strength in the face of Soviet aggression, which was the only language the communist regime recognized. This meant that the United States needed to dedicate itself to the containment of Soviet-style communism anywhere lest it spread everywhere. When adopted as the basis of the Truman Doctrine, this became American policy for the subsequent three decades. Throughout that period, maintaining access to global oil supplies was critical. This was demonstrated by the expansion of operations by American oil companies throughout the Middle East and the concurrent engagement of the United States in Saudi Arabia, Iran, and Egypt.[32]

One of the most important aspects of this containment policy was promoting American-style globalization. President Truman began by championing the Marshall Plan, which funded the reconstruction of war-torn Europe and Japan. As Tyler Priest writes, this included providing "European nations recovering from war with scarce dollars to purchase large amounts of oil flowing from American operations in the Middle East, thereby sealing Western Europe's long-term dependence on this source." The program succeeded terrifically, as evidenced by the strong trade relationships that still exist between the United States and these economic powerhouses. The International Monetary Fund, the General Agreement on Tariffs and Trade, and the Bretton Woods Agreement further advanced Western capitalism. Four major military alliances— the North Atlantic Treaty Organization (NATO), the Southeast Asian Treaty Organization (SEATO), the Australia, New Zealand, United States Security Treaty (ANZUS), and the Central Treaty Organization (CENTO)—built upon this foundation and represented one of the most momentous shifts in American national security policy. Although the process had begun with McKinley, this signaled a formal break with the isolationist policies that had held sway since George Washington's neutrality proclamation. These security arrangements were crucial in the nuclear age and are a major reason why there have been no great-power wars in well over a half century. At the same time, however, the United States began developing institutions that made it far easier for America to engage in more limited conflicts. As he was preparing to leave the presidency, Dwight Eisenhower delivered a farewell address that diagnosed the danger in this changed reality. "Until the latest of our world conflicts, the United States had no armaments industry," he said. But as a result of the Cold War, "we have been compelled to create a permanent armaments industry of vast proportions. . . . Now this conjunction of an immense military establishment and a large arms industry is new in the American experience. . . . We recognize the imperative need for this development. Yet, we must not fail to comprehend its grave implications. . . . In the councils of government, we must guard against the acquisition of unwarranted influence, whether sought or unsought, by the military-industrial complex." This military-industrial complex has never pushed America into war by itself, but there is no doubt that the enduring availability of a massive arsenal of weapons made it far easier for politicians to decide to commit American troops. While presidents had never been hesitant to engage militarily in the past if the national interest demanded it, the ability to do this without waiting for industry to be retooled to provide the necessary arms was unique in the national experience. And in the post-war era the definition of national interest morphed significantly as the United States took on the role of globocop.[33]

The American Empire was never culturally suited for this task, which is probably one reason why the citizenry has never really embraced it in the same way the British public did. The sad reality, however, was that First World (the American sphere) efforts to contain Second World (the Soviet sphere) expansion into the Third World (everyone else) meant that the United States had to have a presence everywhere. This was even truer because of the collapse of the British Empire, which had bankrolled much of the Second World War by selling off military bases across the globe to the United States. Paul Starobin refers to this as America's Accidental Empire, "the global empire that it never sought, as a conscious matter, to possess."[34] The result was that American covert actions and involvement in the Korean and Vietnam Wars began to erode the national image (at home and abroad) and contributed to a relative national decline starting in the 1970s.[35] The country was fortunate that from 1933 to 1981 the INNATE revolutions had resulted in such tremendous social growth that it was possible to weather the coming storm, at least temporarily.

INNATE Revolutions: Electrification
Electrifying the South and West

Two of the most important pieces of legislation passed during the twentieth century are little remembered today. Their impact was largely hidden, unlike high-profile successes like social insurance and civil rights and environmental protections. Yet these bills fundamentally changed the demographic development of the nation and ultimately accelerated the steady demise of an entire way of life. They led to a massive population migration away from the Northeast and Midwest and toward the South and West. In several ways, they created a society that was far less sustainable. The bills were the Tennessee Valley Authority Act of 1933 and the Rural Electrification Act of 1935.

During the early twentieth century rural leaders had been concerned that the communities they represented were being left behind economically, largely because they had limited access to electricity. In 1910 a report on country life found that only 2 percent of farmers had electricity. They had also long suffered from limited access to good schools, roads, hospitals, and communications. Yet whereas many European countries were acting to bring electricity to rural areas, American leaders took a laissez-faire position: if the farmers wanted electricity, they would figure out how to obtain it. For more than half a century, agrarians had struggled with the double-edged sword of more efficient farming techniques. On the one hand, they could adopt innovations (steel plows, mechanical reapers, chemical fertilizers) that allowed them to grow far greater amounts of food; on the other hand, these new technologies created labor difficulties as prices decreased. As industrialization and electrification transformed the nation, farmers were hurting. As David Nye writes, "Poor, isolated, and squeezed by falling prices, farmers had become dependent on the banks and the distant markets that determined the value of their labor."[36]

The problem was that private utilities saw no financial upside in providing electricity to farmers, largely because distribution lines were so expensive and the number of customers so small. This was a classic market failure. Electrification had the potential to provide great benefits to rural Americans, but the costs were too high for market actors. Agricultural communities already along an interurban grid were the exception. In those areas, the arrival of electricity completely revolutionized growing and animal husbandry. For example, cows produced more milk with electric pumps continuously replenishing their water supply, electric milking machines allowed one and a half times as many cows to be milked in a given period, and electrically heated water allowed for better sterilization of equipment necessary to get milk to market. "With electricity farm families had less heavy physical work to do, and they could support larger herds using the same manpower." Unfortunately, this didn't mean greater profits, because as supply increased prices decreased. Regardless, farmers still eagerly sought electricity. For those farmers not situated near an interurban grid, this meant finding decentralized sources like water motors, water dynamos, micro hydroelectric plants, wind turbines, and small gasoline-powered generators. Even with these new technologies, by the mid-thirties only 10 percent of US farms had electricity; other advanced nations had much higher levels.[37]

One of the most interesting developments with regard to rural electrification was regional. The Northeast and the Far West saw much higher levels compared with the Midwest, the South, and the Great Plains. In the Northeast, this was because the region enjoyed population density and high average rainfall, both of which created an incentive to increase productivity. The small farms in these states were also close together, making laying local grids more economically feasible. The much drier Far West, meanwhile, benefitted from a significant national investment in hydroelectric dams and irrigation projects.

Both regions relied on vegetable and dairy farming, whose products needed to reach markets quickly and thus benefitted a great deal from refrigeration. The Midwest, South, and Great Plains had much larger farms that weren't as easy to connect to the grid. They also focused on grain crops, tobacco, and cotton, which didn't need to be immediately transported to market. These regions offered "low and irregular demand spread over a dispersed area."[38]

During the 1932 presidential election, rural poverty emerged as an important political issue. President Herbert Hoover had declared that rural electrification was a state issue. Governor Franklin Roosevelt, who had promoted rural electrification in New York, argued that electrification was vital to bringing modernity and better living to American farms. This was during a period when Jeffersonian notions of an agrarian paradise were experiencing a resurgence, with people blaming industrialization for the financial collapse. As the nation was undergoing one of the most wrenching economic reorientations in history, many argued that the only way to save (or restore) the republic was to return to the land. Electrification was probably not the best means for achieving this, since it dramatically reduced the need for labor on farms. Industrialization and urbanization, however, were easy targets during this era. Given the existing widespread hardship, it is understandable that people engaged in fantasies about how much better things would be if the country returned to its decentralized roots. Ironically, only the national government could provide rural electrification, which was itself a sign of things to come.[39]

Shortly after his election, Roosevelt began to develop a plan to tackle the problem. The first step was the creation of the Tennessee Valley Authority (TVA), which would impact Alabama, Georgia, Kentucky, Mississippi, North Carolina, Tennessee, and Virginia, all states central to the New Deal Coalition. Nye writes that Roosevelt was "excited by the possibilities of creating a series of dams that would transform the economy of a region two-thirds the size of England through inexpensive power, flood control, and improved navigation." To make this happen, the government would have to bypass private utilities and create a government corporation. Not surprisingly, this led to claims of a socialist takeover. Although a national board did direct the TVA, rural cooperatives managed the transmission facilities—hardly the centralized approach skeptics decried. Regardless, in the late thirties the TVA faced several lawsuits arguing that it was unconstitutional for a government corporation to sell power. In *Tennessee Electric Power Company et al. v. TVA* (1939), the US Supreme Court found that the arrangement was constitutional. In the coming decades the TVA would construct nearly fifty dams and eventually diversify into coal plants, gas plants, nuclear plants, and renewables. From an economic perspective, the TVA provided a significant stimulus and essentially delivered modernity to a large region of the nation.[40]

Although less well known, the Rural Electrification Administration (REA) almost certainly had a bigger impact, largely because it served far more of the nation. "Unlike the technical-aid missions of the Cold War, which transferred U.S. technology to so-called Third World nations," writes Ronald Kline, "the REA targeted an 'undeveloped' society *within* the United States: six million American farms." Although bringing power to these farms was initially opposed by private utilities, they got on board once they realized they could sell electricity wholesale to rural cooperatives. Begun during the heart of the Great Depression, the program stimulated local economies both by creating construction jobs and by establishing new markets for electrical appliances. By the start of the Second World War, it had connected nearly a million farms to the grid. REA activities were highly decentralized, with the national headquarters distributing loans to small rural cooperatives (serving on average about eight hundred customers) that built and operated the system. The program slowed considerably during the war years as budgets were diverted to other, more pressing needs, but within a half decade after V-J Day nearly every rural farm had electricity. That almost every dollar invested in this massive infrastructure development was ultimately repaid to the Treasury is an indication of how vital the national government can be in providing macroeconomic security without creating large budgetary deficits. According to Malone, "The R.E.A. is one of the most immediate and profound successes in the history of federal policy-making."[41]

With decades of electrical innovations arriving on farms virtually overnight, it isn't surprising that rural life was instantly transformed. The initial priority for most was buying the home appliances that could already be found in almost any urban apartment or suburban home. Eventually farmers also realized that electricity could modernize the entire production process. Lighting added time to the workday. Electric pumps provided water to livestock for pennies a day. Engines powered energy-intensive gristmills. Electrified wire could easily contain herds. The result was near parity with urban working conditions. Of course this also meant that farms increased production while also reducing labor requirements, eliminating millions of agricultural jobs. Rather than providing a bulwark against the rapid urbanization of the country, it served to decrease the necessity for a large agrarian workforce and thus led to the further depopulation of rural areas. Thus, it was yet another nail in the coffin of the Jeffersonian ideal.[42]

One of the major results of rural electrification was that living outside the urban core became more attractive. As David Nye writes, the program "encouraged urban deconcentration, and Americans moved farther and farther away from the city as rural areas were electrified, attempting to retreat from the city while giving up as few modern conveniences as possible. The result was not a pastoral utopia . . . but rather the sprawling 'crabgrass frontier' of an extended suburbia."[43] It is hard to overestimate the importance of this new reality. Electricity distribution is far more efficient when the population is densely packed together. As cities sprawled, more and more power production was needed to maintain American standards of modernity.

At the same time, regions that had previously been largely inhospitable to human life were increasingly welcoming. Prior to the Second World War, the South had few cities, and the areas in between towns were sparsely populated. James Howard Kunstler argues that three things changed all that: cars, electricity, and air conditioning. Automobiles made it far easier to travel around the vast countryside. The TVA and the REA provided electricity, with suburbanization and industrialization leading to expanded economic opportunities. This in turn began the migration to the South of workers looking for new opportunities. The introduction of air conditioning in the 1950s and 1960s accelerated this movement, as areas that had previously been extremely uncomfortable could be made livable through artificial means. A worker could travel from his or her air-conditioned house in an air-conditioned car to an air-conditioned office or factory. As Daron Acemoglu and James Robinson argue, the success of the civil rights movement also removed the extractive nature of southern economic institutions, making the South a more attractive place to live. All of this meant that a region that had previously been lightly populated was now open for business, and during the postwar years millions moved to this region. In retrospect, this resulted in two significant problems. First, because land was so cheap, the cities that began rapidly expanding in the region developed horizontally rather than vertically. The lack of density meant that these urban and suburban areas used much more energy than did their more vertical northern counterparts. Second, since these cities were being developed in zones with hostile weather that required air conditioning during a significant portion of the year, they used more fossil fuels. Later this same process repeated itself in the Southwest, particularly in cities like Phoenix, Las Vegas, and Denver.

Places that had previously been deemed poor locations for large-scale settlement ultimately joined the Atlantas and Houstons as exciting destinations for a rising middle class. None of this would have happened if not for the large national investment in a rural electricity grid under the TVA and REA. On the other hand, the large negative climate impacts might have been avoided.[44]

* * *

The Electrification Revolution took place during the same period when the American Empire was emerging. The national economy became the largest in the world just as electricity was being introduced. By the time the United States had become the most important nation geopolitically, it had become heavily electrified, with this new power source dominating urban and industrial settings and redefining the expansion of the suburban landscape. It can be persuasively argued that this economic and geopolitical reality wouldn't have been possible without electrification. Coal-fired electricity helped push the nation forward economically. The resulting economic might, combined with geographic isolation, helped Americans weather the storm of two world wars better than any other major power. Thus, the country was able to fill its superpower role in the postwar years.

Electricity also changed what it meant to be an American. In its early days having indoor lights was the ultimate status symbol. Today electricity is everywhere, in our factories and houses and appliances and smart phones. Vaclav Smil writes that "electricity generation, transmission, and use represented unparalleled achievements in energy innovation."[45] It required an entirely new infrastructure that fostered an interconnected society in a way that previous power sources had not. At the same time, the scope of electricity production changed fundamentally over time. While early boilers were relatively tiny, today they are many stories tall and consume fossil fuels at an incredible rate. In 2011 electricity accounted for nearly 40 percent of American energy consumption. Given that 70 percent of electricity is produced by fossil fuels (mostly coal and gas), one-third of national greenhouse gas emissions come from this sector. Therefore, although the Electrification Revolution was clearly critical to the American Empire, today it poses a significant threat to the planet.[46]

INNATE Revolutions: Transportation
The Rise of Highways, Part 2

After years of halting progress in building good-quality roads throughout America, during the Great Depression President Franklin Roosevelt provided major funding to the endeavor for the first time in an effort to put tens of thousands of unemployed men to work and inject cash into the economy. The New Deal transportation projects shifted away from solely financing rural roads, with more money directed toward urban routes and a national highway network. This adjustment occurred largely because so many more people were driving. Henry Ford's pioneering of the mass-production system for automobile construction, which brought down costs significantly, led to an explosion in the number of cars and trucks on the roads. This clogged urban roadways at a time when municipal transit systems were facing nearly insurmountable challenges. Bruce Seely writes that "transit operations were in trouble because bad service, crowded and old cars, weathered stock, corruption in franchise awards, and general mismanagement produced tight regulation of fares at a time when street railways needed capital for expansion or equipment. Automobiles struck many as an apolitical solution to urban transportation and other problems, such as corruption and the need to ease access to the urban periphery from the squalid inner city. In other words, the auto promised both social and moral reform." Ironically, the Second World War saw both the biggest boom in streetcar passengers and the roots of public transportation's demise. Although transit ridership vastly increased during the war years as a result of gas rationing and restrictions on automobile manufacturing, the sudden surge in ridership put too much stress on the already fragile infrastructure, which essentially crumbled in the years after the war.[47] Thus, during the postwar years it was logical for Americans to fully transfer their transportation allegiances back to automobiles. With economic prosperity returned, this led not only to an explosion in automobile ownership but also to an explosion in oil consumption (fig. 3.2). Driven by the transportation revolution, petroleum surpassed coal as the most important hydrocarbon resource, a status it continues to hold.

By the late 1930s President Roosevelt had become an avid supporter of superhighways, which could connect the nation's cities from coast to coast. He actually had a much bigger role in the adoption of the highway network than President Eisenhower. During Roosevelt's second term, Thomas MacDonald's Bureau of Public Roads undertook a series of studies to design a national highway system, with an emphasis on determining how to fund the project. In 1938 the bureau released *Toll Roads and Free Roads*, which proposed a 26,700-mile network of highways. The report argued that toll roads were not feasible, because they would only earn back 40 percent of needed construction costs since most automobile trips were less than twenty miles. Instead the bureau should focus on urban freeways: "In the larger cities generally only a major operation

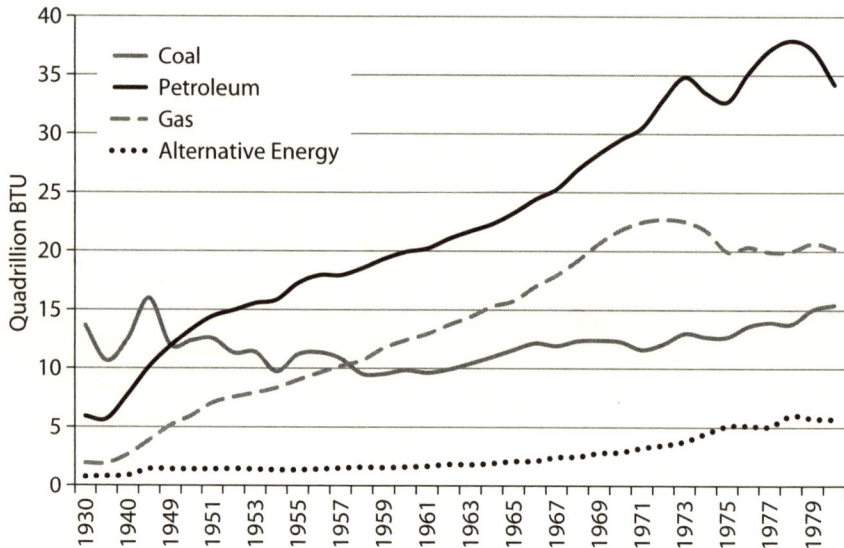

Figure 3.2. US Energy Consumption by Source, 1930–1980. US Energy Information Administration, *Annual Energy Review, 2009.*

will suffice—nothing less than the creation of a depressed or an elevated artery (the former usually to be preferred) that will convey the massed movement pressing into, and through, the heart of the city, under or over the local cross streets without interruption by their conflicting traffic." Earl Swift writes, "To read these passages today, tucked into the dense text of a little-known 1939 report, is to feel a twinge of foreboding, for the urban-renewal formula laid out in *Toll Roads and Free Roads* was exactly that adopted by cities across the nation a few years later—and because, for all of its clarity and comprehensiveness, the document overlooked an important element of the slum areas it targeted: degraded though they might be, they were home to millions of people."[48] This was a quintessential example of the engineering mind-set. Bureau engineers were more interested in finding ways to efficiently move metal beasts out of urban areas than they were in the effect this would have on people living in those cities. Of course at this point in American history no one knew that the combination of urban freeways and long-distance highways would also have such an enormous impact on the very health of the planet. Drawing on economic and national security rationales, FDR accepted the findings of the report without fully understanding its full effect.

As the nation was still battling to defeat fascism and solidify its geopolitical leadership, Congress approved the Federal-Aid Highway Act of 1944. This legislation authorized construction of a continental interstate highway system with routes designed to connect major metropolitan areas. Not surprisingly given the importance of big-city voters in the New Deal Coalition, the bill also directed national funding to urban freeways. As suggested earlier, the postwar period saw a renewed embrace of automobiles as many industrial facilities switched from producing armaments and started manufacturing cars again. Although the roads were in terrible condition after years of wartime neglect, no significant new funding was allotted for maintenance or expansion for more than a decade. "Soaring auto and truck traffic soon overwhelmed plans for sufficient road building. . . . Inflation, war, and changed auto and truck design, all factors outside highway department control, hobbled efforts to build needed mileage." President Dwight Eisenhower stepped into this morass with almost no knowledge of the complexities of the issue, although he had campaigned for modern highways. One of his first acts upon taking office was to replace Thomas MacDonald with Francis du Pont as head of the newly created Public Roads Administration (now housed in the Department of Commerce). An heir to the Delaware chemical fortune, du Pont might have seemed an odd choice to replace the nation's top expert on road construction. But du Pont was a skilled engineer who turned out to know a lot about highways and was eager to employ MacDonald's deputies to move the network forward. The year after Eisenhower took office, his administration announced: "A $50 billion highway program in ten years is a goal toward which we can—and we should—look."[49]

As work began to make this pledge a reality, the special interests began to circle. Federal highway engineers wanted to produce superhighways that would not only relieve congestion but also create jobs and revitalize cities. They also wanted the system to be funded at the national level, guaranteeing their influence over the project. Truckers advocated superhighways that linked urban centers and relieved urban congestion while also reducing taxes and fees. Supporters of rural roads wanted linkages between small towns. Farm-to-market advocates preferred roads that would make it easier to get crops to market. Eisenhower was most influenced by economists, who favored a highway system for its immediate and long-term economic benefits. Interestingly, among all these groups there was a near universal desire to significantly expand the road network. There was no outcry for renewal of the crumbling railroads; for better or worse, the torch had been firmly passed to the highways.[50]

The Federal-Aid Highway Act of 1956 ultimately succeeded not because it chose sides but precisely because it refused to do so; it gave everyone what they wanted. It approved a highway system that would be forty-two thousand miles in length. Federal aid would fund most of the network, with 90 percent of the funding coming from the US Treasury. This funding would be generated by modest tax and fee increases, in particular an excise tax of three cents per gallon of gasoline. The money would be directed to a highway trust fund, which guaranteed that the aid could not be diverted and that anything left unspent could be banked to ease the distress caused by unavoidable future dips in revenue. According to Christopher Wells, this approach of levying essentially invisible taxes "helped the self-fueling system in the political world: the legal linkage of gasoline taxes to road budgets has shielded car-and-highway-centered transportation policy from the regular discussion and debate that typifies the allocation of general funds." As a result, there was no political cost to authorizing $25 billion for road construction over thirteen years, even as the actual cost soon ballooned to $130 billion. Almost eighty years of policy development and special interests" wrangling culminated in this landmark legislation. Although the subsequent decades would see significant challenges in actually constructing this system, particularly from urban activists concerned about the societal impact of the highways, the entire network would ultimately be built. It was arguably the largest public-works project in history. Today it represents one of the key threads in the fabric of the nation, one that requires spending nearly $150 billion *annually* to operate and maintain.[51]

What is troubling from today's perspective is that the highway system provided the circulatory structure that automobiles and trucks would use to transform the national landscape. As Smil writes, "Changes brought by private cars—economic, social, and environmental—rank among the most profound transformations of the modern era." On the positive side of the ledger, the program to build the highway system was a key macroeconomic security policy that promoted growth in both the construction and manufacturing sectors. America would be the world's largest exporter of automobiles for several decades. Open roads and individual cars provided a sense of freedom that fit the cultural and emotional mind-set of many Americans extremely well. Unfortunately, the negative side of the ledger seems to have increasingly prevailed. First, rather than revitalizing cities, highways actually led to the demise of urban centers. They promoted the rapid expansion of suburbia, which James Howard Kunstler has called "the greatest misallocation of resources in the history of the world." Suburbia inherently lacks sustainability, largely because it turns its back on the many advantages of urban density. Second, Americans have "been willing to put up with enormous death, injury, and pollution costs" to preserve car ownership. Third, and most importantly, highways and suburbia elevated the internal-combustion engine to its position of supremacy within the transportation sector. The internal-combustion engine "is a poor choice of power. It dissipates 80 percent of its energy as heat, even before it reaches the rear axle." Rather than relying on much more efficient transportation options, like trains and public transit, America turned almost solely to automobiles and trucks for local and regional travel. As a result, transportation now produces more than ten billion metric tons of carbon dioxide annually, more than one-quarter of all American greenhouse gas emissions. None of this would have occurred if national leaders had not championed highways and written off railroads and mass transit systems as relics of the past.[52]

INNATE Revolutions: Transportation
Emergence of Commercial Airlines, Part 2

Although the 1930s were the most innovative years in aviation history, the technology did not progress fast enough to make the public feel safe. Flying during this period was still very dangerous, with a high number of deaths per passenger mile on commercial airlines. Several high-profile accidents during this decade dramatically influenced the government's role in the aviation industry. On 31 March 1931 a TWA Fokker F-10A crashed near Bazaar, Kansas, killing eight, including the famed Notre Dame football coach Knute Rockne. The crash led to the grounding and eventual commercial demise of the once popular plane. The most significant aviation accident in history occurred a few years later. On 6 May 1935 a TWA DC-2 crashed in Missouri, killing five. Among the fatalities was the Republican senator Bronson Cutting of New Mexico. While a Bureau of Air Commerce report cited the probable cause of the crash as poor weather reporting and pilot error, an independent study conducted by the US Senate cited poorly manufactured navigational aides. The controversy over Senator Cutting's death led to renewed legislative efforts to improve aviation safety. The result was passage of the watershed Civil Aeronautics Act of 1938.[53]

Earlier aviation legislation had focused primarily on standardization in the name of safety. The Civil Aeronautics Act, however, laid the foundation for comprehensive government oversight of the commercial aviation industry. The statute created a new agency, the Civil Aeronautics Authority, whose principal goals were to maintain order in the industry, shield carriers from damaging competition, and settle ticket rates at levels fair for consumers and acceptable for airlines—quite different from what had happened earlier with railroad regulation. One of the most important early tasks of the new organization was promulgation of the Civil Aeronautics Regulations. Since 1926, safety regulations had been issued without any system for clearance through a central office, which hindered enforcement. To resolve this problem, Fred Fagg and John Wigmore, of Northwestern University Law School, were appointed to compile a new set of regulations. These would ultimately transform flight safety. Two years later, Congress acted upon the advice of President Franklin Roosevelt to create the Civil Aeronautics Board, tasked with aviation-safety investigations and economic regulation of the airline industry, including awarding routes and establishing fare structures. These actions set the foundation for the rapid expansion of the airlines, but before this could occur, the nation was engulfed in a worldwide conflagration.[54]

The Second World War fundamentally altered the course of the commercial aviation industry. Prior to the war, there were fewer than three hundred transport aircraft in the entire United States. By war's end, more than ninety thousand planes were being produced annually. Although these were largely fighters and bombers, the American fleet of transports grew considerably. During the war years, commercial airlines had more business (for both passengers and cargo) than they could handle. It was during this period that the Boeing Stratoliner became the world's first aircraft to have a pressurized cabin, allowing it to fly at an altitude of twenty thousand feet and a speed of two hundred miles per hour. This technical advance set the stage for a major expansion in the industry in the postwar years.[55]

The massive government investment in the development of military aircraft during the war enabled rapid progress in aviation in the postwar era. Looming on the horizon was the development of jet aircraft, although serious challenges to making this technology viable for military and commercial airplanes remained. After the war, the United States lagged behind the Europeans in jet-engine technology. Critics blamed the National Advisory Committee on Aeronautics for allowing the nation to fall behind, an argument with some legitimacy considering the obvious lead the British enjoyed in developing jet aircraft. The concept for practical jet propulsion emerged in the late 1920s. Frank Whittle, a twenty-one-year-old student at the Royal Air Force Officers Training College, proposed the revolutionary alternative to the piston engine in his senior thesis. Although Whittle received a patent for the engine, the British Air Ministry initially dismissed his idea and did not provide funding for developing the engine until 1939. The Gloster Meteor, launched in 1944, was powered by a jet-propulsion system, but it did not fly combat missions. In fact, the first jet fighter was the German Me-262. Starting a full five years behind the British, the German airplane designer Dr. Hans von Ohain, working with substantially more developmental funding, fielded a combat-ready aircraft the same year as the British (although its introduction was far too late to sway the outcome of the war). Ohain and Whittle were recognized as the co-inventors of the jet engine.[56]

The US Army Air Forces led preliminary American efforts to develop a jet aircraft during the 1940s. General Henry "Hap" Arnold negotiated an agreement with the British to utilize Whittle's turbojet technology, which during the war years led to an intense competition between various defense contractors to design a jet bomber. This struggle ultimately translated into a battle for preeminence in the construction of commercial jet airliners during the postwar years. The Americans came from behind in the race to create the first practical commercial jetliner and eventually overtook the British with the Boeing 707. By the early 1950s, based on the work it had done on the B-47 and B-52 bomber programs, Boeing had become the world's leading builder of large military jet aircraft. Within a few years the company made the risky decision to develop a commercially viable jet airliner. In October 1958 the Boeing 707 entered service with Pan American Airlines and was an instant success. The 111-passenger, four-engine jetliner could fly at an altitude of forty-one thousand feet and a speed of nearly six hundred miles per hour. The airplane made its first New York–to–Paris flight in under nine hours, vaulting Boeing into the lead among worldwide airframe manufacturers, a position it would maintain for the next three decades.[57]

Two victims of the success of the airlines were trains and ocean liners, neither of which could compete with the speed and cost of air travel. As trains and the telegraph had done in the previous century, airliners made the world a dramatically smaller place. "The speed and range of these planes . . . made it possible to travel among virtually all major cities of the planet in a single day. Naturally, cargo deliveries [could] duplicate this feat."[58] The national government had adopted safety regulations and made critical investments in technology and infrastructure that made all of this possible. And underlying it all was cheap oil, which provided the energy that powered the entire endeavor. Today the aviation sector produces approximately 3 percent of all greenhouse gas emissions.

* * *

America's adoption of cutting-edge transportation technologies starting in the early nineteenth century exemplifies how America attained national greatness. The Transportation Revolution began with the use of steamships and the construction of canals to open the trans-Appalachian West. It accelerated as the population embraced development of a coast-to-coast railroad network that was far larger than any other at the time. The transition to automobiles dramatically bolstered the manufacturing, construction, and hydrocarbon sectors. Finally, the adoption of airliners made America the leader in the other key transportation sector of the twentieth century. What was most important about all these modes of transportation was their synergistic relationship with the national economy. "Systems and methods of moving goods and people have driven the American economy forward," writes Joseph Devine, "and advances in the general economy have propelled improvements in transportation."[59] As with manufacturing, the history of transportation helps us understand how and why the United States increased its demand for energy within this crucial economic sector.

Almost certainly, America could have made different decisions about transportation and still benefitted economically. As a nation, it made a series of decisions that effectively killed railroads in favor of cars, trucks, and airliners. Although there were good reasons for these decisions, more intentional thinking about likely outcomes might have led the county in a different direction. With the benefit of hindsight, it is evident that the chosen path has had negative consequences for the global environment. A critical question as the nation moves into the future is whether it can change direction and adopt a system more suited to the twenty-first century and beyond.

INNATE Revolutions: Agriculture
America's Agricultural Revolutions

Native Americans had been farming for nearly five millennia when Europeans arrived. Maize, beans, and squash underpinned agricultural societies in North America. Many Native Americans used fire to shape the landscape for farming. They learned to use intercropping and multiple cropping to increase crop yields, while maintaining soil health. Particularly in the American Southwest, they built sophisticated irrigation systems and terraces to provide the water needed for farming in a warm climate. The tribe often controlled the land, although farmers with specified plots raised crops to meet the needs of both their own families and the larger community. North America was not an unspoiled wilderness when Europeans arrived. In fact, Native Americans had completely transformed the landscape over the millennia. However, a huge proportion of the native population had been lost to diseases brought ashore by earlier explorers.[60]

Christopher Columbus's rediscovery of the Americas in 1492 set off a scramble for empire that would shape the entire global economy. The Spanish were the early winners in the Western Hemisphere, conquering the Aztecs and the Incas before setting up silver mines that funded conquests throughout Europe and led to tremendous price inflation on the continent. It was a century before the Dutch and the British entered this battle, and their focus was on North America. While the Dutch were largely interested in trade partnerships with the natives, the British initially tried to repeat Spanish successes in South America by locating large precious metal reserves. When those efforts failed, they turned to agriculture. The British established colonies up and down the Eastern Seaboard, ultimately seizing New Amsterdam from the Dutch and renaming it New York.[61]

Settlement of the Americas came at a particularly opportune time for many Eurasians. Evidence suggests that with existing agricultural technologies, by the fifteenth century it was becoming harder and harder to feed European and Asian populations. The Americas provided expansive new lands that could be farmed, although the process of developing these continents would take hundreds of years. Far more importantly, the Americas offered new crops that would ultimately become mainstays of Eurasian diets; at the same time, many Eurasian crops found fertile new territories to conquer. This Columbian Exchange saw "wheat, sugar, rice, and bananas mov[ing] west and maize, potatoes, sweet potatoes, tomatoes, and chocolate mov[ing] east." Crops that had only been cultivated in specific regions of either Eurasia or the Americas crossed back and forth across the Atlantic Ocean and created entirely new trade networks.[62]

In 1630, as he traversed the Atlantic Ocean on the *Arbella*, John Winthrop delivered a speech that would frame America's self-identity. He told his fellow Puritans that they had a covenant with God to colonize the New World and that the Old World would be watching. "For we must consider," he pronounced, "that we shall be as a city upon a hill. The eyes of all people are upon us." With this speech, Winthrop gave birth to an American exceptionalism that would define the future nation. It was rooted in a belief that success was divinely preordained as long as Americans followed a pious path. A flood of immigrants would follow the Puritans, most becoming relatively prosperous farmers. They not only grew enough food to be largely self-sufficient, but also raised crops and animals for sale. Although they sold some crops locally, they also sought to open regional and international markets that would bring in more money. This profit orientation often led to poor land-management practices, but since land was cheap and abundant, there was essentially no economic penalty for this lack of sustainability.[63]

The primary early markets for American crops were in the West Indies. Islands like Barbados provided an extraordinarily good environment for growing sugar. Given the backbreaking work required to grow and harvest sugar, the British needed large numbers of laborers. Because they couldn't persuade most Europeans to emigrate to sweltering, disease-ridden Caribbean colonies, the British turned to the slave trade to provide the needed workforce. Between 1662 and 1807, 3.5 million Africans would be forced into bondage and transported to the Americas. Because it didn't make financial sense to use prime land to produce foodstuffs, the British needed to find a source of provisions for their slaves. The solution was to develop a complex trade network. As Niall Ferguson writes, "By 1770 . . . Britain's Atlantic empire seemed to have found a natural equilibrium. The triangular [slave] trade between Britain, West Africa and the Caribbean kept the plantations supplied with labor. The mainland American colonies kept them supplied with victuals. Sugar and tobacco streamed back to Britain, a substantial proportion for re-export to the Continent. And the profits from these New World commodities oiled the wheels of the Empire's Asian commerce."[64]

In the southern colonies, maize was the most important early crop, necessary to feed the increasing number of settlers. But shortly after arriving in Virginia, John Rolfe raised the first tobacco for export to Britain. This was the ultimate cash crop. Like sugar, it also required a large labor force, which was once again provided by African slaves. Tobacco farmers' heavy dependence on European markets, three thousand miles away, however, meant that it actually returned far less profit than was originally expected. Regardless, huge plantations organized around tobacco monocultures dominated the southern landscape. Tobacco quickly depletes soil nutrients, which ultimately meant that vast tracts of land were left fallow for up to twenty years as farmers pushed westward from the coastal regions onto the Piedmont plateau. Along with tobacco, rice cultivation also became popular in South Carolina and Georgia. It too required intense labor, and so plantation owners bought African slaves to do the work. Finally, a couple decades into the nineteenth century southern plantation owners began growing large amounts of cotton, yet another crop that depended on slave labor.[65]

In the northern colonies, early agriculture focused almost entirely on grain production. This included European grains such as wheat, rye, and barley, but eventually maize became equally important. Although southerners were somewhat dedicated to raising cattle, animal husbandry became a cornerstone of the northern agrarian economy, primarily because raising livestock provided the best opportunities for making profits in a cold climate with rocky soil. Boston and New London soon became major cattle markets, with beef and pork packed for export to the West Indies. Rhode Island became the largest producer of wool, with a larger population of sheep than the rest of New England combined. New York, New Jersey, and Pennsylvania focused largely on growing wheat, so much so that they became known as the Bread Colonies. Philadelphia, New York, and Baltimore became important cities partially because most of the flour from these middle colonies was milled in their urban centers. Since it was so difficult to ship wheat from Europe to the West Indies without it spoiling, the island plantations came to depend on American wheat. Although all but the richest northern farmers worked their own land, they often needed additional labor. They depended far less on slaves, instead relying predominantly on indentured servants.[66]

In the ten thousand years since the beginning of the Neolithic Revolution, almost all increases in global food production have been achieved by expanding the amount of land under cultivation. "Eventually, though," Richard Manning writes, "expansion ran up against the limits of the planet's supply of plowable land. From that point, almost all the increases in total food production have had to be achieved by increasing yield—by harvesting more bushels per acre. This shift set the terms of modern agriculture, and is almost as significant as the development of agriculture itself." During the First Agricultural Revolution in America, these increased yields were produced as farmers began to mechanize. In the colonial era, most farm work had been done by hand. Few farmers used tools beyond axes to clear and hoes to prepare the land. Most farmers could not afford expensive plows from British merchants, and there were restrictions on American manufacturing of such implements. As Peter McClellan writes, the first significant agricultural revolution in America would begin about 1815, when "a new spirit of inquiry" provided the verve needed for a rapid spurt of innovation that transformed American farming methods. While this involved "changes in fertilizer use, crop rotations, pasture designs, and livestock breeding practices," the introduction of new machines was far more important. New or better implements were designed and manufactured for plowing, sowing, harrowing, cultivating, reaping, threshing, and winnowing. During the Civil War, Congress stepped in with supportive government policies. Paul Conkin writes that "these policies included generous terms for the disposal of public lands; federal and state funding of roads, canals and railroads; and land grants for public school systems and colleges. After the

Civil War, this support increased and soon included government-supported research, regulatory legislation, subsidized irrigation projects, and accessible, low-interest credit." This led to big increases in efficiency over the coming decades. "By 1870, one able farmer growing wheat or corn could feed three other families. By 1900, one farmer might feed five other families; by 1930, almost ten." This led to year-by-year decreases in the number of farmers required to feed the nation. During the colonial period, nine in ten Americans had some connection with agriculture, but this would decrease to about 50 percent by the early twentieth century. What is most startling, all this was before the beginning of the second, far more significant agricultural revolution, which began around 1930. Thus, most of these initial advances were made without the massive benefits that would come from using hydrocarbons. While many have called this more recent transformation the Green Revolution, for simplicity's sake and in an effort to avoid confusing and potentially contradictory jargon, it seems wiser to call it the Second Agricultural Revolution. While it is almost always overlooked as a cause of climate change, the primary and secondary impacts of this transformation make it the single leading contributor to this growing global crisis.[67]

* * *

The first significant technological advance during the Second Agricultural Revolution came with the introduction of the tractor. The earliest steam-driven tractors appeared in America during the 1880s, but it wasn't until the following decade that they were widely used. They were operated almost exclusively in the Great Plains, where they pulled plows and combines under expansive blue skies. About the same time, the first tractors with internal-combustion engines were being designed. Shortly after the turn of the century, more than a dozen companies were building these powered tractors, with International Harvester leading the way. These tractors were enormous (weighing at least ten tons) and so expensive that only the wealthiest farmers could afford them. In 1917, Henry Ford started selling a smaller and cheaper tractor called the Fordson; within three years he had produced a hundred thousand. International Harvester soon developed a worthy competitor, and prices quickly fell. These machines were particularly profitable to operate in the Wheat Belt, largely "because the newest models could do all the soil preparation, drilling, and combining, or all the work required for small grains." Still, tractors remained relatively rare for many years, and most farmers continued to rely on horses. It wasn't until the mid-twentieth century that tractors (and self-propelled harvesters) replaced draft animals. The impact was immense. Why? Tractors eliminated the need to set aside millions of acres to feed beasts of burden, effectively increasing the amount of arable land available for producing human food by more than 50 percent.[68]

During the Great Depression, the federal government embarked on the most assertive farm policy experimentation in the history of the nation. Without such an intervention, as many as half of all farmers would have gone bankrupt; as it was, about a quarter did. According to Conkin, "Farmers received payments for cutting production and subsidies to carry out necessary conservation practices; they received price supports for five basic commodities and crop insurance as a form of disaster relief." Although these projects remain controversial, farmers who received assistance were the first to recover from the economic calamity. One other clear outcome was that because government programs focused so much on production control, it became more difficult for small farmers to survive. Price-raising subsidies, in particular, had increased the value of land and placed a significant barrier to entry on small operators, giving a competitive advantage to already affluent farmers.[69]

This trend was amplified as the mechanization of agriculture accelerated, with the introduction of many new types of expensive hydrocarbon-fueled farm equipment. The result was that "American farms became so capital intensive that the cost of entry into farming discouraged aspirants." The primary machine driving the shift toward bigger operations in the post–Second World War era was the combine. Although this technology had been around since the late nineteenth century, new oil-powered versions were larger and far more efficient. Just as important was the introduction of combine heads that could strip and shell corn while harvesting, rather than requiring an extra step to complete these tasks. Ultimately these types of specialized heads were developed for every major seed crop. They allow a single farmer to shell more than five hundred bushels per hour. Conkin writes, "Given the versatility of combines, they are clearly the supreme implements in contemporary agriculture. No other machine . . . has displaced so many farmworkers, so lowered the unit cost of crops, and so directly affected the structure of farming in America. Their level of technological sophistication is almost beyond belief." This explains why large combines cost more than a quarter million dollars. It also explains why fuel for tractors, combines, and many other machines is the single largest contributor to farm energy consumption.[70]

∗ ∗ ∗

Perhaps the most significant technological breakthrough during the Second Agricultural Revolution was the creation of hydrocarbon-based fertilizers.

These were critical because they could replace soil nitrogen, which population demands had drained from farmland. Nitrogen is critical for plant growth because it is a component of chlorophyll (which allows plants to utilize the energy in solar radiation) and amino acids (which are the building blocks for all proteins). The natural process of nitrogen fixation, conducted by microbes, can replace this resource in the soil. These microbes use the enzyme nitrogenase to convert atmospheric nitrogen (N_2) into ammonia (NH_3), which is used by plants. Usually working in a symbiotic relationship with the root nodules of legumes, these microbes can replace lost nitrogen in the soil. One or two seasons of planting legumes like alfalfa or soybeans provide enough nitrogen to grow nitrogen-dependent grain crops. Organic waste such as crop residues and animal manure can be used directly because they contain small amounts of useful nitrogen, but there is never enough to replenish all arable land. As population grew, however, neither of these processes could adequately nitrogenize enough arable land to feed everyone. If crops could not be grown year after year on the same land, it was impossible to keep up with rising demands.[71]

In 1904, the German chemist Fritz Haber engaged in a series of experiments that proved that ammonia (a compound of nitrogen) could be created synthetically but only in small quantities. More experimentation in the coming years, however, showed that larger yields could be achieved if the chemical reactions were undertaken at high pressures. This made it seem that producing synthetic fertilizers might be economically viable. Another German chemist and engineer, Carl Bosch, then took up the problem. Bosch, who was an innovator in the field of high-pressure industrial chemistry, wanted to determine how to manufacture large amounts of both feedstock gases, hydrogen and nitrogen, at a reasonable cost. He succeeded in finding appropriate catalysts and designed steel containers capable of bearing the huge pressures needed for an effective reaction to occur. The process required large amounts of hydrogen, which was initially provided by methane extracted from coal. In 1914 the German company BASF began using the Haber-Bosch process. The company was able to produce more than twenty metric tons of ammonia daily, and that could be converted into thirty-six metric tons of ammonium sulfate fertilizer. By the early 1930s, the consumption of artificial fertilizers consumption had skyrocketed around the world.[72]

At the outset of the New Deal, the Roosevelt administration became interested in finding ways to produce better fertilizers. The Muscle Shoals facility in Alabama became home to the National Fertilizer Development Center (part of the Tennessee Valley Authority). Within a decade, powered by the nearby Wilson Dam, Muscle Shoals succeeded in creating granulated and liquid fertilizers that were far easier to apply. By midcentury, American farmers were rapidly increasing their use of these fertilizers. "In many cases," writes Conkin, "farmers used too many [fertilizers] or used them at the wrong time, with major pollution problems as a consequence." During the coming decades, fertilizer use increased by 4.5 percent annually, from 2.7 million tons in 1960 to 11.4 million tons in 1980. The result was that for the first time, American farmers could grow crops on the same land every year. This had extraordinarily negative impacts on soil health but allowed the US agricultural system to feed billions of humans.[73]

* * *

The arrival of miracle seeds represented the third tectonic shift during the Second Agricultural Revolution. The shift began in the 1920s with the development of hybrid corn. Although other developments with wheat and rice would have significant worldwide consequences, the innovation in corn production had the most substantial effect on America. Hybridization involves allowing plants to only self-pollinate, which reduces genetic diversity. After several plant generations, researchers could isolate desirable traits, such as stronger stalks, that were difficult to uncover normally. As research continued, scientists were able to identify additional desirable traits (e.g., disease resistance, increased yield, and better shipping qualities), which allowed them to introduce new seeds. One result was that the plants did not breed true, meaning that specific traits were not passed on to offspring, so farmers had to buy seeds every year from the large company that had conducted the research and held the patents. At the beginning of the New Deal, hybrid seeds only accounted for 1 percent of the corn harvested, but a decade later that number had increased fiftyfold. The main reason was that yields increased from twenty-five bushels per acre to forty bushels per acre. In the coming decades that figure would more than triple as continuing advances were made, allowing farmers to feed far more people. Hybridization was so successful that the process was soon used to improve most vegetable and orchard crops. Two outcomes of this process were the development of large monocultures and accelerated population growth.[74]

By the mid-twentieth century, the successes experienced with corn had led to new research on wheat and rice. Because both of these grains were already self-pollinators, hybridization wouldn't work. Plant breeders tried to redesign them so that they wouldn't grow as tall and thus the seeds wouldn't have to compete with long stems for solar radiation. Additionally, the new stems were far stronger, which allowed them to bear the weight of seeds that had become

much heavier because of the increased amount of nitrogen supplied by fertilizers. The pioneer in the field was the American agronomist Norman Borlaug,
who starting in the mid-1940s worked on dwarf varieties with support from the
Rockefeller Foundation. His first task was to develop wheat varieties that were
resistant to stem rust, one of the deadliest plant diseases in Mexico (where he was
doing his fieldwork). In 1952 he began breeding his stem rust–resistant varieties with Norin 10, a Japanese variety that "had unusually short, strong stems
... and responded well to heavy applications of nitrogen fertilizer." During the
1960s, when he turned his attention to developing similar varieties for South
Asia, he was able to create plants capable of providing fivefold yield increases,
which were invaluable for nations like India and Pakistan, which were experiencing regular famines. Borlaug was awarded the Nobel Peace Prize for providing bread to a hungry world. Comparable breeding techniques were adopted
by the Philippines-based International Rice Research Institute to similarly increase the yields for rice. Miracle seeds and petrochemical fertilizers had proved
to be an amazing combination. As Borlaug suggested, "If the high-yielding
dwarf wheat and rice varieties are the catalysts that have ignited the green revolution, the chemical fertilizer is the fuel that powered its forward thrust." By the
turn of the century these dwarfs were being used on more than 75 percent of all
cultivated land in Asia and Latin America and on more than 65 percent in the
Middle East and Africa. "In just a single eleven-year period, 1975 to 1986, rice
yields jumped 32 percent worldwide [and] wheat yields by 51 percent," writes
Manning. "Coupled with gains made earlier from corn hybridization, these
quantum leaps created a technical and social revolution in the United States,
especially in the heartland." This had occurred because the capital intensity of
this new agricultural system was steadily driving small family farmers out and
heralding the arrival of large industrial farming operations. In more ways than
one, it was truly a new world.[75]

* * *

Another important player in the Second Agricultural Revolution was pesticides. Three classes of pesticides have contributed significantly to the rapid increases in farming yields: insecticides, fungicides, and herbicides. Insecticides
date from the time of the Romans. By the early twentieth century three natural
sources were being used: arsenic, rotenone, and pyrethrum. These were very
important for American farmers because hundreds of insects had the potential
to destroy crops. During the Second World War, scores of new synthetic insecticides based on chemicals like chlorine, carbon, phosphate, nicotine, sulfur,

and carbonic acid were introduced. Although these provided comprehensive crop protection, over time a variety of safety concerns were uncovered. In 1962, Rachel Carson published *Silent Spring*, in which she described the role of DDT in wiping out large bird populations. The book launched the environmental movement. In time, most insecticides were banned. Those that remain are still considered to be critical because they are relatively inexpensive tools for protecting crops. In addition to synthetic insecticides, bacterial insecticides have become far more common. The most famous of these is a soil-dwelling bacterium named Bt *(Bacillus thuringiensis)*, which is widely used to protect corn.[76]

Fungicides have become important in protecting certain types of crops, in particular corn, wheat, potatoes, and soybeans. These chemical compounds provide protection against mold, rust, scale, blight, yeast infection, scab, and smut. One difficulty with fungicides is that resistance is acquired quickly, so companies manufacturing these products have to change them every year. In the early 1980s the Fungicide Resistance Action Committee was established to bring scientists together to develop strategies for collective efforts to manage fungicide resistance. Because the resulting fungicides are so expensive, farmers generally don't use them prophylactically, instead waiting until an infection has been uncovered.[77]

Herbicides weren't used before the outbreak of the Second Agricultural Revolution. As a result, their discovery had a far larger impact in the past century than did insecticides and fungicides. This was primarily because they essentially eliminated the need to cultivate crops to remove weeds. Not only did this severely reduce the amount of farm labor needed during the growing season but it meant that row crops weren't competing with other plants for scarce resources. Yields skyrocketed as a result. Conkin provides the following example: "For corn, herbicides raised production more than had hybridization. Farmers could now reduce the width of corn rows from three feet or more to as little as twenty inches, in some cases almost doubling production."[78] Over time the introduction of ever better herbicides paved the way for no-till agriculture, which saves time and money, while also protecting the soil. Herbicides have become the subject of a very contentious debate among environmentalists, with some supporting them because they preserve critical soils and others arguing that they violate organic principles.

* * *

Large-scale irrigation projects were the final critical factor in the Second Agricultural Revolution. When Europeans first arrived in North America, much of the continent was not well suited to agriculture, which is why most Native Americans were engaged in hunter-gatherer lifestyles in large sections of the continent west of the 100th meridian. What we now call the Great Plains, which most of us think of as one of the most productive agricultural areas in the world, was referred to on nineteenth-century maps as the Great American Desert. One early American explorer, the geographer Edwin James, wrote that the region was "almost wholly unfit for cultivation, and of course uninhabitable by a people depending on agriculture for their subsistence." Much of the American Southwest was even worse, containing some of the world's most inhospitable landscapes for farming. With the Homestead Act of 1862, however, the federal government hoped to incentivize people to move into these regions, often portraying them as agrarian paradises. As huge numbers began moving westward, fighting over access to the little water that was available became a way of life. This was an obvious problem for a government that wanted to extend the reach of its continental dominion. The solution was irrigation on an enormous scale. As Stephen Grace writes, "Promethean dams allowed the region to be replumbed on a scale unprecedented in world history providing water and power that allowed a new nation hungry for growth to surge westward and develop its economy into a juggernaut."[79]

The efforts had actually begun in earnest in the 1840s with Mormon settlers who worked miracles to irrigate the land around the Great Salt Lake in Utah. Using techniques learned from Hispanic communities on the Rio Grande, they began by successfully diverting a stream in the Wasatch Mountains using nothing but shovels. By the turn of the century they had more than six million acres under cultivation in several western states. "When the U.S. government got into the business of irrigation at the beginning of the twentieth century, it based its program largely on the Mormons' technical achievements turning dry earth to flourishing fields." In 1902 Congress established the Bureau of Reclamation to oversee water resource management in the American West. Over the coming century it would help irrigate more than ten million acres of new farmland, using water from rivers that many had previously assumed could not be tamed—the Colorado, the Columbia, the Sacramento, and the Snake, to name but a few. Although this was a relatively small share of the available land in the Far West, these regions produce roughly 60 percent of the nation's vegetables and 25 percent of its fruits and nuts.[80]

In the western Great Plains, most of the early attempts to irrigate dry lands were modest efforts by individual farmers and ranchers to divert streams.

Windmills were also used to pump water from shallow aquifers. The key breakthrough was the introduction of mechanized pump technologies, which allowed farmers to tap the 3 billion-acre-foot Ogallala Aquifer. One of the world's largest aquifers, it underlies eight states—South Dakota, Wyoming, Nebraska, Colorado, Kansas, Oklahoma, New Mexico, and Texas—and covers approximately 110 million acres. This is the region of the Great Depression–era Dust Bowl, created because the soil was so poor that wheat farmers quickly exhausted the region and created massive dust storms that impacted 100 million acres. Although huge portions of the region never recovered, after the Ogallala was exploited the turnaround in many areas was phenomenal. By the beginning of the twenty-first century there were "over 150,000 pumps [roaring] day and night during growing seasons on the plains. They feed water onto crops planted fencerow-to-fencerow on thousands of farms." The gas-powered pumps soak 7 million acres of wheat, sorghum, corn, and alfalfa fields, more than a quarter of all irrigated land in the nation. These grains are shipped to markets thousands of miles away or added to the feed for cattle being raised in massive livestock operations (which themselves draw eight to ten gallons of water per head of cattle from the reservoir daily). The problem is that the Ogallala is not being recharged anywhere near fast enough to keep up with current usage. Between 1960 and 1990 alone, farmers pumped one billion acre-feet of water. Some experts believe if current trends hold, the aquifer will be exhausted by midcentury. This could result in the amount of irrigated land plummeting to 2 million acres, which would substantially decrease the grain production of the nation. According to John Opie, "Mechanized irrigation may offer only a half century of prosperity for an otherwise harsh and hopeless region."[81]

The development of the Colorado River provides an excellent example of the costs and benefits of damming rivers to create arable land in the American West. With the exception of its amazing thirteen-thousand-foot drop from source to sea, the Colorado is not the most impressive river in the nation. It is shorter than the Missouri and the Yukon and has far less volume than the Columbia or the Ohio, and the Mighty Mississippi is 75 percent longer and has nearly thirty times more volume. But as Marc Reisner writes, the Colorado, called the American Nile, "has more people, more industry, and a more significant economy dependent on it than any comparable river in the world." The world's first great dam was built on the Colorado. The Hoover Dam and the nearly thirty other dams on the river created basin reservoirs capable of holding four times the annual flow of the river; it has been estimated that each drop of water in the watershed is used and reused up to seventeen times. And yet the demands are so significant that only in the wettest years does the river still reach the Gulf of California.[82]

The Colorado River is arguably the most heavily managed and litigated waterway in the world. These efforts began in 1922 with the signing of the Colorado River Compact by representatives from seven states. This governance document divided the river into an Upper Basin (Wyoming, Utah, Colorado, and New Mexico) and a Lower Basin (California, Arizona, and Nevada), each of which received 7.5 million acre-feet of water annually, with the remaining 1.5 million reserved for Mexico. But the river had nowhere near this amount of water. California, the driving force behind the agreement, tried to solidify rights to divert the river via the All-American Canal to thirsty farmland in the Imperial Valley. Although the river doesn't flow through the state and thus it had no legal rights to this water, California used its political muscle to gain access to 4.4 million acre-feet (well over half the Lower Basin allotment). In the years that followed the signing of the pact, as it became clear just how much water the river had, the Golden State's power allowed it to prevail in most of the litigation over who would actually get access to the existing flow.[83]

Another significant problem with the Colorado Compact is that it has resulted in water-use patterns that make very little sense from a cost-benefit perspective. The American system of federalism has led to a situation in which the Colorado River's water is being used extraordinarily inefficiently. One could make a compelling argument that the Upper Basin should have limited access to the river flow for at least two reasons. First, with the exception of Denver, there are few large population centers in the Upper Basin that rely on the river for drinking water. In the Lower Basin, however, the Colorado River provides water for some of the largest cities in the nation, including Los Angeles and Phoenix. Second, the amount of natural evaporation related to cold weather and high elevation make it extraordinarily inefficient to irrigate the Upper Basin. Three times more water is needed to grow low-value crops like alfalfa, hay, and wheat, which is a double whammy from an environmental perspective given that the first two crops are used to support large livestock operations. Farming the Imperial Valley, on the other hand, is far more efficient, because the warm weather is perfect for growing high-value crops like kiwis, nuts, grapes, olives, melons, oranges, tomatoes, and lettuce. "It makes infinitely greater economic sense," writes Reisner, "to build dams and irrigate in warmer regions than in colder ones—even if it makes infinitely greater political sense to do otherwise." Thus, the parochial interests of state officials have trumped the national interest. This has resulted in a failure to holistically plan how the river should be used to benefit the largest number of Americans and Mexicans while also being ecologically responsible. Regardless of the problems that have become increasingly apparent as the nation endeavored to irrigate the Great Plains and the Southwest, one thing is clear: these projects have made it possible for the United States to feed many more people.[84]

* * *

The cumulative impacts of the Second Agricultural Revolution on agricultural productivity have been immense. After the First Agricultural Revolution, annual growth rates in crop yields had already risen to about 1 percent annually, but these had more than doubled by the mid-twentieth century. Whereas labor productivity in all other sectors grew by 250 percent, in the agriculture sector it expanded by 700 percent. Although the number of farmers was halved, overall production rose. "In 1900 it took 147 hours of human labor to grow 100 bushels of wheat. By 1950 this had shrunk to only 14, and by 1990 to only 6. For corn, the number of hours per 100 bushels shrank from 147 hours in 1900 to 16 in 1950 to 3 in 1990." New animal-breeding techniques and electrification led to even greater labor efficiencies in livestock operations. At the same time, the yield per acre of crops accelerated spectacularly. With the benefits of hybridization and herbicides, corn yields went from 25 bushels per acre in 1900 to 120 bushels a century later. While the increases in the yields of other crops weren't as large, there was a more than doubling in many different areas, such as cotton and milk production. During the past half century, many of the innovations that produced these results have proliferated to other parts of the world, saving hundreds of millions from starvation and contributing significantly to the economic and geopolitical resurgence of nations like China and India. Critics contend that this agricultural revolution has also "caused massive environmental damage, destroyed traditional farming practices, increased inequality, and made farmers dependent on expensive seeds and chemicals provided by Western companies." Without question, it has profoundly changed the world.[85]

The use of hydrocarbons was central to nearly every aspect of the Second Agricultural Revolution, with the exception of miracle seeds. In the past century the energy input per acre of American cropland has increased more than eightyfold, more than half of this increase resulting from synthetic fertilizer use. The globe now produces nearly twice as many calories per person, so that the diets of people worldwide have changed; in particular, the quantities of meat and dairy currently being consumed were previously unimaginable.

In the United States today, food-related energy use accounts for 15 percent of the total national energy budget. While much of this energy is expended beyond the process of growing and harvesting food (from production to preparation), growing and harvesting food still represents a large share of the total energy expenditure.[86]

* * *

The increased use of hydrocarbons within the agricultural system represents a significant addition to greenhouse gas emissions, but the real reason for the Second Agricultural Revolution's impact on climate change is that it enabled the tripling of the global population in a very short period of time. That population growth spurred much larger increases in hydrocarbon use in almost every economic sector. The global society requires far more coal, oil, and gas today than it did a century ago. It is true that much of the population growth took place in developing economies, which use less energy than do advanced economies. Still, the sheer number of people in those nations and the fact that many of them are experiencing rapidly rising living standards means that cumulatively they are adding staggering amounts of greenhouse gas emissions to the atmosphere. What's more, the population in advanced economies has also increased tremendously. As just one example, the number of Americans has tripled in the past century. Thomas Friedman is a key popularizer of the term *Americum*, which was defined by Tom Burke as "any group of 350 million people with a per capita income above $15,000 and a growing penchant for consumerism."[87] Friedman focuses a lot of attention on the fact that by 2030 the globe will see an increase from three *Americums* to as many as nine, with up to 3 billion people living an energy-intense American lifestyle. This would not have been possible without the proliferation of Second Agricultural Revolution technologies. This is perhaps not too surprising, given that for the past ten millennia population growth has been closely related to developments within the agricultural sphere. In the coming pages, I rely heavily on a wonderful book by the distinguished plant physiologist Lloyd Evans called *Feeding the Ten Billion: Plants and Population Growth*. In addition to providing a thought-provoking set of ideas regarding the global agricultural future (which I will return to in a later chapter), Evans presents an immensely useful history of agricultural progress and human population growth that dates from before the Neolithic Revolution. His work helps contextualize what has already been discussed in this book, particularly contemporary population growth.

Since *Homo sapiens sapiens* emerged about two hundred thousand years ago, humanity has spent 99 percent of its existence in hunter-gatherer societies. Over tens of thousands of years our forerunners slowly developed technologies that allowed them to increase their numbers, including "stone and blade tools . . . bows and arrows . . . and [fire] not only for cooking but also to manipulate vegetation and facilitate hunting." Demographers estimate that by the time humans began the slow transition to agriculture during the Neolithic Revolution, the total number of people had reached only five million. During the successive six millennia, population grew by about tenfold as the agricultural lifestyle became dominant in most parts of the world. As we saw in chapter 1, the story of this period was largely about the domestication of plants and animals. In Mesopotamia this meant barley, wheat, rye, and oats. In China the key crop was rice. In America the central grain was maize, but other produce included potatoes, sweet potatoes, manioc, beans, squash, peppers, peanuts, tomato, avocado, pineapple, and papaya. At the same time, Eurasians were domesticating goats, sheep, cows, horses, oxen, pigs, and chickens. In sum, this domestication led to a population explosion that was unprecedented.[88]

Humanity experienced another tenfold increase in population between 2000 BCE and 1500 CE, reaching about a half billion at the dawn of the modern age. Most of this growth was associated with the expansion of farming throughout the middle latitudes, although technological advancements such as the heavy plow and the horse collar also increased productivity. This extension of arable land was accomplished mostly by clearing forests but also by cultivating grasslands and terracing hillsides. As agriculture moved beyond Mesopotamia, new plants like oats and rye gained importance. Eurasian farmers were also learning how to better alternate crops and maintain soil nutrients using animal manures. During the medieval period, the three-field system became common. Farmers would plant one field with wheat or rye and another with oats, barley, or legumes and leave a third field fallow, rotating the crops each season. Overall this allowed communities to maintain relatively high yields over long periods, although significant productivity constraints remained.[89]

About 1500, sustained and rapid population growth began, leading in just over three centuries to a doubling of the global population, which reached 1 billion about 1825. One of the most significant contributors to this explosion was a fundamental shift in property rights. While most farmers had previously worked under feudal lords, small planters throughout Europe started to become landowners, with two interconnected consequences. First, farmers had new incentives to ensure that their land was being used as efficiently as possible. Second, they were more likely to break with tradition and get creative.

As a result, the landed gentry developed an entirely new field of learning, with scores and scores of treatises on farming techniques being published and followed. This led to such new practices as replacing fallow fields with a rotation of turnips or other winterfeed (and ultimately potatoes), which both increased the productivity of animal husbandry and generated more manure for grain crops. This in turn led to ley farming, planting fields with grains or cash crops for several seasons before using the fields to grow animal feed or pasture for a number of years while soil vitality was restored. The net result was increased yields and a doubling in the productivity of farm labor, which allowed society to feed more people, while freeing labor for the First Industrial Revolution.[90]

The other major development of this period was the Columbian Exchange, discussed above, in which Eurasians imported plants previously unknown to them, mostly maize, potatoes, sweet potatoes, and sugar. At a time when the existing system was nearing its breaking point, these new plants allowed farmers to bring more land under cultivation and thus feed growing populations. The arrival of maize and sweet potatoes in China saw the population grow from 140 million in 1650 to 400 million in 1850. During this same period, maize, potatoes, and sugar facilitated European growth from 103 million to 274 million people. Tom Standage argues that this exchange, and particularly the huge increase in sugar consumption, was a key precondition for the First Industrial Revolution. Sugar was critical because it provided much needed energy for factory workers, consumed in the form of sweetened tea and jam. Potatoes also played an essential role, because they fed the Irish farmers who were producing wheat for English workers. So, the Columbian Exchange paved the way for the development of a hydrocarbon society and dramatically increased the number of humans who desired to benefit from its advancements. It also began the long process of decreasing the proportion of humanity living in conditions of chronic hunger, largely by providing more overall calories and reducing the number of famines.[91]

During the century starting in 1825, the global population doubled again, reaching 2 billion. While most of this population growth was the result of cultivating ever-larger swaths of land (specifically in America and Russia), the scientific method was the most significant development for long-term population growth. Justus von Liebeg, who discovered that nitrogen was an essential element in plant nutrition, launched the marriage of science and agriculture. His discovery led to the introduction of fertilizers like Chilean nitrate, Peruvian guano, German potash, and superphosphate (this last developed by John Bennet Lawes and considered by some to represent the birth of the chemical fertilizer industry). Used initially in Western Europe and Japan, these fertilizers increased yields dramatically. As these early efforts gave way to the Second Agricultural Revolution, the productivity of farmers continued to improve rapidly.[92]

As the Second Agricultural Revolution truly began to take hold, population growth literally exploded. The third billion was added from 1927 to 1960, yet the technologies weren't even having a significant impact yet; it was still the ability to bring more land under cultivation that drove growth. The fourth billion was added in only fifteen years, 1960–75, when there were rising concerns about the sustainability of the food supply. Yet the arrival of dwarf wheat and dwarf rice, more aggressive use of synthetic fertilizers, and better irrigation systems were changing the rules under the feet of the environmental movement. For the first time in human history, the food supply was easily keeping pace with population growth while actually increasing average calorie intake significantly.[93]

During the next thirty-six years, from 1975 to 2011, the global population increased by 3 billion while continuing to increase the per capita food supply. And this was done without adding any new acreage for farming. This amazing transformation was almost solely connected to the sustained impact of the Second Agricultural Revolution. Average yields of rice (+32%) and wheat (+51%) continued to rise owing to the expanding use of dwarf cereal varieties, synthetic fertilizers, and modern irrigation systems. Genetically modified organisms also appeared during this period; however, their long-term impact on the ability to feed the growing global population remains to be seen.[94]

It was also during this period, in the aftermath of the first oil shock, that many began to fully internalize the deep connection between energy and these impressive gains. Still, although the amount of energy needed to maintain the growing industrial agriculture system was considerable, these gains were accomplished without intensifying the energy input per bushel. This did not prevent many from beginning to investigate ways that agricultural practices could be improved upon to be more energy efficient. One outcome was the development of minimum-tillage farming, which benefitted from the arrival of highly effective herbicides. Interestingly, one side effect of this more energy efficient approach was that it allowed more land to come under cultivation, as farmers could now plant on land with relatively severe slopes.[95]

Because the overwhelming majority of recent growth in global population has been in developing nations, the potential for dramatic future climate impacts is immense. At the time of the negotiation of the Kyoto Protocol, in the mid-1990s, it was estimated that advanced economies had contributed more than two-thirds of all greenhouse gas emissions. The rapid population and economic growth in nations like Brazil, India, China, South Africa, South Korea, Nigeria, Mexico, Saudi Arabia, Turkey, and Indonesia, to name a few prominent examples, has quickly closed that gap. A recent study found that as of 2013, advanced economies were responsible for 52 percent of cumulative emissions, and developing economies for 48 percent. Furthermore, the study team estimated that by 2020 the developing nations would account for a majority of cumulative emissions dating from 1850. These rapidly emerging nations are heavily dependent on hydrocarbons to maintain economic growth. Their national leaders consider it perfectly reasonable to take this route into the ranks of the advanced economies, since that is exactly how all the current members joined that community of developed countries. This explains why these nations will soon close the historical and contemporary emissions gap. As Carl Lopes, UN under-secretary-general, stated in a 2013 speech, "It is only a matter of time before developing nations catch up with the developed world with regards to cumulative emissions." While this will certainly have serious impacts with regard to the continued negotiation of international climate treaties, the more important point is that the combination of rapid population growth and economic growth has greatly increased the amount of greenhouse gases in the atmosphere. This would never have been possible without the ability of farmers to feed this growing population using technologies and approaches developed during the Second Agricultural Revolution. This is a largely untold story within the larger debate regarding both energy and climate policy, but it is fundamentally important to consider as we move further into the twenty-first century and determine whether to expand upon fundamental environmental protections put in place during the Third Political Age.[96]

INNATE Revolutions: Industrial
America and the Third Industrial Revolution

The First Industrial Revolution was characterized by the transition to a new manufacturing process that dramatically improved productivity; it also witnessed the transition from biofuels to coal. The Second Industrial Revolution was characterized by the introduction of steel, mass production, and electrification; it also coincided with the rise of oil. The Third Industrial Revolution was exemplified by the advancement of digital technologies that spawned an information age and set the groundwork for the replacement of human labor in the manufacturing sector; it also saw the addition of gas to the suite of widely used hydrocarbons. This most recent technological transformation has greatly influenced the course of human events. On one hand, it has tremendously improved economic efficiency and eased the path of globalization, which has lifted hundreds of millions out of poverty and created a more peaceful planet. On the other hand, rapid progress in this sector has helped mask the reality that the vast majority of Americans have less economic security and political influence. Digital technology and its myriad platforms have also placed new demands on the global energy infrastructure, augmenting previously existing emission streams. The question going forward will be how to enjoy the benefits of this revolution while countering its negative impacts.

The roots of this revolution began growing in 1837, when Charles Babbage published the first serious article discussing the idea of a programmable computer. Herman Hollerith built upon this work, using punchcard machines to more quickly conduct the census of 1890, later using the same technique when he founded the International Business Machines Corporation (IBM). During the early twentieth century, luminaries like Lord Kelvin and Vannevar Bush developed sophisticated analog computers. In the 1930s, however, it became apparent that digital, binary, electronic, and general-purpose machines would be the hallmarks of modern computing. In 1937, in a key academic paper, Alan Turing used the idea of a "logical computing machine" to consider a contemporary question of mathematical theory. That same year, Claude Shannon wrote "the most influential master's thesis of all time, a paper that *Scientific American* later dubbed 'the Magna Carta of the Information Age.'" As a graduate student working at Bell Laboratories, Shannon had come up with the idea that electromechanical switches could be assembled into circuits to perform specific tasks. At the same time, a Bell Labs mathematician named George Stibitz developed a method for using relay switches to make binary computations. Finally that year, Howard Aiken advanced the conceptual design for IBM's Mark I computer, which would ultimately use a system of wheels and counters to make automatic calculations, with no human intervention required while it was processing.[97]

In 1939, John Vincent Atanasoff built the first electronic digital computer using vacuum tubes to perform computations. He also developed a way to store information in his computer's "memory" using capacitors that could briefly hold an electrical charge, with a system of rotating cylinder drums that both retrieved and stored new data while also recharging the capacitors. When the Second World War intervened, the work was abruptly halted, and a student who didn't know what it was later dismantled the machine for parts. Fortunately, another computer pioneer, John Mauchly, had visited Atanasoff before the war and was able to use what he had learned to push modern computing forward. Mauchly had been collecting ideas from a wide variety of innovators as he designed his own computer. With funding from the War Department, he worked with J. Presper Eckert to develop ENIAC (the Electrical Numerical Integrator and Computer). Their initial objective was to create a machine that could accurately calculate missile trajectories. Although it wasn't binary, relying instead on a decimal system, it was the most advanced computer developed to that point. The massive computer weighed in at thirty tons. Its seventeen thousand vacuum tubes took up the space of a three-bedroom apartment. It was able to do five thousand additions and subtractions per second, one hundred times faster than any other machine at the time.[98]

While these developments were taking place in America, the British were engaged in a secret computer development program intended to crack Nazi codes. Based at Bletchley Park, the effort was led by Alan Turing. Turing and his team determined that to break the German Enigma machine's code, they needed to develop a far more advanced computer. The attempt to develop this machine, to be named Colossus, was led by Max Newman and Tommy Flowers, who realized that previous computers had been inadequate because they could not decipher incoming messages fast enough for military leaders to take needed actions. After a code was broken, what was needed was a machine that not only stored the code in its memory but also could compare it with incoming messages at great speed. By the end of the war the British has built eight of these machines, each with fifteen hundred vacuum tubes, which were critical to war planning in the final year of the conflict. With many of the important technical problems solved by the ENIAC and Colossus teams, the Digital Revolution was ready for the next set of transformations.[99]

For more than a hundred years, the focus of inventors had been almost entirely on hardware, but the next group of innovations were in programmable software. The key early actor was a naval officer named Grace Hopper, who was tasked in 1944 by Howard Aiken to write the first computer-programming manual for the Mark I. She developed the idea for subroutines, "a clearly defined, easily symbolized, often repeated program," and compilers (critical for using the same program on different machines). At the same time, a group of six women working with ENIAC were laboring on similar programming innovations. They were Jean Jennings, Maylyn Wescoff, Ruth Lechterman, Betty Snyder, Frances Bilas, and Kay McNulty. One of the primary early goals was to develop programs that could be stored in a computer's memory, which would make the machines easily reprogrammable. It was one of Turing's former mentors, John von Neumann, who had the key insight that made this possible. While employed on the Manhattan Project, von Neumann had worked with several high-speed computers to assess whether they could be used to solve key equations for nuclear implosion. He determined that stored-memory computers were needed to complete this work in a timely fashion, so he became a consultant to the ENIAC team to help make this needed breakthrough—Eckert and Mauchly had already been thinking about this problem for more than a year. As Walter Isaacson writes, von Neumann's key insight was "realizing the importance of the computer's ability to modify its stored program as it ran and for creating a variable-address functionality to facilitate this." By spring 1945 the collaborative effort had led to the proposal for a new machine called the Electronic Discrete Variable Automatic Calculator (EDVAC). After the war, the Eckert-Mauchly Computer Corporation (which ultimately became Unisys) was founded to develop the first stored-program computers. The corporation built UNIVAC, which was used by a wide variety of clients, including the Census Bureau and General Electric. The computer became a celebrity when it was used to predict the outcome of the 1952 presidential election, which it would continue to do for future contests. "Beginning in the 1950s," writes Isaacson, "innovation in computing shifted to the corporate realm, led by companies such as Ferranti, IBM, Remington Rand, and Honeywell."[100]

Early computers had been expensive because they depended upon costly, delicate vacuum tubes. Beginning in the late 1940s, however, a team working at Bell Labs began work on a new technology that would fundamentally change the world. The "true birth of the digital age," writes Isaacson, "the era in which electronic devices became embedded in every aspect of our lives, occurred in Murray Hill, New Jersey, shortly after lunchtime on Tuesday, December 16, 1947. That day two scientists at Bell Labs succeeded in putting together a tiny contraption they had concocted from some strips of gold foil, a chip of semiconducting material, and a bent paper clip. When wiggled just right, it could amplify an electric current and switch it on and off." The transistor had been born. These devices were capable of controlling the flow of electricity in electronic equipment just as vacuum tubes had done, but they had several major advantages. They required far less power and were much smaller, sturdier, longer lived, and energy efficient. They were invented and advanced by a trio of researchers: the experimentalist Walter Brattain, the quantum theorist John Bardeen, and the solid-state physicist William Shockley. While the idea for the transistor had emerged from thinking done by Shockley before and after the war, it was Brattain and Bardeen who successfully invented the device over a period of about a month. In the months that followed, Shockley, his competitive juices flowing, worked in secret (violating the collaborative culture of Bell Labs) to develop a more straightforward and robust way to produce a transistor. The group jointly revealed their landmark discovery in June 1948, and soon, beginning with the now legendary transistor radio, electronic products using transistors began to appear in the marketplace. The relationship between the three inventors had disintegrated as Shockley tried to take full credit and later formed his own corporate laboratory to manufacture transistors. In 1956, the year he shared the Noble Prize in Physics with Bardeen and Brittain, Shockley brought two brilliant young semiconductor engineers named Robert Noyce and Gordon Moore into the company. The following year, however, as Shockley became increasingly unstable, Noyce and Moore broke away with six others to form Fairchild Semiconductor, with funding from Sherman Fairchild, founder of Fairchild Camera and Instrument. Fairchild Semiconductor was in the perfect position to lead the Digital Revolution's next phase. The company that had built the world's first transistor radio, Texas Instruments, would join it in this quest.[101]

As semiconductors advanced in the late 1950s and early 1960s, a problem emerged. For every increase in the number of components in a circuit, there was a tenfold increase in the number of hand-soldered connections required. The solution was an integrated circuit, which became known as the microchip. An early innovator in this field was Jack Kilby, who joined Texas Instruments in 1958. His "monolithic idea," which came shortly after he arrived at the company, was to place all of the components on one large piece of silicon. As he wrote in his notebook after having the brainstorm, "The following circuit elements could be made on a single slice: resistors, capacitor, distributed capacitor, transistor." In September, he tested a prototype of the idea that worked as expected. This was the first microchip. Yet it remained ungainly and totally impractical to manufacture on a large scale. At Fairchild Semiconductor, Robert Noyce had been toying with a similar concept. A Fairchild physicist named Jean Hoerni had been working on a process to protect the transistors on a circuit board with a thin layer of silicon oxide. This had led Noyce to consider whether there could be other uses for this planar process when he realized that not only could copper lines be printed atop this oxide layer (replacing the hand-soldered wires) but a single silicon chip could hold more than one transistor. In theory, this would make a circuit board far easier to manufacture compared with the comparatively clumsy microchip Filby had patented. Fairchild filed a competing patent and let the lawyers go to work. In the subsequent patent case, *Kilby v. Noyce*, it was determined that Kilby had come up with the microchip concept earlier, but some components had not been included in his patent application. In 1967, with the use of microchips exploding, the patent court ruled in favor of Kilby, only to have this decision reversed two years later by an appeals court. This legal wrangling meant little, however, because as it was proceeding the two companies signed a cross-licensing agreement. In 2010, Kilby won the Nobel Prize in Physics for the development, an award he almost certainly would have shared with Noyce if the latter had not died a decade earlier.[102]

Starting in the 1960s, microchips began to transform society. They were first widely used by the military and by the space program, in ballistic missiles and to help place the first humans on the surface of the moon. Then they began making their way into consumer products, initially in hearing aids and pocket calculators. As this new technology developed, it became increasingly faster and more powerful, while the products shrank in size and became much cheaper. It was at this point that Fairchild's other founder, Gordon Moore, articulated his now famous law suggesting that chip performance would double every twelve to twenty-four months. In 1968, with a change in corporate leadership and after many of their former colleagues had left to form Teledyne, Noyce and Moore decided to create a new company. They worked with the venture capitalist Arthur Rock to put together financing for Intel, which would become the world's dominant semiconductor company. They brought in another Fairchild alumnus, the action-oriented Andy Grove, to serve as the day-to-day manager. As they set about taking control of the semiconductor industry, this Silicon Valley–based company was also critical in the development of a totally new type of business culture that flipped the top-down management style preferred by East Coast firms. Noyce and Moore favored a collaborative and meritocratic approach that promoted innovation, flexibility, and fun—which most of the corporate world believed would result in reduced productivity. As the Silicon Valley culture has proven time and time again, however, the opposite is true.[103]

Shortly after its founding, Intel brought a Stanford professor named Ted Hoff on board. Hoff would make the next big advance in the Digital Revolution: a general-purpose, programmable chip. This would essentially be the general-purpose computer Turing had written about decades earlier, but on a scale that had then been unimaginable. Made public in 1971, this new device was called a "microprocessor." It soon began to appear in a huge variety of consumer products ranging from coffeemakers to medical devices. But most importantly, it made possible the personal computer. In the coming decade, hundreds of companies would join the effort to build a commercially successful machine using Intel's microprocessor. While this struggle was ongoing, however, another technology was being invented that would transform the economy and eventually act to ever more tightly connect the world. And this time the American government would return to lead the way.[104]

* * *

While the federal government had been an actor in promoting science and technology since the founding of the nation, its role started to expand at the beginning of the twentieth century, when the main structural elements of the modern US innovation system took shape. Actors within academia, industry, and the government became heavily engaged in research and development, although there were few efforts to coordinate policy across these sectors. This reality began to change, however, with the outbreak of the First World War. President Wilson and Congress determined that more federal involvement was needed, which led to the creation of the Naval Consulting Board, the National Research Council, and the National Advisory Committee for Aeronautics. Another significant factor that forced the rapid maturation of American industry was the reduction of key imports as a result of the trade war against German interests. As David Hounshell writes, "There is nothing like confiscating domestic manufacturing plants, patents, and trademarks and then implementing a large tariff to help get an industry started." A final important outcome of the First World War I was the swift propagation of corporate laboratories: more than one thousand were created between 1919 and 1936 (and they accounted for two-thirds of all research spending). Interestingly, the Great Depression led to a backlash against technology, with social critics arguing that it had been the primary culprit behind the economic downturn. Although President Roosevelt's National Resources Board took up some of the slack, research funding was slashed by both public and private institutions during this period.[105]

Although a few research organizations were created in the early part of the century, the overall federal role in the national innovation system remained fairly small prior to the Second World War. As war neared, those at the highest levels of the government understood that preparing for a conflict meant mobilizing the nation's scientific and technological resources. To begin mobilizing scientists, Roosevelt approved creation of the National Defense Research Council (NDRC) in early 1940. The brainchild of Vannevar Bush, himself an innovator in analog computing, the NDRC was modeled on the successful National Advisory Committee for Aeronautics. It was a quasi-governmental committee whose purpose was to stimulate innovation in military technology by directing federal funding to projects aimed at addressing national security concerns. As Bush noted in his memoirs, the NDRC determined to more fully integrate university scientists into the national innovation system. This was a decision, he wrote, that "proved to be important, not only for the war years but also for the postwar period. In fact it set a pattern that meant a great deal eventually to advanced education in this country. We decided that we would make contracts for research directly with universities, not with individuals therein. And we decided that, in so doing, we would pay the full costs of the programs. This does not sound like a very radical departure from previous practice, but it was." This created a government-university research partnership and set a precedent for large-scale federal research projects. In 1941 the NDRC became the operating arm of the Office of Scientific Research and Development (OSRD). This new organization provided a steadier, more reliable funding stream for both university and industry research via direct appropriations from Congress. The OSRD funded applied military research, as well as basic research in the physical and life sciences. During the war, the NDRC established university-based labs and project teams that developed sophisticated microwave-based radar systems, proximity fuses, nuclear weapons, and early computers.[106]

During the postwar period, there was widespread consensus that the federal government should continue in peacetime to play an active and direct role in promoting innovation by funding university science as well as contracting with industry for important defense-related research. The resulting system was described in Bush's brilliant report *Science: The Endless Frontier*, which envisioned an independent civilian agency that would have decision-making authority over federally funded basic research. This eventually led to the creation of the National Science Foundation, the National Institutes of Health, and the various national laboratories currently managed under the auspices of the Department of Energy. At the same time, the military was building or expanding its own research capabilities. Each of these efforts was an archetype of what became known as Big Science—large-scale, government-funded scientific endeavors. As the Cold War became the defining geopolitical reality, two characteristics would define the federal role in the innovation process. First, national security and related expenditures dominated government research investments, with military agencies receiving 50–60 percent of all funding, the civilian space program receiving 10–15 percent, and atomic energy garnering 5–10 percent. Second, the "spinoff paradigm" became a key philosophical underpinning of American innovation policy. In *Science: The Endless Frontier*, Bush had posited that the creativity of basic research (as opposed to applied research focused on accomplishing a defined real-world objective) should be maintained whenever possible because it would result in unintended and highly positive discoveries. These findings would then make their way to scientists or engineers working in corporate labs, who would use them to conceive goods or services that could be widely adopted in the commercial marketplace. In this way, the invention would "spin off" from the government to industry. This was exactly how the most important invention of the Digital Revolution came about, with a coordinated effort from research institutions like the Advanced Research Projects Agency, the Lincoln Laboratory, Bolt, Beranek and Newman (BBN), the Stanford Research Institute (SRI), Xerox PARC, and the RAND Corporation.[107]

The key early actor in creating what would eventually become known as the Internet was J. C. R. "Lick" Licklider, an MIT psychology professor who envisioned a decentralized network that would enable the sharing of digital information and the interfaces necessary to make using such a web functional for its users. Together with the artificial-intelligence innovator John McCarthy, he helped develop a system of computers that allowed users to communicate directly with one another. This type of interactive computing was further developed at the Lincoln Laboratory, which Lick had cofounded in 1951. The lab had been tasked with establishing an early-warning system to detect enemy missile or bomber attacks, which it accomplished through a powerful interactive network called SAGE. Lick was brought on board to fabricate human-machine interfaces to increase the system's efficiency. This led him to think about a network that would have much broader applications, which he began working on when he joined BBN in the late 1950s. His work there on a report titled "Libraries of the Future" predicted many of the functionalities of the modern web. In the mid-1960s he was recruited to lead a group at the Defense Department's newly created Advanced Research Projects Agency (ARPA) investigating more effective ways to share data via interactive computers.[108]

Licklider stayed at ARPA's Information Processing Techniques Office (IPTO) for only a short time before being replaced as director by Bob Taylor. Believing there was a great desire for better information sharing and that current approaches were ineffective, Taylor had the idea of creating a network to connect various research centers doing work funded by ARPA. He brought in another engineer, Larry Roberts, to run this project. One of the team's first brainstorms, suggested by Wesley A. Clark, was to connect the supercomputers at various facilities through ARPA-produced microcomputers that would route data. Leonard Kleinrock (ARPA-IPTO), Paul Baran (RAND), and Donald Davies (Britain's National Physical Laboratory) worked together to figure out how to actually transmit data across the network using a highly efficient method that became known as packet switching, with all data, regardless of type or size, broken up into diminutive, uniformly sized units for transmission. Baran, who was attempting to design a fully decentralized communications system that could survive a nuclear attack, was the first to conceive of this approach. He believed the best method was to break messages into small pieces that would travel along disparate paths before arriving at the appropriate network address. RAND recommended the system to both the US Air Force and AT&T, but no one was interested until 1967, when Larry Roberts learned about it. Roberts also became aware of similar work done in Great Britain by Davies, who wrote about a system using "packets" to distribute information. Kleinrock was working on another idea, digital-queuing theory, which was critical to understanding how the data would navigate through bottlenecks in switched-data systems. BBN's Frank Heart and Robert Kahn were contracted to build the microcomputers, known as Interface Message Processors (forerunners of contemporary routers that were the size of a refrigerator), which would actually connect the larger computers housed at various labs. The original network comprised four research centers, at UCLA, SRI, UC-Santa Barbara, and the University of Utah. As Isaacson writes, "Like typical senior professors, the researchers at these centers enlisted a motley crew of graduate students to do the work." UCLA's Stephen Crocker coordinated the group, although his best friend, Vint Cerf, also provided intellectual leadership. Working over a couple years, they established the ARPANET, which went live on 29 October 1969, the precursor to the Internet.[109]

In the coming years, several other packet-switching networks were created, but these systems could not communicate with one another. Two ARPANET alumni, Bob Kahn and Vint Cerf, conducted the key work to solve that problem, in the process earning themselves the moniker "Fathers of the Internet." Over three months in 1973, they worked out the fundamentals for a new, overarching framework. One of their key findings was that such an approach demanded a common protocol so that all computers linking to the network would be able to share data. This led to their creation of an addressing scheme known as Internet Protocol (IP) and a Transmission Control Protocol (TCP) that ensured packets would be reassembled at their destination in the correct order. When they published "A Protocol for Packet Network Interconnection" in the *Transactions on Communications* of the Institute of Electrical and Electronics Engineers the following year, the Internet Age was birthed, with TCP/IP at its foundation. This final step in a public-private partnership that had begun little more than a decade earlier would prove to be the most important spinoff of the Digital Revolution and would transform human existence in ways both good and bad.[110]

* * *

It would be more than two decades before the Internet was widely used by the public. In the meantime, the next big development of the Digital Revolution was the creation of personal computers and the software that gave them purpose. Constructing personal computers had been made possible by the advent of microprocessors, but they also emerged because members of the Counterculture Revolution were seeking to develop technologies that could enhance individuals' ability to challenge societal hierarchies. "Computing went from being dismissed as a tool of bureaucratic control to being embraced as a symbol of individual expression and liberation." Early work on the concept was done by Douglas Engelbart, who focused much of the work at his Augmentation Research Center at the SRI on the challenges of human-computer interaction. His first major breakthrough was the development of the computer mouse, which allowed people to more easily navigate a digital interface. Then he built his oNLine System (NLS), which included on-screen graphics, document sharing, e-mail, formatting, and hypertext. Alan Kay, one of Englebart's protégés, operationalized these early ideas while working at Xerox PARC. Kay collaborated there with Butler Lampson and Chuck Thacker to construct a machine using these technologies. The resulting Xerox Alto was the first desktop computer, but the company was not interested in turning this into a consumer product. Ultimately the first marketable machine was created by Steve Wozniak and Steve Jobs. Before they accomplish this task, however, another young duo would need to provide the world with an operating system that could make a personal computer functional for large numbers of regular users.[111]

Paul Allen and Bill Gates, who had grown up together in Seattle, were both living in the Boston area in the mid-1970s. As the battle to construct the personal computer heated up, they decided that the real profits lay in the operating system and other software that would be run on these new machines. The two had been programming for years, taking advantage of the opportunity to learn coding at their elite preparatory school. Over the period of eight weeks in January and February 1975 they used these skills to engage in a programming whirlwind that would shift the entire digital landscape. The result was BASIC, which, as Isaacson writes, was the "first commercial native high-level programming language for a microprocessor. And it would launch the personal computer software industry." Allen and Gates formed Microsoft and set out to build upon their initial success to become the dominant force within the software sector. As they began building their empire back in Seattle, Wozniak and Jobs were just getting started in their now famous Cupertino garage. Wozniak, who was five years older, was the hardware guy, while Jobs brought a vision and design aesthetic that would help define Apple for decades. Wozniak's breakthrough was to marry a monitor and keyboard in an integrated system—this would become the Apple I. The Apple II was a far bigger commercial success, with the company selling more than one hundred thousand machines in the late 1970s. The personal computer had arrived.[112]

In the mid-1980s, Apple and Microsoft would introduce new products that continued the transformation of computing and helped deliver them into the homes of millions of Americans. In 1984 Apple released the Macintosh, which featured both a graphical user interface and a mouse, as the first mass-market personal computer. Although the company quickly lost the lead in personal computing to rival IBM, it would continue to maintain a core market share with new offerings until it once again rose to the top of the computing industry after the turn of the millennia. In 1985 Microsoft introduced its Windows 1.0 operating system with an innovative graphical user interface (which would be radically improved with Windows 2.0 two years later). Gates's key business innovation was to design this software for the IBM personal computer and any of its clones, while maintaining licensing rights within his company. In the next two decades, it became clear that he and Allen had been correct that this was where the real profits were in the computing industry. As the competition between Microsoft and Apple was evolving, they and many other companies were releasing the thousands of software packages that would make personal computers an essential part of the business world, government, and academia and increasingly common in American homes.[113]

Although the Digital Revolution had been picking up steam for decades, it did not become mainstream until the mid-1990s. Manufacturing computers and producing software had become important economic activities, but it was the arrival of the Internet via the World Wide Web that made the Information Age a force that touched nearly every sector of the broader economy. Although networking emerged in the 1970s, no commercial services were allowed on the Internet, and thus it failed to attract the attention of many people. Things began to change in the early 1980s, when the Hayes Smartmodem provided an inexpensive way to connect with networks via the telephone. Even then, however, relatively few people were interested in connecting regardless of how cool Mathew Broderick made it look in *Wargames*. The use of e-mail to communicate introduced average Americans to the potential benefits of the online world. E-mail and online forums helped enable the development of virtual communities, both professional and personal. With these new tools available and becoming somewhat popular, the window had opened for a new company to step forward. In 1991 Steve Case launched America Online (AOL) as an online community for people with limited knowledge of computers. The service offered a modem, e-mail ("You've Got Mail" became a cultural meme), chat rooms, and gaming. It did not provide a connection with the Internet, which remained inaccessible to most people, but the company attracted millions upon millions of users and is widely credited with popularizing being online.[114]

Although his involvement has become a national joke for political reasons, the person most responsible for providing public access to the Internet was then Senator Al Gore. The "opening up of the Internet, which paved the way for an astonishing era of innovation, did not happen by chance," according to Isaacson. "It was the result of governmental policies, carefully crafted in a thoughtful and bipartisan atmosphere, that assured America's lead in building an information-age economy." And it was Gore who led this effort in Congress. While he championed several important pieces of legislation, the most significant was the National Information Infrastructure Act of 1993 (signed after he became vice president), which opened the Internet to companies and the general public. The two inventors of the Internet, Cerf and Kahn, both credit Gore with fostering its expansion into the public sphere. Once Gore had pushed Congress into taking action, the creation of the World Wide Web provided the structure for the profit-driven and nonprofit endeavors that would create the information economy. Timothy Berners-Lee, who developed a system in which interlinked hypertext documents could be accessed via the Internet using a graphical user interface, invented the web. The person who made it readily accessible to the general public, however, was a software engineer named Marc Andreessen. Together with Eric Bina, Andreessen created the first widely used web browser, which was initially called Mosaic but became famous as Netscape Navigator. This tool allowed users to bypass services like AOL and access the web directly, which made connecting with sites far easier. While inventions such as blogs and wikis further advanced the web, it was the search engine designed by Larry Page and Sergey Brin that really accelerated usage. Google's algorithm allowed people to easily find what they were looking for through access to the millions of sites being created. This combination of developments made the web an effective medium for exchanging information and transacting business.[115]

* * *

The Third Industrial Revolution is still unfolding. Although its full potential has not yet been realized, we have a good sense that it is having a large impact on society. There is little doubt that the Digital Revolution has increased labor productivity and provided new outlets for human creativity. It is far less clear whether it has fundamentally enhanced our quality of life. In the coming decades we will be forced to come to grips with what it means to be human in an increasingly digital world. In the meantime, however, we can consider what impact it has had on our global and political ecosystems. While I do not favor reversing course, it is clear that the Digital Revolution has come with real environmental consequences. These negative results include the costs of extracting billions of tons of precious minerals needed to manufacture the gadgets that are now part of everyday life, the massive energy inputs needed for their production, and the perpetual electrons needed for their daily charging. The end result is often ecological devastation on a local and regional level and increased warming on a global level. Once again, I do not mean to suggest that we should turn our back on modernity. Rather, we need to be more vigilant in safeguarding the natural world as we continue to enjoy the benefits of the Third Industrial Revolution.

Third Political Age
Second Progressive Cycle, 1969–1981

The presidential election of 1968 was one of the most painful in American history. The contest evolved in a year that saw race riots, antiwar protests, and the assassinations of Martin Luther King Jr. and Robert Kennedy (who was himself a candidate for the highest office). The chaotic Democratic Convention led to the fracturing of the New Deal Coalition, which had held sway for three decades. In the election itself, a vehement race baiter and segregationist won five states and forty-six electoral votes. In the end, Richard Nixon emerged victorious by playing the race card with a "southern strategy" that capitalized on regional abhorrence of the Civil Rights Act of 1964 and the Voting Rights Act of 1965. This approach was so effective that Hubert Humphrey won only a single southern state (Texas) in a region that had been loyal to the Democratic

Party since before the Civil War. Yet when he entered the White House the following year, Nixon largely governed as a moderate and sought only to reorient but not reverse New Deal and Great Society programs. And perhaps his most notable achievements were within an entirely new sphere, environmental security. As a result, this minor realignment inaugurated a short-lived progressive cycle in which a great deal of attention was focused on passing legislation to protect public health and the natural environment. As a result, environmental security policy became one of the primary factors distinguishing conservatives from progressives.

When he reached the White House in 1969, Richard Nixon was not an environmentalist. Yet during his more than five years in office some of the most important environmental legislation in US history was passed by a Democratic Congress and signed by this moderate Republican—something that is barely imaginable today. Senator Edmund Muskie of Maine was the driving force behind much of the legislation, but Nixon was happy to go along because he believed that his association with the popular movement would improve his political image among young Americans. In 1969, after a high-profile oil spill off the shores of Santa Barbara, Nixon signed the National Environmental Protection Act. This bill mandated that government agencies submit environmental-impact statements for most federally funded projects and created the Council on Environmental Quality to coordinate environmental policy in the executive branch. In December 1970, after mass demonstrations associated with the first Earth Day, Nixon formed an independent regulatory body to oversee the implementation of environmental policy, the Environmental Protection Agency (EPA). Later that month he also signed Muskie's Clean Air Act Extension, which tasked the newly created EPA with generating and enforcing regulations to protect the public from airborne contaminants known to be hazardous to human health (e.g., sulfur dioxide, nitrogen dioxide, particulate matter, carbon monoxide, ozone, and lead). This has proven to be the most important environmental act in American history. During the remainder of his term in office Nixon also signed the critically important Marine Mammal Protection Act of 1972, the Endangered Species Act of 1973, and the Safe Drinking Water Act of 1974. Although he also approved the Alaska Pipeline and vetoed the Clean Water Act of 1972 (an action that was overridden by Congress), it is fair to say that the statutes and regulations fashioned with his support made providing some level of environmental security a new role of the federal government.[116]

During the Oil Embargo of 1973 (discussed more fully in chapter 4), Nixon

announced an initiative he called Project Independence. "Let us set as our national goal," he said in a major presidential address, "in the spirit of Apollo,
with the determination of the Manhattan Project, that by the end of this decade
we will have developed the potential to meet our own energy needs without
depending on any foreign energy source." Although its goals had little to do
with environmental policy given that climate science was barely in its infancy,
it would have had dramatic impacts on the development of the global crisis.
Unfortunately, little was done to make this goal a reality. After Nixon's resig-
nation from office, his successor, Gerald Ford, championed a similar plan that
would have led to the construction of 200 nuclear power plants, 150 coal-fired
power plants, and 20 synthetic-fuel plants. This idea also went nowhere. It
wasn't until the election of Jimmy Carter that real efforts were made to change
American energy-consumption patterns. Unfortunately, this helped lead to his
political downfall four years after taking office.[117]

Despite rising petroleum prices in the aftermath of the 1973 embargo, oil
imports had jumped 65 percent by the time Carter took office four years later.
Americans remained profligate energy consumers, using more than twice as
much as citizens in other advanced economies. The new president was committed to reversing the nation's dependence on foreign energy, as well as to
creating a sense of stewardship for natural resources. During his presidential
transition, Carter announced that he favored a comprehensive plan that would
foster conservation, reduce dependence on imported hydrocarbons, and promote alternative energy. His key adviser in this area was James Schlesinger,
who had directed the Atomic Energy Commission and served as secretary of
defense under both Nixon and Ford. Upon taking office, Schlesinger kicked
off a secret ninety-day effort to develop the wide-ranging energy strategy the
president favored. Two days before the new plan was revealed, Carter delivered
a groundbreaking national address on 18 April 1977; it was the first of several
important speeches in which he attempted to speak truth to the American
people.[118]

"Tonight I want to have an unpleasant talk with you about a problem unprecedented in our history," he told the television audience. "With the exception of preventing war, this is the greatest challenge our country will face during our lifetimes. The energy crisis has not yet overwhelmed us, but it will if we
do not act quickly. . . . We simply must balance our demand for energy with our
rapidly shrinking resources. . . . This difficult effort will be the moral equivalent
of war—except that we will be uniting our efforts to build and not destroy." He
warned his listeners about declining domestic hydrocarbon supplies and the

danger of relying on ever-greater imports from abroad, particularly given the order-of-magnitude increase in spending for imported petroleum from $3.7 billion to $37 billion in just six years. He spoke about past energy transitions from wood to coal and from coal to oil and gas, suggesting that it was time for yet another transition, from oil and gas to conservation and renewable energy. He argued that conservation could be achieved without harming the American way of life, pointing to the fact that nations with similar standards of living used half as much energy. He endorsed a shift to smaller cars and the revitalization of public transit systems. "Other generations of Americans have faced and mastered great challenges," he concluded. "I have faith that meeting this challenge will make our own lives even richer. If you will join me so that we can work together with patriotism and courage, we will again prove that our great nation can lead the world into an age of peace, independence and freedom."[119]

The plan the Carter administration drafted reminds us of two things. First, when it comes to governing, there are few new things under the sun. Humans have been governing themselves for several millennia. During that time they have tried just about every way to accomplish societal goals. Perhaps the single best tools policymakers have are various types of taxes. Second, many of the policy alternatives being debated today as solutions to climate change emerged decades ago. The Carter plan included an expansion of the gas tax (pegged to inflation), a tax on the purchase of gas-guzzling automobiles, tax credits to promote upgrading residential and commercial buildings to make them energy efficient, a carbon tax on domestic oil aimed at raising prices to decrease consumption (which would have funded a rebate for consumers), deregulation of intrastate gas markets, tax incentives aimed at fostering more coal-based manufacturing (an idea that most likely wouldn't have been included if climate science had been better understood), promotion of new nuclear power plants, and inducements to install solar heating. This was an audacious plan, submitted to an American public that over a century had been increasingly enthralled by petroleum. Even though the administration did not consult Congress during the development of the proposal, Speaker Tip O'Neill was able to shepherd the bill through the House of Representatives largely unchanged. Lacking such leadership in the Senate, however, oil-state legislators forced the White House to water down significant portions of the legislation. Still, the National Energy Security Act of 1978 was a remarkable accomplishment. Carter had created the Department of Energy, under the leadership of Schlesinger, to regulate the energy industry and fund research on alternative energy (particularly wind and solar power). It was tasked with implementing many of the bill's provisions, while additional agencies were created to take on specific tasks. By themselves, the creation of a cabinet-level department and passage of this sweeping legislation would have cemented Carter's environmental legacy. But he wasn't done.

On 23 May, Carter delivered a special message to Congress outlining his broad agenda for environmental and energy security. The impressive message contained action items to control pollution and protect health; assure environmentally sound energy development; improve the urban environment; protect natural resources; preserve America's national heritage; protect wildlife; affirm concerns for the global environment; and improve implementation of environmental laws. In the coming years the Carter administration would work faithfully to achieve these objectives, with a great deal of success. Among the major pieces of legislation passed were the Clean Air Act Amendments of 1977, the Soil and Water Conservation Act of 1977, the Surface Mining Control and Reclamation Act of 1977, the Antarctic Conservation Act of 1978, the Endangered American Wilderness Act of 1978, the Alaskan National Interest Lands Conservation Act of 1980, and the Superfund Act of 1980. When a new oil crisis struck in the third year of his presidency, Carter took the opportunity to seek approval for a second wave of energy legislation. He also delivered one of the greatest and most misunderstood speeches in American history (discussed at length in chapter 4). One repercussion of its ultimate failure was the collapse of the short-lived second progressive cycle of the Third Political Age. With this breakdown, environmental security would soon be downgraded severely as a national priority, as hydrocarbon companies and their political allies reestablished themselves within the economy and the halls of government.[120]

PART TWO

SUSTAINABILITY AND AN AMERICAN REBIRTH

CHAPTER FOUR

Energy Insecurity and the American Decline

Peak oil has become one of the more contentious theories in modern times. On one side are oil-industry diehards who claim that petroleum is essentially infinite and should be used regardless of any negative consequences. On the other side are radicals who argue that not only have we passed the peak in global oil production but it will cause a societal meltdown, a prediction in which some from this camp seem to revel. As with most things, the truth lies somewhere in the middle, and it depends to a great degree on how particular terms are defined. In his book *The Quest*, Daniel Yergin attempts to define the concept as a whole: "The peak theory, in its present formulation, is pretty straightforward. It argues that world oil output is currently at or near the highest level it will ever reach, that about half the world's resources have been produced, and that the point of imminent decline is nearing." Yet in suggesting that the ideas behind the peak oil concept are so simple, Yergin glosses over much of the complexity. Describing how extremists on both sides of the debate employ the word *oil* can solve some of this difficulty.[1]

The key sticking point is the distinction between conventional and unconventional oil. Generally speaking, conventional oil is liquid petroleum found in large reservoirs that can be pumped to the surface. It comes in a variety of forms, ranging from pure sweet crude to heavy oils that contain far greater impurity levels. In making their arguments, peak-oil radicals almost always refer to these varieties of conventional oil. Whether these conventional sources have already peaked is still being debated, but both camps agree that this is likely to happen sometime in the first quarter of the twenty-first century. The hitch is that there are large global reservoirs of unconventional oil, found most prominently in shale oil and oil sands. Shale oil occurs in sedimentary rocks in the form of a waxy solid known as kerogen. The world's largest reserve is the Green River formation underlying Utah, Wyoming, and Colorado. The generally high price of contemporary conventional oil (which has been prevalent except during the short period when Saudi Arabia took action to reduce prices and secure its market share), combined with the introduction of new technologies like horizontal drilling and hydraulic fracturing, made production from shale-oil reserves in places like North Dakota economically viable. Shale-oil producers in the United States have essentially become the swing suppliers of the global petroleum markets. Oil sands consist of rock bonded by heavy bituminous materials, with usable oil extracted by excavating huge amounts of soil and baking it to separate the petroleum by-products. The largest known accumulation of oil sands is in the Athabasca area of northern Alberta, although some suspect there might be even larger reserves in Siberia. Extracting petroleum from both shale oil and oil sands requires larger energy expenditures for each recovered barrel of oil, meaning that these sources have a relatively poor EROI (energy return on investment). When producers first started pumping conventional oil out of the ground, the pressure built up in these reservoirs was so great that the EROI was about 100, meaning that the energy found in one barrel of oil was sufficient to extract one hundred barrels from the ground. Over time this EROI slowly decreased, to about 20 by the 1950s. As unconventional oil began to be produced in large amounts in places like Canada and the United States, this number plummeted even further, to about 10, and it continues to decline as more petroleum from shale oil and oil sands enters the market. Since these resources also have far more impurities, this represents an environmental double whammy.[2]

On the positive side of the ledger, unconventional oil is abundant. Since the 1870s the global economy has burned about 1 trillion barrels of conventional oil. By most estimates there are about the same number of barrels remaining, though some diehards say 1.5 trillion is more likely. Some assume that because it took 140 years to burn through half of the conventional oil, it should take another 140 years to use the remaining amount. The problem is that both the world population and the energy intensity of that population have climbed sharply, so that at current usage rates this petroleum would be used far more quickly. Estimates of how much unconventional oil is available vary widely, but most experts would agree that it is very likely more than 5 trillion barrels. Given the long-term trend of higher oil prices, at current EROI levels this means that oil production is very likely to rise in the coming decades. This is particularly the case in nations like the United States and Canada, which pioneered the technologies to leverage these hydrocarbons. "It appears," Yergin writes, "that world [oil] production capacity should grow from about 93 million barrels per day in 2010 to about 110 mbd by 2030. This is a twenty percent increase." So even if conventional supplies have peaked, the addition of unconventional supplies has led to increasing worldwide production. From an economic perspective, once again, this appears to be a positive development. But major impediments remain to maintaining economic modernity and sociopolitical stability while relying so heavily on petroleum.[3]

To begin with, rapidly declining EROI levels pose a significant threat to macroeconomic security. The geographer David Murphy writes that "as the EROI of the average barrel of oil declines, long-term economic growth will become harder to achieve and come at an increasingly higher financial, energetic and environmental cost." Paul Roberts echoes Murphy, arguing that we have reached the end of cheap petroleum. Even if large supplies remain, he says, it is unclear whether they will be economically viable to extract: "Although we will not run out of oil tomorrow, we are nearing the end of what might be called the easy oil. . . . This fact means not only higher prices, but more volatile prices." This price volatility was borne out when Saudi Arabia took action to lower prices, a tactic that would be expected as conventional producers like the Saudis seek to protect market share for as long as possible. Regardless, most experts believe that in the long term prices will continue to rise. Another significant challenge, one that is often overlooked by those who are bullish about future growth in oil production, is the expansion of global demand. The International Energy Agency (IEA) predicts that as a result of population growth and economic globalization, overall energy demand will increase by one-third from 2011 to 2035. While the share of hydrocarbons in the energy mix is expected to decrease, the IEA calculates that they will still represent three-quarters of all demand in three decades. Petroleum will still be providing 27 percent of world energy needs, with 90 percent of the demand growth in emerging economies. The big question, therefore, is whether new oil production by itself can keep pace with mounting demands from the nations that are driving global economic growth, which is directly linked to sociopolitical stability. Finally, and perhaps most troubling, the contribution of petroleum to global climate change will not diminish if we maintain a relatively high level of consumption well into the twenty-first century. Providing climate security will be nearly impossible without a major decrease in petroleum utilization. For a variety of reasons, then, there are serious questions to be answered before we determine whether we are in fact on the verge of a new era of oil abundance. The fact that there is oil in the ground doesn't mean that we will be willing or able to extract and consume it.[4]

* * *

Regardless of the shape of future hydrocarbon developments, one thing is clear historical fact: America's production of conventional oil peaked in 1970.

The geologist M. King Hubbert was the first to seriously consider that there were limits to the supply of conventional oil that could be produced within the United States. (He later made similar computations for global output.) At the time when he began to discuss these boundaries he was considered to be a giant in the field, having taught at Columbia University and worked as the chief of research at Shell Oil before moving to the US Geological Survey. Clearly brilliant, Hubbert had melded physics and mathematics with geology in his graduate work at the University of Chicago. At the same time, he also had an abrasive personality that easily offended colleagues, which may have contributed to his not being tenured at Columbia (forty years later the same university would award him the Vetlesen Prize). His most controversial work emerged from mathematical models he ran in the mid-1950s, while still working for Shell Oil. His key insight was that production would parallel oil field discoveries, with an approximately four-decade lag between the two. Given that the unearthing of new reservoirs in the United States had crested in the 1930s, he calculated that American production would peak between 1965 and 1970. Other geologists roundly ridiculed Hubbert until it became clear that America had indeed peaked in 1970, at 11.3 million barrels a day. This roughly coincided with "multiple Middle Eastern nationalizations of Western oil interests (Iraq 1972, Libya 1975, Iran 1979) and the last surge of new OPEC members (the United Arab Emirates 1967, Algeria 1969, Nigeria 1971, and Ecuador 1973)." The combination of domestic peaking and these developments abroad severely threatened American energy security, with many far-reaching impacts for the nation, many of which have never been fully recognized. As David Painter writes, "The availability of inexpensive oil encouraged the United States to adopt patterns of socioeconomic organization premised on high levels of oil use. Understandable when oil was inexpensive and access secure, this way of life has become less sustainable as economic, strategic, and environmental conditions . . . changed."[5]

In Daniel Yergin's view, one of the most immediate effects was a "major geopolitical rearrangement. The United States could no longer largely go it alone. All through the 1960s, even with imports, domestic production had supplied 90 percent of demand. No longer. To meet its own growing needs, the United States went from being a minor importer to a major importer, deeply enmeshed in the world oil market." American supply peaked at a time when global demand was high; demand doubled in little more than a decade, from nineteen million barrels a day in 1960 to forty-four million barrels in 1972. A primary reason for this surge was the reemergence of Europe and Japan as major industrial powers, posing a significant challenge to American postwar economic dominance. At the same time, controlling the Middle East had become a proxy battle in the Cold War, with both the United States and Soviet Union seeking influence in the region. After years when their existing petroleum reserves developed slowly, nations like Saudi Arabia had finally come into their own. These countries were considering throwing off the corporate colonialism that had reigned supreme in the industry for several decades. OPEC had emerged in the previous decade as a competitor to the seven large petroleum companies that had previously set the global price for oil—Anglo Persian, Gulf, SoCal, Texaco, Royal Dutch Shell, Esso, and Socony. This was a precarious moment for the United States to begin importing large amounts of oil from unstable nations. The result was a threat to the very existence of industrial capitalism, which had provided unprecedented national and economic security to the American people. The country had no choice but to reorient its entire foreign policy to ensure the continued flow of the modern era's "black gold."[6]

This new reality was not lost on either the Soviets or OPEC. Tensions had been rising for years, but they reached a boiling point when the Yom Kippur War broke out in 1973. In October, Egypt and Syria (both Soviet client states) commenced a surprise attack against Israel aimed at regaining territory in the Sinai and Golan Heights lost during the Six Day War. President Nixon quickly decided to provide military aid to the Israelis, which allowed them to gain the upper hand in the conflict. In response, Arab oil ministers launched an embargo on oil exports to the United States, Western Europe, and Japan. The price of oil increased by 70 percent overnight. Combined with the devaluation of the dollar following the American withdrawal from the Bretton Woods Agreement, this devastated the economy. Although the war ended after three weeks, the embargo endured as a Saudi-led coalition took actions to ensure that the price of oil would remain high and that OPEC nations would retain far more of the profits. According to David Painter, "Increased oil earnings allowed producing countries to buy back company-owned concession rights and establish national oil companies." Within a decade, oil-producing countries increased their ownership share from 10 percent to 70 percent, which gave them far more control over the development of their own reserves. By the time the embargo ended in March 1974, the base price of petroleum had quadrupled. Because of America's dependence on oil, the economic strain shattered national confidence. The US automobile industry suffered tremendous trauma since it could only produce large gas-guzzlers. This provided an opening for Japanese automakers, who were able to manufacture small, efficient, and better-engineered cars that instantly began to grab market share. The embargo and easy-money policies set off persistent national inflation, with the annual rate rising to 14 percent by 1980. At the same time, higher oil prices led to stagnant economic growth. The resulting stagflation was a drag on the American economy for years. None of these outcomes would have been nearly so dire without a peak in domestic oil production. America never imported a large percentage of oil directly from the Middle East, relying far more on imports from nations within the Western Hemisphere (e.g., Canada, Mexico, and Venezuela). Yet as John Duffield writes, "Because so much economic activity now crosses national boundaries . . . the health of the U.S. economy is closely bound up with the economic well-being of a number of other countries that are themselves large oil importers." Therefore, because the Saudis had proven that they could control international pricing, key American leaders believed that the only option was to increase diplomatic, economic, and military engagement in the Middle East. The costs of this decision have been immense.[7]

In the decade after domestic oil supplies peaked, there was a major acceleration in the shift of American force deployments toward the Middle East. This included the creation of the US Central Command (CENTCOM) to oversee the region, the establishment of rapidly deployable sealift and airlift forces, and the prepositioning of significant, varied naval forces. The primary goal of this new national security orientation was to ensure the free flow of oil through the Straits of Hormuz, chiefly to US allies throughout Eurasia. American foreign policy became focused on four objectives: to persuade oil-producing nations in the region to increase production and avoid embargoes; to provide military aid to these countries to ensure their internal and external security; to encourage the discovery and development of new oil fields; and to send US military forces to the region if required to guarantee the flow of oil to America's defense and economic partners. This shift in US foreign policy was pursued even though it conflicted with longstanding national values, such as the "promotion of democracy, good governance, respect for human rights, and compliance with international law." As a result of these changes in America's geopolitical strategy, it found itself engaged in two oil wars over a quarter century—the Persian Gulf War and the Iraq War. Similarly, the nation's involvement in the Global War on Terrorism (which included the Afghanistan War) was directly related to its need to guarantee the free flow of petroleum. As former New Mexico governor Bill Richardson, who also served as UN ambassador and energy secretary, has written, "Every American can make the intuitive connection between global dependence on Middle Eastern oil, dissent in Middle Eastern societies, and terrorist attacks on the United States." Thus, in the modern era, American dependence on imported oil has resulted in significant hidden outlays that don't show up at the pump. Fighting three petroleum-related wars cost at least $4 trillion. (Interestingly, only $100 billion of this was spent on the Persian Gulf War, the only conflict with a clearly positive outcome for American forces.) The country spends an estimated $50 billion annually policing the Persian Gulf. Amazingly, as Roger Stern writes, "On an annual basis, the Persian Gulf mission now costs about as much as did the Cold War."[8]

The military costs of domestic peaking were bad, but the direct economic losses were perhaps more severe. As David Greene and Sanjana Ahmad write, between 1970 and 2005 the necessity to rapidly increase exports and rely on foreign markets cost the economy more than $10 trillion (in 2014 dollars), approximately $285 billion annually. This price tag was the result of wealth transfers to oil-producing nations, disruptions resulting from significant price movement, and squandered economic potential during periods of artificially high purchasing costs. These were new losses for a nation that had previously enjoyed the fruits of enormous internal reserves that provided 90 percent of its petroleum. That had changed with the domestic peak, which produced significant challenges that the system was ill equipped to handle. "Oil dependence is not simply a matter of how much oil we import," say Greene and Ahmad. "It is a syndrome, a combination of factors that together create economic, political and military problems. It is comprised of the concentration of the world's oil supply in a small group of oil producing states that wield monopoly power, together with the demand-side vulnerability of the U.S. economy to higher oil prices and price shocks. Our vulnerability depends on how much oil we consume, the lack of ready substitutes for oil, and also how much we import."[9]

Since the domestic oil peak, American consumers and businesses have also been paying more for petroleum, with significant long-term economic and fiscal consequences. In 1970 the inflation-adjusted price of a barrel of oil was approximately twenty dollars (in 2014 dollars). Since that time it has experienced an inexorable rise. While real prices fell substantially from 1986 to 2003, as the Saudis flooded the market, nominal prices were higher. The same was true of recent price declines, as Saudi Arabia temporarily sought to protect its market share. The clear trend in petroleum prices from 1970 to 2015 is an upward rise, and most energy analysts admit that this movement will continue in the future regardless of attempts by petro states to game the system. Because coal and gas prices have long been linked with petroleum prices, overall the nation is spending more for energy then was true four decades ago. The Energy Information Agency suggests that energy expenditures as a percentage of GDP in 2008, at the end of the Third Political Age, were greater than those in 1970, 9.9 percent compared with 8.0 percent (this share decreased from 1986 to 2003 but began rising again thereafter). And these data don't include all the social costs associated with America's oil addiction, many of which have accelerated since the domestic oil peak.[10]

Lester Brown, of the Earth Policy Institute, writes: "The many indirect costs to society—including climate change, oil industry tax breaks, oil supply protection, oil industry subsidies, and treatment of auto exhaust-related respiratory illnesses—total around $12 per gallon ($3.17 per liter), marginally more than the cost to society of smoking a pack of cigarettes. If this external or social cost is added to the roughly $3 per gallon average price of gasoline in the United States, a gallon would cost $15. These are real costs. Someone bears them. If not us, our children." I discuss climate change at length below, but it is worthwhile to briefly look at the social costs associated with direct subsidies, tax breaks, and health care. Over the past century, the federal government has guided approximately $470 billion in subsidies and tax breaks to the oil industry. This began in 1916, when Congress passed a tax provision that allowed extraction companies to write off dry holes and intangible drilling costs during their first year of exploration. A decade later, another tax provision provided an over-blown "depletion allowance," which permitted producers to deduct 27.5 percent of their annual revenues, a provision later panned by treasury secretary Henry Morgenthau as the most glaring loophole in the tax code. At midcentury, *Fortune* referred to it as the "greatest single source of modern American wealth." It wasn't until 1969, during the early days of the environmental movement, that action was finally taken to fix the problem. Yet Congress only approved a decrease from 27.5 percent to 23 percent of revenues. Six years later the legislature eliminated the depletion allowance for big companies, although it was left in place for smaller exploration outfits. Over the coming three decades Presidents George H. W. Bush, Bill Clinton, and George W. Bush would slowly expand the number of companies eligible for the allowance. Meanwhile, on the subsidies front, funding for fossil fuel–related research had begun a rapid rise during the OPEC oil embargo, jumping 1,000 percent in just five years, and it continued to increase over time. An indirect subsidy was added in the mid-1990s, when Clinton signed legislation that allowed companies to drill in the Gulf of Mexico without paying royalties. In recent years, the total cost of these various direct subsidies has been about $10 billion per year.[11]

For all the concern about the societal costs of these subsidies, there is far less discussion of a much bigger harm. A 2009 report from the National Research Council, which had been requested by Congress, found that an estimated $120 billion in annual costs were associated with American consumption of hydrocarbons (not including damages from climate change, ecosystem destruction, air pollutants such as mercury, and risks to national security). Damage from pollutants released by coal-fired power plants, which provide electricity to the nation, accounted for $62 billion of this estimate. Damages from gas-fired power plants and heating systems accounted for less than $3 billion, which is striking given their growing prevalence. Damages from the use of petroleum, for both transportation and industry, accounted for the remaining $56 billion. While these hydrocarbon-related costs are not necessarily greater then they would have been if American oil supplies hadn't peaked in 1970, the decision to increase petroleum imports rather than find cleaner alternatives should certainly be considered in assessing the overall impacts of the domestic peak on American politics.[12]

Although roundly ignored by national leaders and most scholars alike, the peaking of American domestic oil supplies was a pivotal turning point in US history. The resulting societal damage, which included grave harm to the political system, was considerably magnified when a conservative cycle began a decade later. Since the peak, the country has spent trillions of dollars protecting the free flow of petroleum. Direct percentage expenditures on energy have increased substantially, and experts concur that they are likely to increase in the future. Direct subsidies, tax breaks, and research funding intended to prop up the ailing petroleum industry have added billions in costs. Finally, some proportion of the vast annual energy-related health-care expenses paid by Americans can be added to the tally. In sum, the costs to the nation of peak oil and the continued propping up of the hydrocarbon system since 1970 easily surpass $500 billion annually. Obviously, these costs have placed dramatic fiscal pressure on government. As figure 4.1 shows, total energy consumption and overall economic growth slowed substantially—from 1930 to 1980 the economy grew sixfold, but from 1980 to 2008 it didn't even double—after the domestic peaking of oil production. At the same time, the government budget expanded at an even more sluggish pace, while the national population continued to swell and a large share of resources were dedicated to maintaining the Hydrocarbon Age. This combination of factors heralded a drastic rise in political dysfunction and the failure to make the big decisions needed for the country to thrive

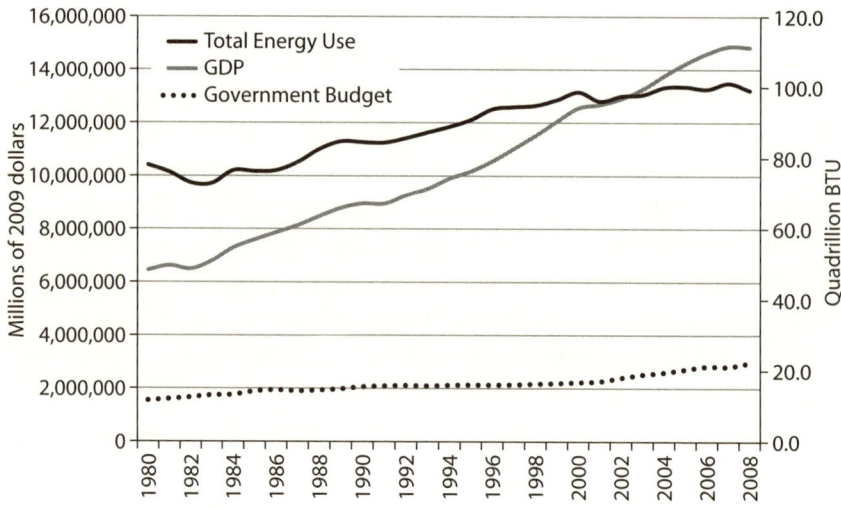

Figure 4.1. US Total Energy Use, GDP, and Government Budget, 1980–2008. US Energy Information Administration, *Annual Energy Review, 2009; Budget of the US Government, Historical Tables, 2017;* Louis Johnston and Samuel H. Williamson, "What Was the U.S. GDP Then?," 2017.

in the twenty-first century. When these shortcomings were combined with an increasingly interventionist foreign policy, the result was a markedly weakened country struggling to maintain its national image abroad.[13]

Close of the American Century?
The American People Fail Jimmy Carter

Although few Americans recognized the fact, by the late 1970s their nation had begun its relative decline. It had been a decade of monumental transitions, of which peaking conventional-petroleum production was simply the most important. The period also saw the reemergence of Europe and Japan as global powers. It saw a distinct thawing in the Cold War as the policy of détente held sway, with the United States shifting its geopolitical focus toward the Middle East—coinciding with a dramatic increase in global terrorist incidents. It saw an American president resign his office after covering up a political scandal and then lying about it to the entire nation. It saw stagflation temporarily upset a widely held belief in national exceptionalism. It saw the global population reach four billion as the impacts of the Second Agricultural Revolution began to take hold. It saw the initiation of policies that would signal the rise of new economic behemoths that would come to upset the global balance of power.

It saw the coming of age of the Third Industrial Revolution, which included the beginning of a conversion to digital technologies. And it saw the end of a three-decade period of slight global cooling, during which the high concentration of sulfate aerosols pumped into the atmosphere by coal-fired power plants had temporarily masked the warming effect of carbon dioxide; from this point forward climate trends became ever clearer. Given the influence of these shifts on the ability of the federal government to continue providing national and economic security to its people, there is good reason to believe that the decade signaled the close of the American Century.

The events of 1979 made clear to President Carter and his advisers that some of these transformations were eroding American strength. The troubles the nation faced found expression in the Iranian Revolution. Almost four decades earlier, Mohammad Reza Shah Pahlavi had become the sitting monarch in Iran. In 1953, more than ten years into his reign, a US-backed coup had removed the country's democratically elected prime minister, Mohammad Mosaddegh, because he had nationalized the country's petroleum reserves. The shah supported the continued exploitation of Iranian oil by foreign companies, largely because it was making him extraordinarily wealthy. By the late 1970s, however, his people had grown tired of his continued rule. Even though the nation enjoyed relative prosperity, protests calling for the shah's removal began in 1977, and the following year general strikes crippled the national economy. In January 1979 the shah was forced into exile. Three months later a national referendum declared the country an Islamic republic with a theocratic-republican constitution, and the Ayatollah Khomeini was selected as the supreme leader. One of the immediate results of the revolution was a 4 percent decrease in daily petroleum production, which spurred a global energy crisis that witnessed oil prices rising to nearly $40 per barrel. As Kevin Mattson writes, "The age of limitless, low-priced gas—what better symbol of American power than that?—had ended." This set off a new wave of inflation increases and seriously eroded the economic abundance that had been enjoyed during the Great Compression. The Three Mile Island accident in March only added to the sense of crisis. Yet a people that had come to expect that the riches provided by hydrocarbons would never end were ill prepared for calls to action. They would rather close their ears and enjoy the cultural excesses of the decade: "disco dancing, roller skating, hot tubs, mood rings, television shows like *Charlie's Angels* and *Three's Company*, a whole slew of fads and mindless diversions."[14]

On 5 April 1979 President Carter spoke on television in an attempt to quell fears of the energy crisis, but few watched. Two years earlier his energy speech had garnered eighty million viewers, but faced with a new crisis, only thirty million tuned in. A Democratic pollster named Pat Caddell persuaded Carter to make another attempt several months later. He believed that the nation was facing an identity crisis and that only by facing it head on could the president get through to the American people. Caddell worried that unless Carter could make a connection with the electorate, his chances of reelection were slim. Caddell pointed to reports that Senator Ted Kennedy was thinking about making a challenge for the Democratic nomination. Such a serious test from within the party could have dangerous repercussions in the general election, even if Carter won the nomination. The president's poll numbers had been sliding precipitously all year as frustration mounted about the growing energy crisis, finally dipping to a record low of 25 percent by July. Gas lines reappeared, and long-haul truckers were striking in protest of rationing and high prices, as if these weren't simply the market realities of a nation faced with reduced domestic oil supplies and increased dependence on unstable foreign exports. The situation had worsened at the end of June, when OPEC ministers decided to hike prices for their oil by 24 percent, largely to protest continued American support for Israel.[15]

Carter had planned to deliver a major address on the energy crisis when he returned from a trip to East Asia in early June. He abruptly canceled it, however, and departed Washington for the presidential retreat at Camp David, in the Catoctin Mountains of Maryland—where ten months earlier he had negotiated an accord between Israel and Egypt that represented the high point of his presidency (and for which he would later be awarded the Nobel Peace Prize)—to redraft the speech and issue a call to arms that would truly connect with the American people. Carter spent eleven days secluded at Camp David, bringing in a select group of advisers, governors, mayors, legislators, businesspeople, academics, and members of the clergy to discuss the problems facing the country. He was told the nation faced a moral and spiritual crisis. He was told Washington had become too disconnected from the people. He was told the nation couldn't continue to import 40 percent of its energy supply. He was told that while mistakes had been made in recent years, it was time to experiment to find solutions. He was told that if he led, the people would follow. He emerged from the meetings believing he had no choice but to speak truth to the citizenry, being clear about the contemporary shortcomings of the American way of life but also speaking to the country's inherent greatness. On 15 July Carter returned to the White House to deliver the defining address of his presidency. Networks broke into their regularly scheduled programming, which gave him the largest audience of his presidency.[16]

Carter's brilliant and prophetic words, which bring energy security front and center as perhaps the most important concern of government, deserve to be quoted at length:

> I want to speak to you first tonight about a subject even more serious than energy or inflation. I want to talk to you right now about a fundamental threat to American democracy. . . .
>
> The threat is nearly invisible in ordinary ways. It is a crisis of confidence. It is a crisis that strikes at the very heart and soul and spirit of our national will. We can see this crisis in the growing doubt about the meaning of our own lives and in the loss of a unity of purpose for our nation.
>
> The erosion of our confidence in the future is threatening to destroy the social and the political fabric of America.
>
> The confidence that we have always had as a people is not simply some romantic dream or a proverb in a dusty book that we read just on the Fourth of July.
>
> It is the idea which founded our nation and has guided our development as a people. Confidence in the future has supported everything else—public institutions and private enterprise, our own families, and the very Constitution of the United States. Confidence has defined our course and has served as a link between generations. We've always believed in something called progress. We've always had a faith that the days of our children would be better than our own.
>
> Our people are losing that faith, not only in government itself but in their ability as citizens to serve as the ultimate rulers and shapers of our democracy. As a people we know our past and we are proud of it. Our progress has been part of the living history of America, even the world. We always believed that we were part of a great movement of humanity itself called democracy, involved in the search for freedom, and that belief has always strengthened us in our purpose. But just as we are losing our confidence in the future, we are also beginning to close the door on our past.
>
> In a nation that was proud of hard work, strong families, close-knit communities, and our faith in God, too many of us now tend to worship self-indulgence and consumption. Human identity is no longer defined by what one does, but by what one owns. But we've discovered that owning things and consuming things does not satisfy our longing for meaning. We've learned that piling up material goods cannot fill the emptiness of lives which have no confidence or purpose.

For the first time in the history of our country a majority of our people believe that the next five years will be worse than the past five years. . . . There is a growing disrespect for government and for churches and for schools, the news media, and other institutions. . . . We were taught that our armies were always invincible and our causes were always just, only to suffer the agony of Vietnam. We respected the presidency as a place of honor until the shock of Watergate. . . . We believed that our nation's resources were limitless until 1973, when we had to face a growing dependence on foreign oil. . . . These wounds are still very deep. They have never been healed. . . .

Washington, D.C., has become an island. . . . What you see too often in Washington and elsewhere around the country is a system of government that seems incapable of action. . . .

What can we do? First of all, we must face the truth, and then we can change our course. We simply must have faith in each other, faith in our ability to govern ourselves, and faith in the future of this nation. Restoring that faith and that confidence to America is now the most important task we face. It is a true challenge of this generation of Americans.

One of the visitors to Camp David last week put it this way: "We've got to stop crying and start sweating, stop talking and start walking, stop cursing and start praying. The strength we need will not come from the White House, but from every house in America."

We know the strength of America. We are strong. We can regain our unity. We can regain our confidence. We are the heirs of generations who survived threats much more powerful and awesome than those that challenge us now. . . . We are at a turning point in our history. . . . All the traditions of our past, all the lessons of our heritage, all the promises of our future point to another path of common purpose and the restoration of American values. . . .

Energy will be the immediate test of our ability to unite this nation, and it can also be the standard around which we rally. On the battlefield of energy we can win for our nation a new confidence, and we can seize control again of our common destiny. . . . The energy crisis is real. It is worldwide. It is a clear and present danger to our nation. These are facts and we simply must face them. . . . I am tonight setting a clear goal for the energy policy of the United States. Beginning this moment, this nation will never use more foreign oil than we did in 1977—never. From now on, every new addition to our demand for energy will be met from our own production and our own conservation. . . . To give us energy security, I am asking for the most massive peacetime commitment of funds and resources in our nation's history to develop America's own alternative sources of fuel. . . . I propose the creation of an energy security corporation to lead this effort. . . . Moreover, I will soon submit legislation to Congress calling for the creation of this nation's first solar bank, which will help us achieve the crucial goal of 20 percent of our energy coming from solar power by the year 2000. . . . I'm proposing a bold conservation program to involve every state, county, and city and every average American in our energy battle. This effort will permit you to build conservation into your homes and your lives at a cost you can afford. . . . To further conserve energy, I'm proposing tonight an extra $10 billion over the next decade to strengthen our public transportation systems. . . .

So, the solution of our energy crisis can also help us to conquer the crisis of the spirit in our country. It can rekindle our sense of unity, our confidence in the future, and give our nation and all of us individually a new sense of purpose. . . . we can succeed only if we tap our greatest resources—America's people, America's values, and America's confidence. . . . In the days to come, let us renew that strength in the struggle for an energy secure nation. . . . I will do my best, but I will not do it alone. Let your voice be heard. . . . Let us commit ourselves together to a rebirth of the American spirit. Working together with our common faith we cannot fail.[17]

Four decades later, Carter's words echo hauntingly in one's ears. He spoke about a threat to democracy, but a continued lack of action to secure the nation's form of government has led to the virtual destruction of We the People. Americans now live in a nation where energy companies, which have the most to gain from continued inaction, wield the ultimate veto power. As a result, the country remains heavily dependent on imported energy (the recent uptick in domestic production notwithstanding) and unwilling to make the sacrifices necessary to develop abundant renewable energy sources. The crisis of confidence that President Carter pointed to as modern America's fundamental problem has never been truly addressed, although those who profit from continued apathy have successfully veiled this dysfunction at the heart of the national existence. Years ago I heard someone pose the following question, which has always stuck with me: "Did Jimmy Carter fail the American people, or did the American people fail Jimmy Carter?" I think the answer is mostly the latter.

The initial reaction to Carter's speech was overwhelmingly positive. Editorialists and pundits lauded the remarkable performance, praising the audacity of a president who was, after all, faced with historically low poll numbers. Positive calls and letters flowed into the White House. *Time* magazine stated that "in the whole history of American politics, there had never been anything quite like it." A *New York Times* headline read: "Speech Lifts Carter Rating to 37%; Public Agrees on Confidence Crisis; Responsive Chord Struck." President Carter's approval rating jumped twelve points almost overnight. In the weeks and months to come the administration would begin to join the fight by successfully gaining passage of key policies such as the Crude Oil Windfall Profits Tax and by taking executive action to implement CAFE (corporate average fuel efficiency) standards mandating more efficient automobiles. Carter set a personal example by turning down the White House thermostat and placing solar-thermal panels on the roof to heat water.[18]

Unfortunately, however, the call for a change in the culture of energy usage never took hold. For too many decades the American people had luxuriated in the lifestyle provided by cheap energy, and few were willing to make the sacrifices necessary to change course. It was one thing to support efforts to clean up the environment; it was quite another to forgo personal comforts. As a natural technocrat, President Carter lacked the requisite skills to persuade the citizenry that these actions were required to maintain national greatness, and he never took the necessary political steps to counter critics who argued the opposite. It might be said that he put far too much faith in the American people. Some have argued that he should have pushed for more aggressive actions by the government to force a change in culture. The reality, though, was that even though many agreed that the United States faced huge problems, they would probably have rebelled had he tried to press the issue. This point was essentially proven in the coming fifteen months as the nation turned against Carter. The news media had quickly changed its tune, accusing the president of being overly pessimistic in suggesting the nation had descended into a moral "malaise," a term he had not used in the address. "The speech, not as a text or a set of actual words but as an imagined event," Mattson writes, "provided a script for Carter's end. The script's lines generated fuel for the fires burning in the minds of those who hoped to depose the president."[19]

Even though Carter had properly identified many of the problems facing the nation, members of his own party quickly smelled blood in the water. Primary among those was Senator Ted Kennedy, who for all his strengths as a legislator never had a concrete reason to run for president; it was simply his birthright. His decision to challenge the sitting incumbent from his own party, while ultimately unsuccessful, would deliver a body blow from which Carter never recovered. It was into this opening that former California governor Ronald Reagan strode. The "happy warrior" slyly decided that the best way to win the presidency was not to challenge the sitting president on the facts but rather to paint a rhetorical fantasy about the nation's position on the global stage. He countered Carter's supposed pessimism with sunny optimism. As the Iran hostage crisis seemed to bolster the perception of presidential incompetence, the American people would ultimately buy what Reagan was selling hook, line, and sinker. The result was the end of a progressive cycle and the beginning of the most damaging conservative backlash in the nation's history. The decision of a generation to reject national progress in favor of individual comfort significantly deepened the gathering climate crisis, while also leading to the triumph of plutocracy and the creation of structures that strikingly reduced the possibility that the nation could avoid future calamities.

Third Political Age
Conservative Cycle, 1981–2009

In 1971 the Harvard professor John Rawls published the most important work of American political philosophy of the twentieth century, *A Theory of Justice*. In this book he argues that the key to a fair civilization is a just social contract between the people and the government. This justice would be produced by a system in which all citizens were treated equally, which would require the creation of institutions (e.g., electoral, education, health-care) that would correct imbalances between the advantaged and the disadvantaged. He advanced the idea that redistributive institutions should be created behind a veil of ignorance, with the framers of legislation making structural decisions without knowing their own place in a future society. Not knowing whether they might find themselves in a poor social position—belonging to the wrong class, gender, race, or sexual orientation, for example—they would be more likely to create a fair society. Rawls was not suggesting a society in which all differences would be eliminated but one in which those differences would be the result of merit rather than privilege. Rawls's book was inherently a defense of the modern welfare state; he believed that only just institutions could generate a fair society. Unfortunately, a decade after the book's publication a conservative backlash began that at its core sought to erode policies aimed at providing economic and environmental security.[20]

It is ironic that a generation committed to returning "power to the people" elected an entire political class that joined the financial elite in taking a wrecking ball to American democracy and creating an entrenched plutocracy. While the Greatest Generation and the Silent Generation were still the dominant force in American politics when Ronald Reagan swept into office in 1980, in the decades that followed, the Baby Boomers would elect leaders who took the vision of government that he advocated to unconscionable extremes. For decades, economic conservatives had been striving to gain control of the Republican Party and win the White House. They had suffered repeated defeats to moderates within their own party and to progressives from the Democratic Party. So how did they reverse this trend? The answer was a new partnership. The fusion of economic and social conservatives was the defining development in modern American politics, because it created a new, extraordinarily powerful coalition. The idea had first emerged when Frank Meyer, editor of the *National Review*, began writing in the mid-1950s about the benefits of such an arrangement. He believed that economic and social conservatives were natural allies. The latter could provide the party with moral order, while the former could destroy the big government social conservatives believed was responsible for moral decay.[21]

As the nation faced its crisis of confidence in the 1970s, social conservatives emerged as a political force in a way that America had never experienced. The instrument of this transformation was the Reverend Jerry Falwell's Moral Majority, which sought to politically mobilize Christians. This organization was created in reaction to the *Roe v. Wade* decision, the ongoing debate about an Equal Rights Amendment, and the Carter administration's failed attempt to take away the tax-exempt status of Christian schools if they failed to enroll minorities. The Moral Majority's entry into politics resulted in a far more proactive approach on the part of key religious leaders. Rather than opposing sex education, they began to advocate for preaching abstinence in American schools. Rather than opposing the teaching of evolution in schools, they advocated also teaching about intelligent design. Rather than wanting government to leave them alone, they demanded that federal funds be directed to local churches and religious groups. And most importantly, rather than taking a back seat when it came to vetting federal-court appointees, they focused on becoming the single most important interest group in this process. Kevin Phillips, the former Republican strategist who was responsible for crafting the party's "southern strategy," has more recently argued that this development signaled "the transformation of the GOP into the first religious party in U.S. history." The benefits to economic conservatives, the champions of the rising oligarchy, were readily apparent. Forcing this brand of social politics onto the national stage made it easier to obscure the nation's economic and geopolitical decline and the slow erosion of Americans' most deeply held democratic principles. And this allowed conservatives to accomplish the only political objectives they really cared about: significantly lowering taxes on the wealthiest

Americans and slashing government regulations. All that remained was to find a leader whom both economic and social conservatives could rally behind. It was in this role that a former Hollywood actor and California governor, Ronald Reagan, was cast.[22]

In 1981 Reagan took office advocating a return to laissez-faire economics and hoping to engineer a partisan realignment capable of reversing the economic security and environmental security gains made during the previous progressive cycles. Although he was never able to fully achieve either goal, he was able to work with congressional Democrats to make headway on both. In his inaugural address, President Reagan delivered an ode to individualism and a rejection of the communitarian principles that had been responsible for the Great Compression. In speaking about the inflation that had afflicted the nation for most of a decade, rather than making clear that the energy crisis and easy-money policies were responsible, he told the American people: "In this present crisis, government is not the solution to our problem; government is the problem. From time to time we've been tempted to believe that society has become too complex to be managed by self-rule, that government by an elite group is superior to government for, by, and of the people." What he didn't share was that in his vision of the nation's future, democratically elected leaders would be replaced by a plutocracy. A supposed governing elite would be replaced by a very real financial elite that had no responsibility to safeguard the public welfare. "If we look to the answer as to why for so many years we achieved so much, prospered as no other people on Earth," he said, "it was because here in this land we unleashed the energy and individual genius of man to a greater extent than has ever been done before." While it is certainly true that the United States has been blessed with fine minds throughout its history, many highlighted earlier in this book, Reagan's historical interpretation ignored the role the government played in providing national, economic, and environmental security over a period of more than two centuries. He concluded: "We're not, as some would have us believe, doomed to an inevitable decline. . . . So, with all the creative energy at our command, let us begin an era of national renewal. Let us renew our determination, our courage, and our strength. And let us renew our faith and our hope. . . . [We must] believe in our capacity to perform great deeds, to believe that together with God's help we can and will resolve the problems which now confront us. And after all, why shouldn't we believe that? We are Americans." This passage shows why Reagan was called the Great Communicator. He could sway public opinion, even if he mischaracterized the beliefs of the long chain of national leaders who had guided the country to international preeminence. There is no doubt that Reagan and his ideological successors were sincere in their belief that government had failed to serve the American people. There is also no doubt that history has proven them disastrously incorrect, a fact that the country is still incapable of fully grappling with.[23]

The minor realignment election of the previous year gave Reagan the political capital necessary to advance a significant legislative agenda in Congress. His two main goals were to cut taxes on the wealthy and expand the military. He sought to pay for the first by making significant cuts to social insurance programs. The problem was that he was still faced by an opposition legislature that resisted those cuts. In the end, as Sidney Milkis and Michael Nelson write, "Reagan persuaded Congress to approve a dramatic departure in fiscal policy: more than $35 billion in domestic program reductions; a multi-year package of nearly $750 billion in tax cuts; and a three-year, 27 percent increase in defense spending." This approach saw the commencement of a new era in which government determined to pull out the proverbial credit card rather than make the tough decisions necessary to address the economic and budgetary challenges that had begun with peak oil a decade earlier. Thus a president who had made a pledge to the American people that he would eliminate existing budget deficits did the exact opposite: he ran up the largest peacetime national debt to that point, tripling it to $3 trillion in just eight years. Reagan was able to sell this to the American people by employing the concept of supply-side economics, which argued that large tax cuts would stimulate so much economic growth that tax revenues would actually rise to balance the budget. Unfortunately, Reaganomics never came close to achieving this objective.[24]

As he was making this domestic argument, Reagan was also suggesting that a large increase in defense spending was needed to bankrupt an economically tottering Soviet Union. The argument was that an arms race would work in America's favor because the US economy could handle the added pressure, while the Soviet economy could not. Although there was certainly a limited amount of truth to this argument, it was not the primary reason for the fall of the communist bloc. The real reason for the fall was plummeting oil prices. In the mid-1970s the Soviet Union had replaced the United States as the global leader in oil production. Early in the next decade, OPEC nations found themselves facing a crisis as their global market share dipped below 50 percent for the first time in many years. "OPEC was in trouble," writes Yergin. "The market confronted it with an unpalatable choice: cut prices to regain markets, or cut production to maintain price." The cartel decided to flood the markets with cheap oil. This was an economic disaster for the Soviets, who depended on the export of hydrocarbons for the hard currency needed for food imports. With oil at lower prices, it could not afford to expand oil production, which would hamper the Soviet petroleum industry until prices began to climb again more than a decade later. Reduced exports and reduced profits on what was sold made the collapse of the Soviet Union inevitable.[25]

The increasing economic instability within the Soviet Union helped accelerate a transition that had already begun to take hold with regard to America's national defense posture. In 1980, in his last State of the Union address, President Carter had announced a new policy aimed at protecting US petroleum interests in the Middle East:

> The region which is now threatened by Soviet troops in Afghanistan is of great strategic importance: It contains more than two-thirds of the world's exportable oil. The Soviet effort to dominate Afghanistan has brought Soviet military forces to within 300 miles of the Indian Ocean and close to the Straits of Hormuz, a waterway through which most of the world's oil must flow. The Soviet Union is now attempting to consolidate a strategic position, therefore, that poses a grave threat to the free movement of Middle East oil.
>
> This situation demands careful thought, steady nerves, and resolute action, not only for this year but for many years to come. It demands collective efforts to meet this new threat to security in the Persian Gulf. . . . Meeting this challenge will take national will, diplomatic and political wisdom, economic sacrifice, and, of course, military capability. . . .
>
> Let our position be absolutely clear: An attempt by any outside force to gain control of the Persian Gulf region will be regarded as an assault on the vital interests of the United States of America, and such an assault will be repelled by any means necessary, including military force.[26]

As discussed earlier, this new Carter Doctrine required shifting military resources to the region. Carter began this process by developing the Rapid Deployment Joint Task Force, which became the US Central Command under Reagan. This force structure included various Air Force fighter and reconnaissance wings; the First Marine Division; multiple Army airborne, infantry, and mechanized divisions; and most importantly, three Navy aircraft-carrier groups. When the Cold War ended, CENTCOM not only remained in place but became the single most important regional command within the armed services. This new geopolitical focus resulted in continued military spending that was, at its core, intended to ensure that cheap oil would continue to flow into global markets. The sole reason for the Reagan Corollary to the Carter Doctrine, which pledged American support for Saudi Arabia if the ongoing Iran-Iraq War spilled over into its territory, was to ensure that the millions of barrels being produced in Saudi Arabia every day would reach global markets.[27]

While tax cuts, rising deficits, and expanding military budgets were all cause for concern, deregulation in the social sphere and especially in the economic sector threatened continued macroeconomic and microeconomic security. Over the course of his presidency, Reagan worked systematically to enfeeble the government agencies tasked with safeguarding the nation and its citizens in these areas. For much of its history, as discussed earlier, the American economy had few regulations, and laissez-faire attitudes prevailed. Beginning with the Progressive Era and accelerating with the New Deal, the government recognized that regulating capitalism was a core function. True believers, however, clung to the notion that markets were always right and government was always wrong. They found their greatest advocate in President Reagan. Particularly in the economic sector, he and his staff worked to rapidly erode a regulatory framework that had resulted in the longest period of market stability in US history. This despite the fact that the first attempt at financial deregulation resulted in the savings-and-loan crisis, which Seth Allcorn and Howard F. Stein describe as the "largest theft in the history of the world at that time." Deregulation allowed the owners of these financial institutions to lend money to themselves, which they invested in highly speculative financial instruments. When these led to insolvency, the federal government intervened on behalf of the plutocrats with a $1.4 billion taxpayer bailout. Rather than pausing their efforts, Reagan and his acolytes determined to go further. They were so successful that they also convinced moderate members of the opposition party, including the only Democratic president during this cycle. "The Reagan deregulation," writes Bruce R. Scott, "was the underpinning for the rapid rise of the financial services sector, with rapidly rising incentive compensation, inducing increased leverage, and an increasing concentration of wealth in this one sector." This would lead to the worst economic disaster since the Great Depression. This was Reagan's truest legacy; he made the nation less economically secure.[28]

As vice president, George H. W. Bush chaired two task forces aimed at "regulatory relief." Once he was elected to the presidency, however, he governed as the moderate he had always been, as previously evidenced by his suggestion during the 1980 presidential nomination fight that Reagan's policy plans were "voodoo economics." One of Bush's first tasks upon taking office was to gain control of what he viewed as runaway deficits. Although he had famously said in his nomination speech, "Read my lips: no new taxes," in an act of tremendous political courage he broke this pledge in reaching a compromise with congressional Democrats to start reducing deficits. This decision was partially responsible for the economic boom of the 1990s, burnishing his legacy of modestly advancing economic security (although he took no actions to reverse the regulatory policies of his predecessor). Although he also signed the Clean Air Act of 1990 and the Civil Rights Act of 1991, Bush's signature achievements came in the national security sphere. This shouldn't be surprising given his foreign policy credentials; he had, after all, been an envoy to China, ambassador to the United Nations, CIA director, and a key actor on Reagan's national security team. As president he masterminded a nimble response to the collapse of communism in Eastern Europe and the eventual fall of the Soviet Union, which involved cheerleading from the sidelines without actively intervening. He led the nation into the Persian Gulf War in an effort to dislodge Saddam Hussein from Kuwait, brilliantly assembling a global coalition and prudently pulling ground troops out once the objectives of Operation Desert Storm had been accomplished, while persuading partners to pay nearly 90 percent of the costs. He also removed the Panamanian dictator Manuel Noriega from power, dispatched twenty-five thousand troops to Somalia to restore order and provide humanitarian assistance, signed the North American Free Trade Agreement, and negotiated the START II Treaty with Russian president Boris Yeltsin. Thus, while his domestic achievements were somewhat lackluster, his accomplishments in providing enhanced national security place him among the best foreign policy presidents of the past century. Unfortunately, however, in the coming years it would become increasingly clear that the end of the Cold War had threatened American geopolitical preeminence. This was because removing the Soviet threat cemented the US military's role as a "global oil-protection force."[29]

Bill Clinton's presidency had mixed results. When he was elected, many progressives had high hopes that he would continue Lyndon Johnson's expansion of microeconomic policies. Initially he tried. He started by setting a macroeconomic foundation, passing a budget deal that raised taxes aimed at eliminating the federal budget deficit; combined with actions taken by his predecessor, this set the nation on a path that would eventually result in a budget surplus. Although he faced a hostile news media and was hampered by inexperienced staffers, he daringly waged a battle with Republicans and members of his own party to gain passage of universal health care. His failure to engage congressional leaders in the process of drafting the legislation, however, not only resulted in an extremely complicated bill but also disenchanted many within Congress. After the failure of the health-care initiative, the Democrats suffered a crushing defeat in the midterm elections, giving the Republicans control of the national legislature for the first time in more than five decades. In order to maintain his relevance, Clinton quickly pivoted, moving to the ideological right and adopting many conservative ideas as his own. After he skillfully exploited a government shutdown in 1995 to bolster his standing with the American people, he announced in his 1996 State of the Union address that "the era of big government is over." Michael Beschloss has described this passage as "a skywritten acknowledgment by someone who in another age might have liked to govern as a liberal Big Government Democrat that the Age of Reagan was so overwhelming that even a Democratic President had to work within its limits." In the coming year this led to the passage of the Welfare Reform Act of 1996, which gutted longstanding welfare programs aimed at propping up low-income families and helped Clinton cruise to reelection. By far his most damaging decision was to support the repeal of the Glass-Steagall Act. As discussed earlier, this legislation had created a firewall between investment and commercial banking and set reserve requirements for commercial banks. The legislation had been an overwhelmingly success, helping smooth the boom-bust financial cycle that had long plagued the nation. The few opponents to its repeal argued that removing needed restraints would lead to reckless speculation and create financial institutions that would be considered "too big to fail." How right they were. Combined with the erosion of regulations during the Reagan administration, this opened the door for an economic meltdown that resulted in widespread suffering. This is one of the key ironies of Bill Clinton's legacy: he helped create the most prosperous peacetime economy of the postwar period yet vigorously supported legislation that would ultimately cripple the country in the future. Although there is much nostalgia for his presidency, on balance he likely made the nation less economically secure.[30]

Even though his vice president had been one of the most outspoken environmentalists in Congress, Clinton had a surprisingly modest record of advancing climate security. By far his biggest accomplishment was tasking Vice President Gore with actively participating in the negotiations that resulted in the Kyoto Protocol. The problem, however, was that Clinton chose to avoid what surely would have been a vicious fight in the Senate to gain ratification of the treaty. While this effort likely would have failed, it would have signaled to the American people how seriously the Democratic Party took climate security, a position that would have been difficult to walk back. Instead the party remained frustratingly uncertain about how to position itself on this issue. When two oil executives were elected to replace Clinton and Gore in the White House, these newcomers proved the political system's inability or unwillingness to take this issue seriously. As Robert Engler wrote nearly four decades earlier, "The presence within [government] of men whose direct origins have been in oil [threatens] political democracy.... A corrosion of democratic principles and practices pervades wherever the interests of private oil and public policy meet."[31]

After reaching the White House in one of the closest elections in American history, George W. Bush pursued an aggressive legislative and regulatory agenda. Despite some reservations about his political acumen, early in his tenure he was surprisingly successful in gaining passage of key bills. The problem for the nation was that many of these policies further endangered economic security, particularly because he presided as a Big Government Conservative, cutting taxes and increasing spending. Rather than making tough decisions, he chose to run the country on a credit card, even in a time of war. In his early years, the now infamous "Bush tax cuts" were adopted, providing tremendous benefits to the wealthiest Americans. Nearly 90 percent of the cuts went to the top 25 percent of wage earners, meaning that 75 percent of the population shared just over 10 percent of the cuts. This exacerbated already troubling levels of income inequality within the country. The tax cuts also drove federal budget deficits to new highs, helping to more than double the national debt during Bush's tenure alone. When President Bush arrived in Washington, the national economy was already on the precipice; his policies helped push it over the edge.[32]

Although his economic policies were poorly conceived, it will be his national security record that most severely tarnishes George W. Bush's legacy. And tragically, for both him and the nation, this was simply the result of poor planning and decision making. In the weeks and months after the attacks of 9/11, Bush had the full support of the American people and Congress. He used this unprecedented popularity to gain approval for an invasion of Afghanistan, whose Taliban leaders harbored those responsible for the first attacks on US continental soil in nearly two centuries. From a military perspective, this conflict was initially successful in removing the Taliban, although it failed to capture Osama bin Laden and thus decapitate Al Qaeda. As a result, maintaining the peace turned out to be a far tougher assignment for the military. It became exponentially more difficult when troops were siphoned off to support President Bush's broader foreign policy adventure. Within days of the 9/11 attacks, his advisers had begun formulating a narrative that linked Iraq and Saddam Hussein with Al Qaeda and Osama bin Laden, an idea that on its face seemed preposterous to many outside experts. This narrative was created by falsifying evidence to suggest that Hussein had large stockpiles of weapons of mass destruction, including fissile materials, which represented a threat to regional and international security. The administration commenced a brilliant campaign to mislead the American people about the level of the existing peril. Meanwhile, Congress abdicated its constitutional authority to decide when the nation should go to war by giving President Bush carte blanche in October 2002. Five months later, he used this power to send US troops to topple the Iraqi regime. The initial military campaign was very successful. Unfortunately, then the wheels came off.

President Bush made two significant strategic errors before American troops were even deployed to Iraq. First, unable to assemble an international coalition willing to provide soldiers, peacekeepers, and financial support, he determined to pursue his aims on a largely unilateral basis. Second, he decided to send a relatively small number of ground troops to the Middle East for the invasion. This was critical to the eventual outcome of the war. In doing so, the Bush administration ignored the advice of US Army Chief of Staff Eric Shinseki, who previously had managed American operations in postconflict Bosnia. Appearing before the Senate Armed Services Committee in February 2003, Shinseki had been asked how many troops would be required to maintain order after combat operations ceased. He responded: "I would say . . . something on the order of several hundred thousand soldiers are . . . required. We're talking about post-hostilities control over a piece of geography that's fairly significant, with the kinds of ethnic tensions that could lead to other problems. And so, it takes a significant ground force presence to maintain a safe and secure environment, ensure that people are fed and water is distributed, all the normal responsibilities that go along with administering a situation like this."[33] The opinion of the military's leading expert on managing this type of operation was ignored. Less than a month later the invasion of Iraq was commenced with only 145,000 ground troops, fewer than half as many as Shinseki had recommended. This decision essentially set Bush's legacy as a war president in stone.

Six weeks after the quick and decisive invasion, President Bush infamously delivered a speech aboard the USS *Abraham Lincoln* in which he declared, "Major combat operations in Iraq have ended. In the Battle of Iraq, the United States and our allies have prevailed. And now our coalition is engaged in securing and reconstructing that country."[34] At that point only 139 American service members had been killed. In the coming seven years, another 4,352 died in a war the commander in chief had declared to be over. What happened? At a fundamental level, the decision to send too few troops meant that when Baghdad fell there simply weren't enough soldiers to prevent civil chaos. This unrest resulted in the dismantling of most parts of Iraq's physical and political infrastructure, with the exception of the oil ministry, which was protected. This initial disorder was made worse by the fact that there had been no serious planning for peacekeeping because it was assumed that, in the words of Vice President Dick Cheney, "we will, in fact, be greeted as liberators."[35] This notion was quickly dashed when it became clear to the Iraqis that US forces had no idea how to stabilize the country. Then two more disastrous decisions were made, this time by L. Paul Bremer, who had been appointed by Bush as the administrator of the Coalition Provisional Authority of Iraq. The first decision was the De-Ba'thification of the Iraqi government, which meant that all civil servants and teachers who had previously served under Saddam Hussein were prohibited from returning to their jobs. Not only did this result in great unrest but it barred those persons most qualified to knit the nation back together from offering their services. The second decision had even larger impacts. Working with senior White House and Pentagon advisers, he disbanded the Iraqi army. This was done without consulting retired Army general Jay Garner, who had been leading reconstruction efforts on the ground. It was done without consulting General John Abizaid or Lieutenant General David D. McKiernan, the forward commanders of American ground forces. It was done without consulting Secretary of State Colin Powell, who had been Chairman of the Joints Chief of Staff during the Persian Gulf War. The result was that overnight a half million soldiers, most of whom were the primary breadwinners for their families, were out of work and angry. Many joined the insurgency that would tear Iraq apart during the remainder of the American occupation and beyond.[36] In addition to the lives lost on both sides, ultimately this decision led to the destabilization of a region of the world considered critical to American national security. It also drove up the global price of oil, endangering energy security and thus compromising economic security. Therefore, this conservative cycle ended with arguably the worst foreign policy disaster in American history. As with the impending financial crisis, it would be left to the first president of the Fourth Political Age to clean up the mess. And he would have to do this while also trying to generate political consensus for a major government push to take climate security seriously.

Political Dysfunction
Energy Insecurity + Extreme Conservative Backlash
+ Constitutional Flaws = Political Dysfunction

A truth about the modern age that has long been overlooked is that it is far easier to govern a nation when it has access to inexpensive and abundant hydrocarbons. It is easier for governments to take on additional responsibilities. It is easier for governments to make the big decisions that are necessary to push a nation forward. The reason is simple. Hydrocarbons were at the foundation of each of the INNATE revolutions, which combined to accelerate economic growth to levels never previously seen in human history. Hydrocarbons created modernity. When a society with inclusive institutions has access to inexpensive and plentiful hydrocarbons, it has the central resources needed to build a dynamic economy, which in turn can provide it with significant geopolitical weight. Because America had each of these ingredients, it was able to rapidly grow its economy while also providing greater security for its people. In fact, it was fairly easy to accomplish these goals. Progressives were able to expand the role and size of government; conservatives found it difficult to show that this was harming the nation. The latter's self-appointed function, therefore, became to keep markets relatively unfettered and to ensure that government did not overextend itself. They did both jobs effectively, forcing key progressive leaders to seek out pragmatic policies that allowed them to maintain course without providing an opening for a broad-based conservative backlash. Over time the nation moved gradually toward the combination of policies that created the Great Compression, which was America's golden age.

There is a flip side to the reality about inexpensive and abundant hydrocarbons. When domestic supplies run short and prices (both observable and hidden) increase, it becomes ever more difficult to maintain the institutions that had become the norm. When the growing costs associated with maintaining energy security coincided with an extreme conservative cycle, the result was a fractured American political system. The national government lost the ability to make the decisions that were necessary to meet the challenges of a quickly changing world, because this would have required true compromise between the two political parties. The reality of American history, however, is that the constitutional system has never been particularly well suited to making hard choices. Big decisions were generally made with widespread support and without the need for compromise between the parties. During the First Political Age, there was widespread approval of the objectives of a national security policy that sought to keep the United States isolated from Europe while gaining continental preeminence. During the Second Political Age, there was widespread approval of a suite of policies intended to provide the nation with greater macroeconomic security. During the Third Political Age, there was widespread approval of adding microeconomic security to the list of government responsibilities. These major policy punctuations gained such support because it was within America's means to expand the government as it enjoyed the fruits of the INNATE revolutions. The country failed to make other big decisions, most notably to eliminate slavery and end southern apartheid, because these issues were far too difficult to solve in a constitutional system with so many veto points. It is not a coincidence that after more than a century of economic and geopolitical advance, significant new challenges emerged in the decade after the peaking of US domestic oil production. The problem was that the system was not equipped to face these contemporary trials, because it had so little experience making hard choices that required real sacrifices. Energy insecurity unmasked constitutional weaknesses that have existed since the founding. While certainly not the only factor in the crippling political dysfunction that the nation still faces, it played a critical role in pushing the country toward the abyss.

* * *

Thomas Mann and Norman Ornstein have identified two sources of dysfunction within the American government. First, they point to deep political polarization, which has created fervently antagonistic parliamentary-style political parties. This is problematic because the United States doesn't have a parliamentary system. Instead, Democrats and Republicans operate in a system with anti-majoritarian rules that make it difficult to gain approval for big policy changes without a significant punctuation in the country's political equilibrium. Second, Mann and Ornstein argue that this partisan extremism is asymmetric: "One of the two major parties, the Republican Party, has become an insurgent outlier—ideologically extreme; contemptuous of the inherited social and economic policy regime; scornful of compromise; unpersuaded by conventional understanding of facts, evidence and science; and dismissive of the legitimacy of its political opposition." They point to the midterm election of 1978 as the beginning of this Republican insurgency.[37]

The 1978 election saw an important class of conservative freshmen elected to the House of Representatives, including Newt Gingrich and Dick Cheney. The former entered Congress with a plan to retake the lower chamber. As Mann and

Ornstein write, his strategy was to "destroy the institution in order to save it, to so intensify public hatred of Congress that voters would buy into the notion of the need for sweeping change and throw the majority bums out." Cheney was a wholehearted advocate for the approach. Starting in the mid-1980s, during the heart of the Reagan Revolution, Gingrich and his acolytes began implementing their strategy by regularly delivering late-night speeches on C-SPAN that bashed the Democratic majority. This raised the profile of their efforts, particularly after Speaker Tip O'Neill told the House that this behavior was the worst thing he had encountered in his more than three decades in Congress. A few years later this was followed by a ruthless flood of ethics charges levied against O'Neill's successor, Jim Wright, which eventually forced him from office. The chance to actually retake Congress finally offered itself in 1994, after President Clinton's bungled attempt to gain approval of universal health care. The midterms saw Republicans pick up fifty-two seats in the House and eight seats in the Senate, giving them control of both chambers for the first time in four decades. The new members were a different breed. They were far more radical than any incoming cohort, from either party, before them. In fact, they were far more zealous than their leader. They were unhappy with Gingrich's bipartisan turn in 1996. After his failed attempt to impeach Clinton backfired politically, they quickly turned on Gingrich and forced him from the Speakership. In his place they elected Dennis Hastert, under whom conservative fundamentalists increased their sway in the years to come.[38]

Mann and Ornstein suggest that while this extremism had deep roots, for more than a century "parties were less internally unified and ideologically distinctive, and more coalitions in Congress cut across parties than is the case today."[39] But Mann and Ornstein do not explain why things changed so dramatically starting in the 1970s. The potential for the government to descend into deep political dysfunction had existed since the founding. In fact, it had been readily apparent during the long debate over slavery in the decades leading to the outbreak of the Civil War. So what had changed? The answer on both ends of this timeline is hydrocarbons. Starting in the 1860s, as the use of coal began to transform the industrial sector, the federal government benefitted from exceptional economic growth, which saw pragmatic progressives from both political parties guiding the nation for eleven of the next twelve decades. Political dysfunction returned in the 1970s not because the United States was running out of hydrocarbons but because of the increased costs to society incurred when domestic oil supplies peaked. It was no longer easy to maintain the progressive path the nation had long enjoyed. When this coincided with a

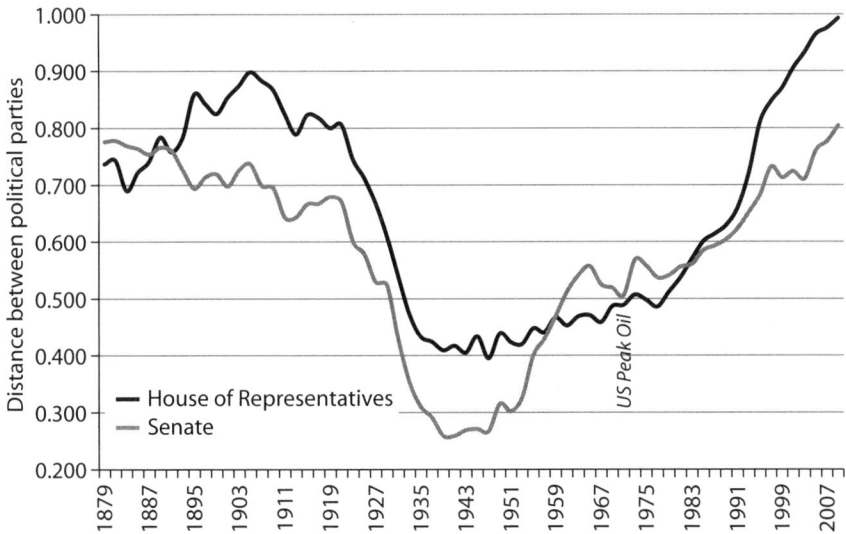

Figure 4.2. Party Polarization in Congress, 1879–2008. Nolan McCarty, Keith T. Poole, and Howard Rosenthal, *Polarized America: The Dance of Ideology and Unequal Riches* (Cambridge, MA: MIT Press, 2016).

conservative cycle, albeit one that saw divided government for all but six years, insurgent Republicans tried to turn back the clock and the Democrats refused to budge. The only solution was to put on a happy face, give both parties what they wanted, and pull out the proverbial credit card to pay for the collective failure to accept that President Carter had already accurately assessed both the problem and the solution.

As we can see in figures 4.2 and 4.3, an initial fall in the partisan divide began just as hydrocarbons were playing an ever-larger role in powering the American economy. During the Great Compression, partisanship remained at historically low levels as broad-based energy security made it relatively easy to find compromises. Partisanship began a steep climb in the late 1970s, after the peaking of domestic petroleum production, and coincided with the jump in American dependence on imported oil (fig. 4.4). At the same time, total energy use began to slow fairly significantly (although this was somewhat offset by the adoption of conservation measures). The resulting fiscal and economic pressures laid bare the existing fissures between the two political parties. One result was increased economic and environmental insecurity, as the federal budget began to fall as a share of the overall economy. Although oil prices remained relatively low, the increased cost of maintaining the global flow of petroleum

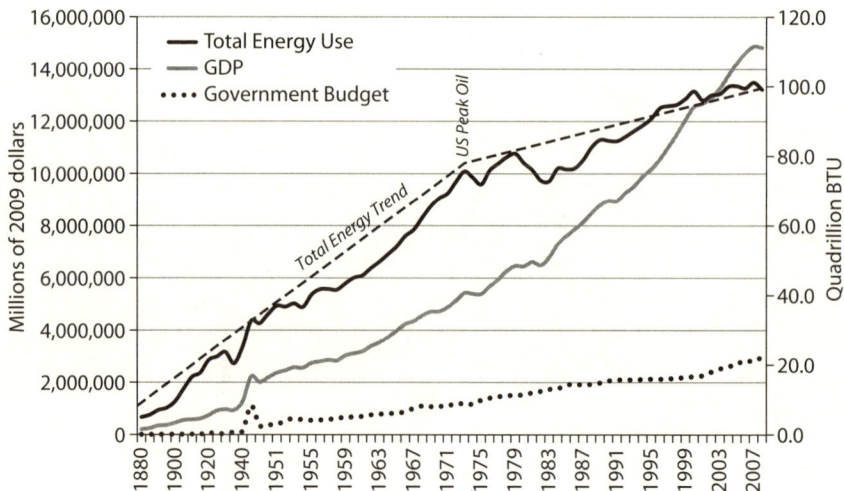

Figure 4.3. US Total Energy Use, GDP, and Government Budget, 1880–2008. US Energy Information Administration, *Annual Energy Review, 2009*; *Budget of the US Government, Historical Tables, 2017*; Louis Johnston and Samuel H. Williamson, "What Was the U.S. GDP Then?," 2017.

had made it more difficult to find funding to maintain existing programs or invest in new endeavors aimed at addressing emerging societal problems. This shift also paralleled a rise in conservative extremism after progressives championed both civil rights and women's rights. Finally, there was a generational aspect to this partisan swing, as an ideological sorting began that saw the more conservative members of the Lost and Greatest Generations retiring to the Sunbelt (which, by the way, was made possible by the miracle of air conditioning) and younger generations moving to neighborhoods, cities, and states that matched up with their political beliefs. With all of these factors taking hold as the nation entered a conservative cycle, it is hardly surprising that it led to a dramatic rise in extremism within the Republican Party.[40]

* * *

While it wasn't appreciated at the time, 1 June 1980 was a sad day for American democracy. On that date, CNN inaugurated the era of twenty-four-hour cable news. In the early years of the Republic, there had been newspapers that were essentially mouthpieces for the Federalists and the Democratic-Republicans. Although they did some damage to the country, their reach was so limited that they had little impact on the health of democratic institutions. Over time

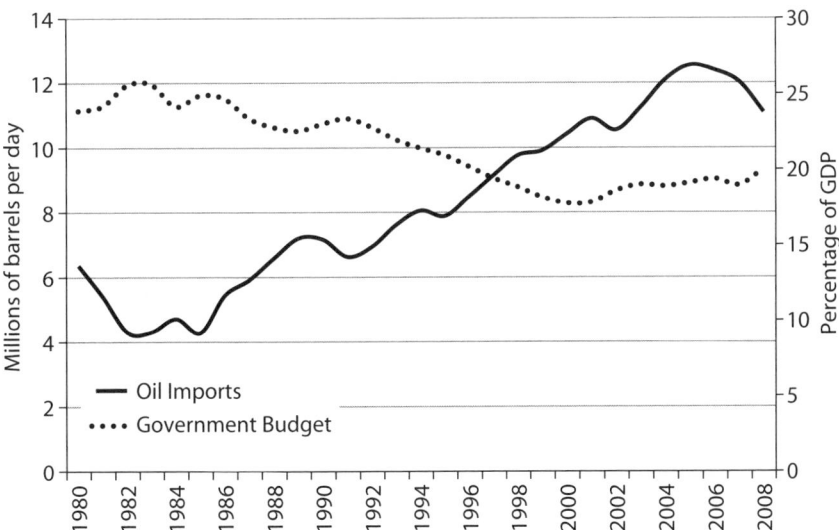

Figure 4.4. Federal Budget as a Share of GDP and Rising Oil Imports, 1980–2008. US Energy Information Administration, *Annual Energy Review, 2009; Budget of the US Government, Historical Tables, 2017;* Louis Johnston and Samuel H. Williamson, "What Was the U.S. GDP Then?," 2017.

newspapers became much more professional, with journalists considering it their duty to act as referees between the two parties. Radio broadcasters like Edward R. Murrow and early television news anchors like Walter Cronkite expanded upon this tradition, earning the public trust. CNN initially did little to change this dynamic, although its very existence began to transform the relationship between the media and politicians. When MSNBC (15 July 1996) and the Fox News Network (7 October 1996) were added to the mix, however, the entire medium became toxic. The primary reason was that it was nearly impossible to find twenty-four hours' worth of real news that would draw a large enough audience to make the network money. This became far more difficult when three networks were competing for the same viewers. The result was that rather than sustaining their role as truth tellers, the networks became purveyors of entertainment. Rather than reporting actual events, they created "debate" shows with partisan hacks from each party battering one another. Rather than relying on journalists, they turned to pundits. Rather than gaining the public trust by speaking truth to both the public and the powerful, cable news anchors sought to become ideological stars by reading the talking points fed to them by the political parties and tearing apart the opposition. Fox

News was the true champion in this area, attracting far more viewers than its competitors by adopting truly outrageous tactics that sought only to skew the public's perception of reality. As the cable networks became more omnipresent and the Internet became a new media force, newspapers that had sought to maintain journalistic integrity slowly went out of business. Sensationalism now trumps truth. While the web seemed to provide hope for a rebirth of candor, the medium is far too fragmented, and many outlets, ranging from the Drudge Report to the Huffington Post, have probably made things worse. And now the Twitter phenomenon has provided the ultimate weapons of mass political destruction by requiring that complicated policy issues be distilled into 140 characters. The end result is a media environment so venomous that it does not bring Americans closer together but pushes them further apart—to the detriment of all.[41]

As the role of television has become more critical for candidates for higher office, the impact of money in politics has evolved into the greatest threat to American democracy. Money has long played a role in electoral politics, but its importance increased significantly as the amount of funding needed to win office skyrocketed when television advertising became the primary avenue for campaigns to communicate with the electorate. Although the Federal Election Campaign Act of 1971 and the Bipartisan Campaign Reform Act of 2003 placed some limitations on the influence of money in politics, the US Supreme Court opened the floodgates with its ruling in *Citizens United v. Federal Election Commission* (2010). Contravening a century of precedent and going far beyond the remedies sought by the petitioner, the conservative wing of the court found that corporations were free to spend unlimited amounts in support of individual candidates. The decision in *Santa Clara County v. Southern Pacific Railroad* (1886) to create juristic persons had fully come home to roost. Justice John Paul Stevens wrote a ninety-page dissent in the case, arguing: "In a democratic society, the longstanding consensus on the need to limit corporate campaign spending should outweigh the wooden application of judge-made rules. . . . The Court's opinion is . . . a rejection of the common sense of the American people. . . . While American democracy is imperfect, few outside the majority of this Court would have thought its flaws included a dearth of corporate money in politics." Within weeks of the *Citizens United* ruling, so-called super PACs were created to pour huge amounts of money into electoral politics.

As Mann and Ornstein point out, campaign finance is not the only area in which the US electoral system falls short of the democratic ideal in the selection of leaders. In recent years, as the share of younger and nonwhite voters has increased, conservatives have launched a number of efforts aimed at restricting voting among these groups. This included the Supreme Court decision in *Shelby County v. Holder* (2013), in which the conservative wing gutted the key provisions of the Voting Rights Act of 1965, which had attempted to ensure that nonwhite voters would be granted their full electoral privileges in jurisdictions that had actively restricted them. In an impassioned dissent, Justice Ruth Bader Ginsburg wrote that the Civil War Amendments had given the power to make such decisions to Congress, not the courts. She further argued that with the 2006 reauthorization of the legislation, Congress had reached the considered opinion "that 40 years has not been a sufficient amount of time to eliminate the vestiges of discrimination following nearly 100 years of disregard for the dictates of the 15th Amendment." To this case have been added actions by state governments to suppress votes, particularly among nonwhite populations and younger demographic groups. Another problem with the electoral system is low voter turnout, among the lowest in advanced nations. There are many reasons for this. One is that the United States holds most elections on Tuesday even though polls have shown that as many as one-quarter of all voters cannot participate because they cannot get leave from work. This effectively disenfranchises a huge number of Americans. These are hardly democratic outcomes.[42]

With all trace of legislative independence and honor removed from the policy process, political dysfunction descended to previously unimagined depths. The situation has become so bad that a recent study by Martin Gilens and Benjamin Page concluded that the United States is no longer a democracy. Rather, it is a plutocracy, with the wealthy controling both financial and political institutions. The authors conducted a multivariate analysis of 1,779 policy issues to measure how much influence specific actors have over policy outcomes. They found that "economic elites and organized groups representing business interests have substantial independent impacts on U.S. government policy, while average citizens and mass-based interest groups have little or no independent influence." The chance of a policy proposal being adopted if supported by modern plutocrats is approximately 45 percent, but the chance falls to 20 percent if the plutocrats are opposed. Thus, even if a majority of Americans favor a policy outcome, the plutocrats have a considerable veto power. Gilens and Page concluded that "Americans do enjoy many features central to democratic governance, such as regular elections, freedom of speech and association, and a widespread (if still contested) franchise. But we believe that if policymaking is dominated by powerful business organizations and a small number of affluent Americans, then America's claims to being a democratic society are seriously threatened."[43]

* * *

As political dysfunction became endemic and led to the erosion of past progressive accomplishments, the antidemocratic flaws in the US Constitution became ever more painfully apparent. In the eighteenth century that document was truly revolutionary, even if there was a limited view of who was included in We the People. It would be foolish not to understand that the framers were, at their core, the ultimate pragmatists. They believed that all thirteen states needed to stay within the same union to secure their collective future, so they were willing to make unpalatable compromises for the cause of union. While slavery has rightly been considered the most disastrous of these concessions, the Constitution was flexible enough to allow subsequent generations to fix this most intractable defect. In modern times, however, it has become ever more apparent that the charter has significant structural shortcomings that must be addressed to meet the challenges of the twenty-first century. These deficiencies have not caused the rise in political dysfunction in recent decades, but they have provided mechanisms by which extremists can prevent the government from taking actions to keep the nation both economically and geopolitically strong in a rapidly changing world. Several contemporary writers, most notably Francis Fukuyama, Moisés Naím, and Thomas Friedman, have argued that the American government is no longer a democracy but rather a vetocracy; no political party can acquire enough power to adopt and implement needed changes. This, of course, plays right into the hands of the plutocrats who are deciding who gets elected and the direction of policy.

Beyond campaign finance, there are two other reasons why the American election process is profoundly broken. First, partisan gerrymandering of seats in the House of Representatives has distorted the election process for that chamber. Under the Elections Clause, states have been granted the authority to draw district lines. Congress previously regulated how this process was to be carried out, mandating contiguous and compact districts representing equal populations, but much of this oversight has been allowed to lapse. Today redistricting has become the subject of one of the most partisan fights in government, with huge sums now being spent to swing state legislature elections to ensure that new maps will be drawn to support a given party. Alexander Aleinikoff and Samuel Issacharoff write that

In a democratic society, the purpose of voting is to allow the electors to select their governors. Once a decade, however, that process is inverted, and the governors and their political agents are permitted to select their electors. Through the process of redistricting, incumbent office holders and their political agents choose what configuration of voters best suits their political agenda. The decennial redistricting battles reveal the bloodsport of politics, shorn of the claims of ideology, social purpose, or broad policy goals. Redistricting is politics pure, fraught with the capacity for self-dealing and cynical manipulation.

The result of this pervasive gerrymandering has been to create mostly safe seats, seats for which there is no real contest between the two major political parties. According to Nate Silver, in 1992 there were as many as 103 swing districts, but by the 2012 election this number had plummeted to 35. The number of seats won by more than twenty percentage points had doubled to approximately 242 seats. Because there is no true contest between the parties, the actual fight for a seat takes place during the nomination process. In these competitions, in which a much larger proportion of the voters are hard-core activists, the winner is almost always the more radical candidate. The end result is representatives who speak for a small majority of the most extreme voters in their districts. Not surprisingly, they turn out to be partisan fundamentalists who aren't willing to work with the opposition party. Given that the Republicans controlled far more state legislatures during the recent conservative cycle, it also isn't surprising that the impact has been to skew the House in their direction. As evidence of this reality, in the 2012 elections Democratic candidates won more votes (48.8% vs. 47.6%), yet the Republicans won a 34-seat majority, hardly a democratic outcome.[44]

Second, in some respects, the way we select presidents is even more undemocratic. The Electoral College was created to ensure that the election of the chief executive was federal in nature, by giving each state a number of electors equal to its number of senators and representatives in the national legislature. States would have some leeway in how to select those electors, but the key result was that the president would not be chosen by the popular vote of the whole people. As with many structural decisions made by the framers, the primary reason for the creation of the Electoral College was to keep southern slave owners happy. Northern states would have had far too much power in the new nation if slaves hadn't been counted for voting purposes, but of course those slaves could not be allowed to vote, because they weren't citizens. The Electoral College and the Three-Fifths Compromise were the solution.

It is curious that although slavery ended more than 150 years ago, we have maintained this archaic system. One fundamental problem with the system in modern times is that a large number of states have almost no impact on presidential elections, because one candidate is expected to win by such a comfortable amount. The campaigns essentially ignore these states, focusing their time and resources on winning battleground states, where the vote is expected to be tight. Thus, big states like California and Texas get almost no attention because everyone knows who will win those contests. This is contrary to the notion of one person, one vote, which has been a guiding principle since the Supreme Court decision in *Baker v. Carr* (1962). Instead, once again the plutocrats enjoy far more power, because they are able to concentrate their campaign funding in a limited number of contested media markets, providing them with a much greater ability to influence the election outcome. The biggest problem with the Electoral College, however, is that presidents have regularly been elected without receiving a majority of the vote (e.g., President Bill Clinton, who won only 43% of the popular vote in a three-way election), and five times the president has been elected even though his opponent received more votes. This latter, most disturbing outcome has happened twice in just the past five elections. President George W. Bush lost by more than a million votes to Al Gore but won in the Electoral College by three electoral votes. President Donald Trump lost by nearly three million votes but won in the Electoral College by a truly shocking seventy electoral votes. Because their legitimacy was challenged from the outset by the opposition, it was easy for partisan extremists to reject offers to compromise and instead wage "holy wars" against their administrations. This is a terrible outcome for American democracy. Elections should have consequences; those elected should be able to take the actions for which the people selected them. But the Electoral College is one of several constitutional flaws that have turned this natural objective into a fallacy. This is hardly a democratic outcome.[45]

While the American election process is clearly broken, in some respects the legislative process is even less well suited to meet the challenges of the modern age. At a basic level, the problem is that there are far too many veto points in the American system of governance. This might not have been as consequential in the eighteenth and nineteenth centuries, but it became a more significant problem in the twentieth. Because of the numerous economic and geopolitical challenges facing the nation in the early twenty-first century, the resultant paralysis is outright dangerous. Federalism itself is the source of many of these difficulties. Under the Constitution, power is divided and shared between the national government and the state governments. The framers believed that federalism would prevent tyranny, make available greater political participation, and finally allow the states to act as policy laboratories. They created various enumerated powers for the national government, including coining money, maintaining armies and navies, and declaring war. Their inclusion of both the Necessary and Proper Clause and the Commerce Clause also provided the national government with broad implied powers, most importantly over the economy. All other powers were nominally reserved to the states under the Tenth Amendment, although because these other powers were not actually spelled out, there is often confusion about who has authority in a given policy area. This, in turn, has stymied efforts to take actions at the national level that have broad support among the electorate. This system of federalism is very expensive given the large number of subnational governments (nearly ninety thousand, according to the Census Bureau). With so many divisions, there is a great deal of inefficiency because of the duplication of services. There are also more access points where the plutocrats can sway policy outcomes, which has become quite clear as the amount of spending for election to state legislatures and state judgeships has risen steeply in recent decades. Finally, federalism is inequitable. This is particularly acute with regard to paying for access to public education, one of the most important macroeconomic security policies, for which more than 90 percent of funding comes from state and local governments. This means that far less money is spent on students from poor states and municipalities with a low property tax base. Not only does this create an inequitable system but it severely reduces the economic potential of the country by failing to adequately develop human capital. This is hardly a democratic outcome.

America's bicameral tradition dates from when it was thirteen separate colonies, when settlers mostly chose to follow the British practice of having two legislative chambers. There was a belief that an upper chamber would be an effective check against the people; thus the Senate would represent aristocratic interests and the House would represent democratic interests. Things haven't changed very much: the Senate today has been labeled the "millionaires' club." Many other advanced economies have also retained bicameral legislatures, although about two-thirds of all nations have unicameral systems. What makes the United States unique is the strength of its bicameral approach. Nations like Britain and France have weak systems; the lower chamber can generally prevail against the wishes of the upper chamber. Other nations with a robust federal character (e.g., Germany, Switzerland, Canada, and Australia) also have strong bicameral systems. Still, a lower chamber and upper chamber with nearly parallel powers, as in the United States, are rare. Modern advocates of a bicameral system laud its "consensualism," which makes it more difficult to gather support for radical change. Of course this type of antimajoritarianism has long been used to stymie change, the most egregious case being the continuation of slavery and southern apartheid. Only in times of great national crisis could a coalition be fashioned to support the progressive changes that were required to advance economic security. This is by far the main criticism of bicameralism. It makes passing legislation too difficult, especially when the chambers have majorities from different parties. Once again, elections should have consequences. Bicameralism almost always makes this an unfulfilled dream. This is hardly a democratic outcome.[46]

While a bicameral system has fundamental weaknesses that call into question its desirability, the far more troubling aspect of the American constitutional approach is how the Senate is constituted and operated. Article 1, section 3, clause 1, provides that "the Senate of the United States shall be composed of two senators from each state . . . and each Senator shall have one Vote." The problem is that senators are apportioned equally among the states. Yet again, it is important to remember that the primary reason for this provision of the Constitution was to protect the slave power of southern states. This anachronism serves little purpose today, except to protect the plutocrats and political interests in small, mostly rural states. In a nation that has become increasingly urban, as the citizenry transitioned from farming to industrial and service jobs, it seems strange to be holding on to this feature of the Constitution. The simple reality is that enormous political power is placed in the hands of voters in small states, just as is the case with the Electoral College. Given the character of bicameralism, this means that senators representing a very small proportion of the population can veto all legislation. As Sanford Levinson writes, "The equal-vote rule in the Senate makes an absolute shambles of the idea that in the United States the majority of the people rule." This is hardly a democratic outcome.[47]

The inequality represented by the antimajoritarian nature of the Senate is exacerbated by internal mechanisms such as the filibuster. This potent parliamentary mechanism is used to delay the passage of legislation, which has come to mean that nonbudgetary legislation requires the support of 60 percent of the body to bring a bill to the floor for a vote. The Constitution provides each chamber with the ability to make its own rules, which the Senate did when it adopted its first rulebook in 1789. This document included the previous-question rule, which allowed a bill to be brought to a vote of the full body with a simple majority vote. In 1805, Vice President Aaron Burr (who had recently been indicted for the murder of Alexander Hamilton) suggested that the Senate rules be streamlined and the previous-question rule removed because it was common sense that a majority vote could bring legislation to the floor for a final decision by the full body, an action taken the following year. In the late 1830s and early 1840s this led to the first filibusters whereby the minority took control of the Senate floor to delay passage of several pieces of legislation. Nearly a century later the cloture motion was added, initially intended to provide a way to stop a filibuster. It required a vote of two-thirds of the body to call the question and bring the bill to a vote. It was later extended to motions to proceed, meaning that two-thirds of the senators needed to sign off on bringing legislation to the full body for a vote. This essentially meant that the body no longer operated on majority rules but instead required a supermajority. The tactic was famously employed to protect southern apartheid from national civil rights legislation. In 1975 the threshold for cloture motions and motions to proceed was lowered to three-fifths of the Senate, which still means that senators representing approximately 20 percent of the nation's population can prevent the national government from taking any action. Perhaps more importantly, in that same year the rules were amended to allow other Senate business to proceed while a filibuster is ongoing. Plutocrats have long supported the filibuster, because it means that fewer elected officials have to be swayed to prioritize the financial interests of the wealthiest Americans. Ironically, this Wall Street agenda increases the economic insecurity of western libertarians and southern social conservatives, who are the Republican Party's key voting blocks.

Numerous legal scholars argue that the filibuster is unconstitutional, because it requires a supermajority for the passage of normal legislation, contradicting the intentions of the framers. The Constitution, after all, requires a supermajority in several specific instances, such as treaty ratification, impeachment, member expulsion, presidential-veto override, and passage of a constitutional amendment. If the intent had been to require a supermajority for regular business, why would there have been a need to include clauses for these specific circumstances? The answer is that there would have been no need. Both Madison (in *Federalist 58*) and Hamilton (in *Federalist 22*) make clear that the Constitution does not require a supermajority in other situations, because it would interfere with the energy of the national government and reverse the fundamental premise of the entire framework of government. In

United States v. Ballin (1892), the Supreme Court found that while "the Constitution empowers each house to determine its rules of proceedings [it] may not by its rules ignore constitutional restraints or violate fundamental rights." Yet this reasoning has never been used to find the filibuster unconstitutional. In 2012, Common Cause and several members of Congress jointly filed a suit requesting such a finding. The case, however, was dismissed on procedural grounds by lower courts, and the Supreme Court refused to hear the case. This despite the fact that more than four hundred cloture motions have been filed since 2001. This despite the fact that what was once considered an exceptional action is now seen as a purely routine method of obstruction. This despite the fact that the government has come to the point where the mere threat of a filibuster halts needed action on Capitol Hill.[48] This is hardly a democratic outcome.

* * *

In 2014 Mark Warren conducted interviews with ninety members of Congress to find out what they thought about the American political system. He discovered widespread, bipartisan agreement that it is broken. Said one Republican representative, "I didn't get elected to Congress to not get things done—most people here want to get things done. I didn't get elected to Congress to make meaningless speeches on C-SPAN and tell lies about people. I didn't get elected to Congress to scare the hell out of the country and drive the sides further apart. I didn't get elected to Congress because I love politics—I hate politics, to be perfectly honest, and if I didn't before I got here, I do now." Many members are worried about the impact of the redistricting process. Aaron Schock, a Republican representative from Illinois, said, "You know, if I had a magic wand, one thing I would love to change—which you can't do unless you're king—is the redistricting process by which our boundaries are drawn[, which] creates hyperpartisan districts." Democratic Congressman John Lewis of Georgia concurred: "The country is changing and change makes some people uncomfortable. But our congressional districts don't reflect that change, and there are so few competitive districts remaining that people only fight for or speak up or speak out for the narrow base of people who reelect them." "We're seeing the political equivalent of segregation going on in the country," said Republican representative Tom Cole of Oklahoma. Independent senator Angus King of Maine pointed out: "You can't be moderate. Who votes in primaries? You have a 10 percent turnout in a primary election in Georgia, and Republicans are 30 percent of the population. So 10 percent of 30 percent—that's 3 percent of the population voting to choose the nominee, and then if it's a multiperson race, and the winner gets 35 percent, that's one third of 3 percent—1 percent of the population chooses the nominee, who in a gerrymandered district will be the eventual member of Congress. That is bizarre, and it has completely polarized Congress." Many members are worried about the impact of the modern media. The media don't care about party cooperation, said Republican representative Morgan Griffith of Virginia: "People . . . getting along, compromising, doing their jobs like adults doesn't have the sizzle of conflict that the media demands in order to hold your interest." Democrat representative Joaquin Castro of Texas put it this way: "It's the coliseum. And in the coliseum, people get hurt for sport." Many members worry about the lack of real human contact between elected officials. Democratic senator Patrick Leahy of Vermont remembered, "When I first came to the Senate, people in both parties went out of their way to have personal relationships. [We would] talk about where our kids went to school, the vital business of daily life, which then enabled us to work together on the vital business of the United States. Those relationships don't happen so much anymore." Many members worry about the impact of the fund-raising imperative. Democrat representative Donna Edwards of Maryland stated, "When you look at the cost of a House seat now—which is about $1.6 million or something—you've got to raise that money. . . . And they have to do it every two years. It's a never-ending hustle. You get elected to this august body to fix problems, and for the privilege, you find yourself on the phone in a cubicle, dialing for dollars." According to Republican representative Adam Kinzinger of Illinois, "There's an entire industry in Washington that makes money on conflict. Some of these outside groups . . . go out and they fundraise by saying that Republicans aren't sufficiently conservative. Or they pick an issue to go to war on because they can stir the base and raise money on it and pay their big salaries. And what that does in the long run is it takes what would be a solid Republican agenda and causes chaos. And they do the same thing on the Democrat side, you know?"[49]

Almost none of this is merely ideological. It is structural. The political process will remain broken until these representatives are willing to accept this fact and take action to fix it. Yet to this point very little has been done to pass significant reforms to overhaul the governing framework and bring it into line with the needs of a great power in the twenty-first century. Instead, elected officials have chosen to remain loyal to their parties rather than do what is in the best interest of the nation. This is a serious threat to both economic security and the ability of the nation to retain its geopolitical edge at a time when many rapidly expanding nations are making strides in both areas. It is well past time to begin speaking about how to solve these problems, and the first target is clearly a series of constitutional reforms. "The public may still revere the Constitution and support the system of government that it shaped," Mann and Ornstein write, "but this is more a measure of patriotism—love of country and pride in being an American—than of satisfaction with how it is working in practice." If Americans were educated about the true failings of the governing document, they might be prepared to challenge these institutional arrangements. In so doing, they would also be laying the political groundwork for the policy effort required to renew and expand American energy security within the constraints of the climate crisis.[50]

Financialization
Energy Insecurity + Extreme Conservative Backlash
= Financialization

Financial crises are not new. They have helped shape global economic and geopolitical history. In their insightful eight-century analysis of past financial crises, Carmen M. Reinhart and Kenneth Rogoff write that

> the most commonly repeated and most expensive investment advice ever given in the boom just before a financial crisis stems from the perception that "this time is different." That advice, that the old rules of valuation no longer apply, is usually followed up with vigor. Financial professionals and, all too often, government leaders explain that we are doing things better than before, we are smarter, and we have learned from past mistakes. Each time, society convinces itself that the current boom, unlike the many booms that preceded catastrophic collapses in the past, is built on sound fundamentals, structural reforms, technological innovation, and good policy.

Reinhart and Rogoff's study revealed one fundamental truth: too much debt is bad. Massive public debt is bad, and massive private debt is bad. Even though the periods when this debt is being accrued are subject to widespread magical thinking, the end result is always a financial crisis. Housing bubbles, rising indebtedness, climbing energy prices, and mounting current-account deficits often precipitate these calamities. The ensuing meltdown takes many forms, ranging from sovereign debt crises to inflationary crises and banking crises to currency crises. Irrespective of their nature, however, the biggest losers are always average citizens. It is equally true that the boom-to-bust cycle, which has persisted for centuries, is bad for the overall health of national economies.

The clearest proof of this fact was the Great Compression, in which government regulations were implemented to smooth out the boom-to-bust cycle and provide long-term stability. The result was the longest period of economic flourishing in American history.[51]

This began to end after the peak in domestic oil production and the 1973 oil crisis. Over the next eight years, billions in American wealth left the nation as oil imports rose. Much of it would ultimately return to the United States in the form of huge arms deals and real-estate investments, and the dollar remained the global reserve currency. As a result, as Tyler Priest writes, the nation was immune to "balance-of-payments crises. The United States could simply print money to buy crude oil, while other countries had to borrow or trade for dollars to do the same." The problem was that this also fueled growth in domestic deficits and debts. The acquisition of such large peacetime debts was new in America's history. The nation had previously accrued large debts during wartime, as can be seen in figure 4.5. Public debt reached about 30 percent of GDP after the American Revolution, the American Civil War, and the First World War. It went through the ceiling in the final year of the Second World War, when it exceeded 110 percent. After each of these events, however, the nation expeditiously paid down the debt, although levels remained somewhat high in the post–Second World War era because large postconflict military forces were maintained during the Cold War. This all changed with the rise in the peacetime national debt, which continued during the Reagan presidency and accelerated under George W. Bush. Political dysfunction fueled by energy insecurity was another reason for the risings debts, as both political parties determined that rather than make hard choices it was far easier to give everyone what they wanted. As David Leonhardt writes, "In the simplest terms, Republicans have won the debate on taxes, and Democrats have won the debate on benefits. We, the voters, have chosen the winner of each. In exchange, we have a federal government facing enormous deficits in coming decades." Neither party gave up anything, because American voters wanted to keep all of the benefits they had enjoyed when the country had unquestioned energy security, and they didn't want to pay higher taxes. This was something new in the life of the country.[52]

Of course the rise in public debt was concerning, but there is reason to believe that it wasn't by itself catastrophic. Reinhart and Rogoff suggest that national debts don't become a hindrance to economic growth until they approach 90 percent of GDP, a level not reached in the post–Second World War era until after the beginning of the Great Recession in 2007–8. But rising public debt wasn't the entire story. Total American indebtedness, which included corpo-

Figure 4.5. US Public Debt as a Percentage of GDP, 1790–2008. US Department of Treasury, Bureau of the Fiscal Service.

rate and personal obligations, was far more problematic. As figure 4.6 shows, total debts began to slowly increase in the early 1950s. This was primarily the result of the transition to broader-based homeownership, which was spurred by government institutions and programs that included the Federal Housing Authority (1934), Fannie Mae (1938), the Veterans Loan Program (1949), and Ginnie Mae (1968). These made homeownership possible for far more Americans, with the rate increasing from less than 45 percent in 1940 to more than 65 percent in 1980.[53] More people took on long-term mortgages, which increased total indebtedness. Starting in the early 1980s, however, total liabilities quickly rose to previously unimagined levels.

Two things combined to bring this about. First, peaking oil production weakened the century-old foundation of American economic growth, which had been constructed by providing access to cheap hydrocarbons. Yet, the entire American economic system is based upon continued growth. And, for good reason. One of the great advantages of modernity is that the consumption of manufactured goods (and services) makes life easier, providing individual security. Access to these goods meant that average people enjoyed a standard of living previously limited to the wealthiest members of society. A great goal of society is to expand the advantages of the modern age to more people, and this inherently requires economic growth. This was made far more difficult

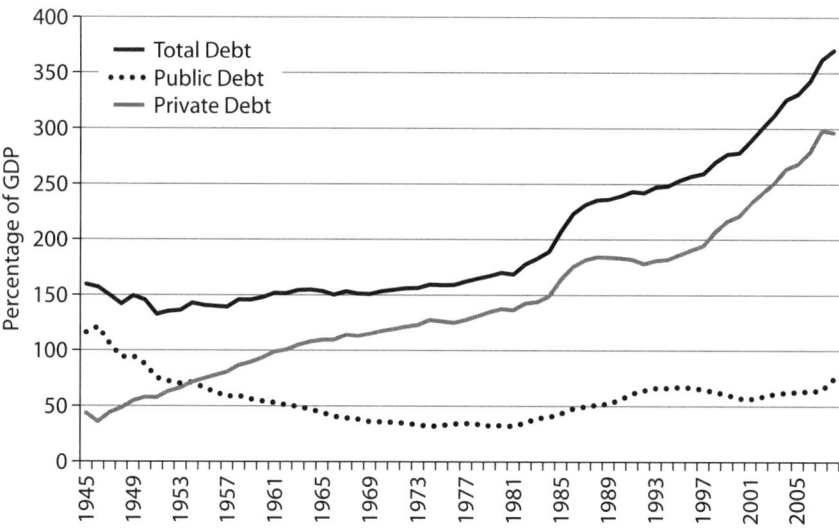

Figure 4.6. US Public and Private Debt as a Percentage of GDP, 1945–2008. US Department of Treasury, Bureau of the Fiscal Service; Federal Reserve.

to accomplish without secure domestic sources of cheap and abundant petroleum. Second, the conservative movement had an idea about how to replace the cheap energy foundation in the American growth model. Through tax cuts and deregulation, conservatives empowered the plutocratic class to increasingly financialize the country. If the nation could no longer rely on cheap energy, the economy could still be juiced with waves of debt mechanisms, some public but most of them private. The rising indebtedness hit the working and middle classes hardest, with the upper and governing classes benefitting from the embrace of personal greed and an apparent disregard for the national good.

As a rising conservative leader in the 1960s, Ronald Reagan became the most persuasive advocate for lowering government spending, slashing taxes on the wealthiest Americans, and deregulating the financial industry. He tried and largely failed to accomplish these goals as governor of California, so he sought a larger stage, eventually bringing his vision to the presidency. He based many of his ideas on the work of Nobel Laureate Milton Friedman, who believed that government programs and regulations were undermining the ability of the market to efficiently grow the economy. Despite the lack of any basis in economic history, the conservative movement gratefully adopted his anti-Keynesian message. According to Jeff Madrick,

Almost every suggestion Friedman made in [his book] *Capitalism and Freedom* was taken up by serious proponents over the next forty-five years. Some were adopted by the nation. The call to privatize Social Security has frequently been sounded since the 1980s and 1990s; increases in minimum wage were successfully avoided for long periods of time; a version of a negative income tax for the poor rather than direct outlays . . . was backed by both Republicans and Democrats; international currencies were floated in the financial markets in the mid-1970s; and there has been a sharp reduction in progressive income taxes, beginning with John F. Kennedy, and still more aggressively by Ronald Reagan and later by George W. Bush.

While some of these ideas had clear merit, on balance they took America in a disastrous direction.[54]

Meanwhile, key bankers and investors were laying the groundwork to take advantage of the Reagan Revolution. After decades of suffering in a staid industry that was heavily regulated by the government, they were ready to bring innovation to the entire sector—the economic health of the nation be damned. This began in the early 1960s, when Citibank's Walter Wriston introduced the negotiable certificate of deposit (CD), which allowed large banks to rapidly raise funds from investors to increase money for lending. This instrument technically violated the Federal Reserve Act, which prohibited payment of interest on demand deposits. However, the Federal Reserve, which was more concerned with banking illiquidity that could slow economic growth, decided to look the other way. As Madrick writes, "Wriston relished his role in diminishing the federal government's traditional role as overseer and regulator of finance, knocking down barriers through audacity and attracting competitors to join him in his freewheeling ways, and thus turning his bank into the biggest in the world." Unfortunately, this meant that the institution had become "too big too fail," which later led to a $1.5 billion aid package to prop up the Mexican government, which had borrowed heavily from the American bank.[55]

The next big financial innovation was the hostile takeover, pioneered in the 1970s by Joe Flom. There was nothing intrinsically wrong with mergers that made good business sense, but Flom and his ilk where pushing acquisitions that primarily sought profits by dismembering companies. Although this was bad, the real problem was how much this changed the management of corporations. "The mere threat of takeovers changed corporate values. Vulnerable companies were desperate to raise the value of their stock to make them less attractive and avoid a takeover, which usually required focusing on improving profits in the short run, often by cutting wages and jobs, just as if they had been taken over. Others bought entities they did not necessarily want or need in order to use up their idle cash in the bank, which otherwise made them tempting targets for hostile acquirers. . . . Thus, American business did indeed become lean and mean—far too much so. It became a narrowly focused revolution and the gains, when made, were short-term." With the vultures circling Wall Street looking for the next victim of a hostile takeover, the temptation to seek out insider information in hopes of making a big score was huge. This led to the adoption of risky behaviors that led to the jailing of arbitrageurs like Ivan Boesky; more importantly, it increased the desire of the less scrupulous to seek out entrepreneurial ways to take advantage of the increasing financialization of the American economy. Another impact of the change in corporate strategy was the new objective of growing ever larger, which protected companies from being acquired but seriously undermined competition within the vaunted marketplace.[56]

As President Reagan was gutting regulations that had maintained financial stability for decades, two management consultants from the consulting firm McKinsey, Tom Peters and Robert Waterman, were trying to explain to the American corporate elite why the nation was in decline. In 1982 they published *The American Challenge*, which was in essence a warning about the impacts financial innovations were having on American business. "The best companies, they found, were those with lean staffs, not top-heavy. They fostered entrepreneurship at many levels of management, from the worker to the foreman to the general manager and higher. . . . They avoided business in which they had little experience or about which they had little knowledge. They acted quickly to fix problems." Unfortunately, corporate management teams were steadily moving in the opposite direction, adopting the bigger-is-better model, which promoted quick profits and high debt levels. They were assisted by financial gurus like Michael Millken, who made billions selling extraordinarily risky "junk bonds" and used the proceeds to underwrite some of the biggest takeovers of the 1980s. (Eventually jailed for insider trading, he remains one of the richest people in the world.) Not only had the era of high risk arrived but even the worst actors faced limited repercussions for destabilizing the American economic system.[57]

The reason for this was that key government leaders had gotten into bed with the plutocrats. The key architect was Federal Reserve chair Alan Greenspan, who for nearly two decades used his free-market ideology to persuade American presidents to prime the pump regardless of the risks posed by over-speculation. Hand in hand with Treasury Secretary Robert Rubin and Deputy Treasury Secretary Larry Summers, he supported the Gramm-Leach-Bliley Act of 1999, which removed regulatory barriers between banks, investment firms, and insurance companies. The act was signed into law by President Bill Clinton. The results were utterly predictable. Huge financial institutions were created that marketed a variety of commercial products, while countenancing highly leveraged gambles. As they began to operate in this unfettered environment, the amount of risk existing within the economy reached untenable levels. During this period the hedge-fund manager reached the heights of the financial pyramid, with men like George Soros and John Meriwether making billions by borrowing aggressively to make sweeping investments in stocks, commodities, options, currencies, and precious metals. This by itself might not have been so terrible if their success hadn't encouraged conventional Wall Street firms to outdo them in leveraging risky behavior. This was paralleled by increasing corruption within the business sector by companies ranging from WorldCom to Adelphia to Enron. As Madrick writes,

> The 1990s through 2002 was the most corrupt 'decade' since the 1920s—and one of the most corrupt in American business history. An infrastructure of corruption, whose seeds blossomed with Mike Millken, Ivan Boesky, and the savings and loan entrepreneurs in the 1980s, spread in the 1990s across the American establishment into elite professions, including commercial and investment banks, accountants, lawyers, consultants, rating agencies, and mutual funds as well as newer hedge funds. The nation did not fully comprehend or reflect upon the excessive degree of business corruption in the 1990s. . . . [Thus, the] uproar over Enron, WorldCom, and the stock market debacle passed, and extreme free market ideology retained its hold for another decade.[58]

The financial vehicles that finally pushed the American economy into free fall were mortgage-backed securities and collateralized debt obligations. These mechanisms pooled mortgages into asset-backed securities. The problem was that they included a large number of subprime loans, which far outweighed the number of solid loans within any given security. Wall Street funneled trillions of dollars from pension funds, mutual funds, and insurance companies into these risky investments. As the market heated up, unscrupulous mortgage underwriters accelerated the number of subprime loans they were approving. Since these transactions were almost entirely unregulated, companies like Countrywide, led by Angelo Mozilo, were making billions, yet no one took notice of the developing imbalance in the mortgage-backed securities and collateralized debt obligations. With so much activity in the housing market, prices rose rapidly, which gave homeowners the sense that they had newly acquired wealth. They took advantage of this by taking out second mortgages, which sped up the ongoing national borrowing spree by providing consumers with artificial purchasing power. "Never before," writes Kevin Phillips, "have political leaders urged such large-scale indebtedness on American consumers to rally the economy."[59]

Another part of the story was the increased financialization of the oil market itself. After the passage of the Commodity Futures Modernization Act of 2000, speculative capital flooded into a "paper oil" market, which over the coming six years led to an enormous price bubble. At the same time, American petroleum imports were funding a spike in petrodollars, which were being steadily reinvested into risky securities. By 2007 both the housing and oil bubbles began to burst, thrusting the highly leveraged American economy into the Great Recession. In Madrick's words, "It was the house of cards built on Wall Street greed, unchecked by Washington regulators, that created the nation's credit crisis—the sudden drying up of lending to consumers, business, and homeowners—and caused the most severe recession in the United States since the Great Depression." Federal Deposit Insurance Corporation chair Sheila Bair wrote that "during the run-up to the crisis far too much money was directed towards booming, oversupplied property markets. The bust that followed is clear evidence that capital was misallocated and could have been put to better use in areas such as energy, infrastructure, or the industrial base." Fortunately, the Keynesian response championed first by President George W. Bush and even more forcefully by President Barack Obama prevented a complete economic collapse. Yet the nation's people had been dealt a severe blow, although the plutocrats had gotten richer still.[60]

In the decades following the peak of domestic oil supplies and leading up to the Great Recession, there was a fantastic transformation of the American economy. As the costs to the nation of maintaining its dependence on hydrocarbons began to increase, political leaders needed a new way to grow the economy. The answer was to encourage the government, corporations, and households to pull out the credit card and debt-finance future growth. As Bair points out, this wouldn't have been the worst idea had the money been directed into solid investments aimed at transitioning away from hydrocarbons, renewing and expanding critical infrastructure, and shoring up the flagging industrial sector. This, of course, was not what the plutocrats chose to do. Instead they took a much easier route. They supported two substantial expansions in the size of the armed services during the safest period in human history, channeled huge amounts into Internet companies with unproven business models, and backed risky mortgage-backed securities chock-full of subprime loans, all the while making it possible for the American people to take on unprecedented levels of personal debt. The result was predictable, yet incredibly damaging to the US democracy. As Phillips writes, "Debt was a critical enabler. Its huge expansion . . . paralleled—and helped to bring about—the growth in U.S. financial services." The upshot was that finance became the nation's largest economic sector. At the outset of the Great Compression, financial services accounted for less than 10 percent of corporate profits, while manufacturing accounted for nearly 60 percent. During the lead-up to the Great Recession, this essentially flipped: financial services rose to 40 percent, and manufacturing fell to 10 percent. This meant that while 30 percent of Americans had previously had industrial jobs, offshoring and automation had decreased this number to less than 10 percent. Only 5 percent of all jobs are in finance, although the sector accounts for nearly 20 percent of GDP (manufacturing fell below 10 percent but is now resurgent). Thus, energy insecurity during the past conservative cycle saw America transition from a nation that made things to one that financed things and accumulated massive debts. There was also a shift in the location of political power in the nation, with the plutocrats firmly reassuming the complete supremacy they had held during the Gilded Age and the Roaring Twenties. By spending billions on lobbying efforts and election contributions, they have solidified their authority to such a degree that their culpability for nearly tearing down the American economy went almost entirely unpunished. With the plutocrat- and hydrocarbon-backed Republican Party now firmly in control of the White House and both chambers of Congress, conservatives continue to weaken the nation and make it less able to secure a safe and prosperous future for everyone.[61]

Income and Wealth Inequality
Energy Insecurity + Extreme Conservative Backlash
= Financialization = Income and Wealth Inequality

Various benefits are ascribed to the capitalist economic system. Pure capitalism has long been considered synonymous with political freedom, because a large and powerful state controls neither the economy nor other areas of human existence. Likewise, capitalism is thought to be efficient, because there are incentives for companies not only to produce goods and services that consumers demand but to do so at the lowest possible cost to fend off challenges from other potential suppliers. Moreover, these incentives are thought to create an environment conducive to innovation and economic growth, which leads not only to greater wealth acquisition but also to overall increases in the standard of living. Since no other economic system is nearly as capable of promoting societal well-being, capitalism has become the preferred economic approach of nearly all the existing advanced nations. Yet none of these nations allows capitalism to have free reign. Governments step in to provide regulation aimed at improving market efficiency and protecting critical social values by correcting market failures such as monopoly, externalities, public goods, and information asymmetries. Governments have also sought to promulgate regulations and policies intended to reduce the tendency toward boom-and-bust cycles, which result in damaging recessions, mass unemployment, and widespread suffering. Finally, modern governments have intervened to reduce the income and wealth inequalities that are fundamental characteristics of capitalist societies.

In his recent widely acclaimed book *Capital in the 21st Century*, Thomas Piketty writes that a historical analysis of capitalism reveals that it will always result in massive wealth disparities, even in countries with so-called welfare-state capitalism. This is the case because over time the return on capital *(r)* has been greater than the rate of nominal economic growth *(g)*. Piketty's time-series analysis reveals that *r,* after tax and capital losses, has long been about 5 percent. Thus, if *g* is less than 5 percent, the wealth of the rich will grow faster than the overall economy, which means that the capital income of the upper class will increase more rapidly than the labor income of the working and middle classes. Given that growth rates rarely exceed 5 percent, especially in advanced economies, the natural result is that the rich get richer and gain ever more political power. Adding to this trend in America is the rise of "supersalaries," whereby a very small percentage of the population is awarded monumentally large compensation. Unlike the rest of the world, where capital income has a bigger impact, in the United States it accounts for only about one-third of the wealth gap. The other two-thirds is the result of this highly inequitable income gap. As can be seen in figure 4.7, the top decile's share of income has reached levels not seen since the 1920s, just before the stock-market crash that plunged the nation into the Great Depression. Similarly, as figure 4.8 shows, wealth concentration in the top decile is now higher than at any time since the Progressive Era, and America now has more wealth concentrated at the top than in Europe. The big question is why income and wealth inequality have returned to levels that haven't been seen since the macroeconomic and microeconomic policies of the New Deal were implemented? Some conser-

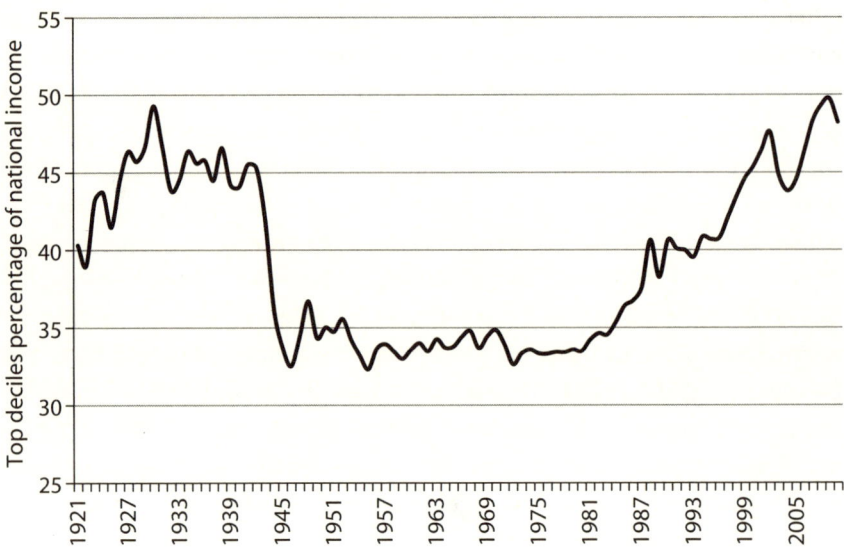

Figure 4.7. Income Inequality in the United States, 1921–2008. Frank-Sommeiller-Price Series, accessed 28 August 2017, tinyurl.com/jtuhd98.

vatives believe that supersalaries are earned and that it would be wrong for the government to penalize people for their personal merit. There are significant holes in this argument, primarily because we know that the United States now has the lowest social mobility of any advanced economy. This suggests that those who succeed because of their "merit" very likely have benefitted from advantages not enjoyed by most Americans. If large incomes were simply earned by the most meritorious, the nation would exhibit much higher levels of social mobility. Piketty puts forward another argument, known as the executive power hypothesis, which states that the reason for large salaries is that under the dominant corporate structure executives essentially set their own salaries. It is easier to set high salaries that ignore social norms, he contends, because it is hard to measure what executives are actually worth. While this explanation is more compelling than the standard conservative mythology, as Paul Krugman points out, its one weakness is that it doesn't account for the concentration of high salaries in the financial-services industry, where there are solid metrics to determine effectiveness. The gap in the executive power hypothesis can be plugged when one considers the reasons for the financialization of the US economy: energy insecurity magnified growing political dysfunction, and the resulting extreme conservative backlash led to the gutting of critical financial regulations and the rise of debt financing to power economic growth.[62]

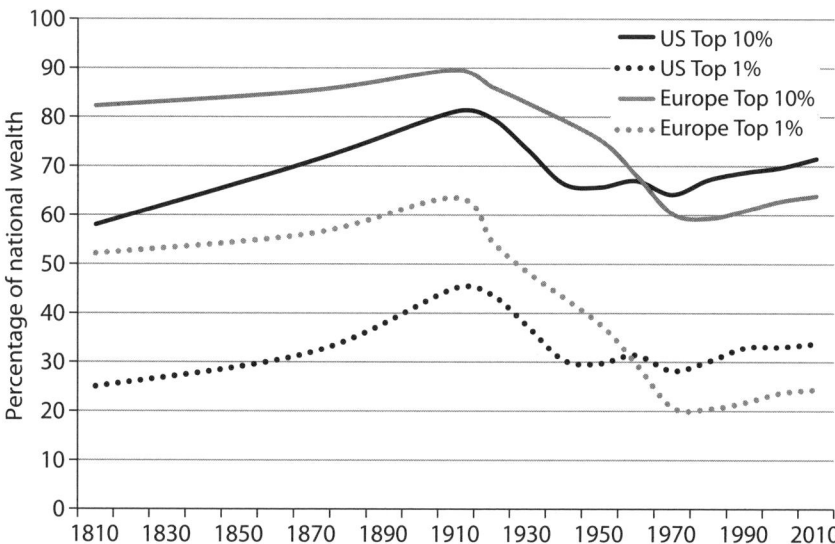

Figure 4.8. Wealth Inequality in the United States and Europe, 1810–2010. Thomas Piketty, *Capital in the 21st Century* (Cambridge, MA: Harvard University Press, 2014).

When hydrocarbons arrived during the Gilded Age, the energy security they provided made it relatively easy to increase the standard of living for most Americans despite growing inequality. Although the rich were getting richer, median salaries were increasing, so most people were relatively happy with the direction of the nation. By the 1890s, however, despite these positive trends, a majority of the citizenry supported progressive policies that would reign in some of the worst excesses. Although there was a short-lived conservative backlash in the aftermath of the First World War, these forces gained much more strength in the wake of the Great Depression. Continued energy security provided resources to alleviate some problems associated with inequality via macroeconomic and microeconomic security policies. The subsequent Great Compression was in many respects the golden age for average Americans, in that it saw the wealth and income gaps decrease. Piketty argues that this was largely the result of wealth destruction during the Second World War and high tax rates implemented to pay down the large postwar debt. According to Krugman, there were two other important factors. First, high wages mandated by the government during the war years had largely been continued. Second, strong labor unions were able to continue to negotiate better wages for their members. Although conservatives would have predicted that higher taxes and more generous wages for labor would lead to poor economic performance, the result was the opposite. As Krugman writes, "America in the 1950s was a middle-class society, to a far greater extent than it had been in the 1920s—or than it is today. . . . None of the bad consequences one might have expected from a drastic equalization of incomes actually materialized. . . . And the era of equality was also a time of unprecedented prosperity, which we have never been able to recapture." This is ironic, given that many conservatives have great nostalgia for the fifties.[63]

Things began to change when domestic petroleum supplies peaked, resulting in comparative economic insecurity. One outcome was wage stagnation. If inflation is taken into account, nonmanagement workers have seen flat or even negative wage growth. A variety of reasons have been given for declining wages, including rising benefit costs, labor-market slack, flagging educational outcomes, and a shift toward low-wage industries. While there is some merit in several of these notions, all but the last fundamentally ignore one of the central truths of our time, namely, the increased cost of maintaining energy security. The average salary peaked in 1973; not surprisingly, this was the same year as the first post-peak oil crisis. Within a few years a conservative backlash created a broad financialization of the economy, which greatly enriched plutocrats, who could participate in an arena from which the working and middle classes were barred because of a lack of wealth to invest. The combination has resulted in the income and wealth disparities that exist today, which represent an existential threat to American democracy.[64]

Although many journalists and commentators have recently reported on the rise in income and wealth inequality in America in stark terms, they almost never explain why these are bad societal outcomes. Actually, there are some positive aspects of income inequality. These include incentivizing high educational achievement, hard work, investment, and entrepreneurship. As long as the level of inequality is properly managed by government to maximize constructive societal outcomes, making it possible for strivers to benefit from their talents and diligence seems quintessentially American. The problem is that government no longer provides either the tax structure or the regulations needed for this to happen. Consequently, the negative aspects of income inequality have come to characterize the economic lives of US citizens. One important argument against extreme inequality is that it actually hinders economic growth. According to Barry Cynamon and Steven Fazzari, wage stagnation among the "bottom 95 percent" of earners destroys their ability to maintain high levels of consumption. Given that consumer spending drives more than two-thirds of the American economy, this is troubling. As discussed earlier, the solution in the period between the domestic oil peak and the Great Recession was simply to expand household debt to preserve the accustomed standard of living. Since the recession, however, the ability and willingness of most consumers to continue profligate borrowing has diminished significantly. This was a key reason for the slow economic recovery.[65]

Another notable consequence of extreme inequality is a lack of social mobility. This is a key focus of Piketty's work. Because *r* usually exceeds *g*, he worries that in the absence of government interventions the advanced economies will return to the high levels of inequality seen in Europe in the nineteenth century. This would lead to the destruction of the entrepreneurial advantages that should be enjoyed in a capitalist system, because large fortunes will be obtained not by the meritorious but instead by those who inherit great wealth. He worries that this form of patrimonial capitalism is damaging to the social order. Alan Krueger, chair of the White House Council of Economic Advisers under President Obama, has shown that nations with extreme inequality also have low mobility. Because America currently has the most extreme inequality, it shouldn't be surprising that it also has the lowest levels of social mobility. This is contrary not only to the country's founding principles but to the ideal of equal opportunity, an ideal its citizens still hold today. Those who believe equal opportunity still exists in the United States are living in a fantasy world. As Piketty writes, the children of wealthy parents have huge advantages over the children of working- and middle-class parents. They have access to a better K–16 education. They are better prepared to excel in both school and the workplace because of their ability to participate in critical extracurricular activities and unpaid internships. And they begin their adult lives not weighed down by large debts, because their parents can pay for higher education and perhaps even buy them a first house. These advantages, along with many others, have destroyed the fiction that the American Dream is equally available to all citizens.[66]

One of the clear results of the rise of extreme inequality has been the takeover of democratic institutions by the plutocrats. The wealthy have the ability to dictate the outcome of political elections by providing large infusions of campaign cash, which gives them an outsized role in determining the outcome of policy debates. This is contrary to the democratic ideals that have long been the basis of the entire political system. Federal Reserve chair Janet Yellen has said about the rise of extreme inequality: "I think it is appropriate to ask whether this trend is compatible with values rooted in our nation's history, among them the high value Americans have traditionally placed on equality of opportunity." William Forbath goes even further, arguing that "you can't have a republican government, and certainly not a constitutional democracy, amid gross material inequality. That's because gross economic inequality produces an oligarchy in which the wealthy rule. Insofar as it produces a lack of basic social goods at the bottom, gross inequality also destroys the material independence and security that democratic citizens require to participate on a roughly equal footing in political and social life. And access to such goods is essential to standing and respect in one's own eyes and those of the community." The only way to restore democratic institutions is through government action. But as the Federal Reserve's Rajashree Chakrabarti and Matt Mazewski have found, there is a strong correlation between income inequality and debilitating political polarization. The resulting political dysfunction perpetuates existing policies supporting the status quo, thus further strengthening the power of the plutocracy.[67]

Most Americans no longer receive anything like a fair share of national income and wealth. Eighty percent of citizens control just 5 percent of private national wealth, and the top 20 percent control the remainder. The top 1 percent control 42 percent; the next 4 percent control 30 percent; the next 5 percent control 13 percent; and the next 10 percent controls 11 percent. In the past four decades, while wages for the working and middle classes have stagnated, the top 1 percent has seen an increase of 165 percent, and the 0.1 percent has seen an increase of 362 percent. While the self-interested can argue all they want about the rewards of merit, it is hard to imagine that most people would support such outcomes if they truly understood what was happening. Although there is reason to hope that books like Piketty's *Capital in the 21st Century* will lead to a fruitful discussion about how to overcome truly un-American levels of inequality, one of the main reasons to fear the growing income and wealth gap is its potential to lead to political instability. Despite the fact that the nation is still enjoying a period of unprecedented domestic tranquility, even today's financial elites need to begin questioning whether the plutocratic takeover of both the political and economic life of the nation may change that dynamic. In the depths of the Great Depression, many observers at the time considered a working-class rebellion against the existing plutocracy a distinct possibility. Some believe there are similar stirrings inside America today, represented in protest movements ranging from the Tea Party to Occupy Wall Street, which came to life in the candidacies of Donald Trump and Bernie Sanders. As the venture capitalist Nick Hanauer wrote in an open letter to his fellow one-percenters, "I see pitchforks. At the same time that people like you and me are thriving beyond the dreams of any plutocrats in history, the rest of the country . . . is lagging far behind." Although he contended that some inequality is a natural part of any market economy, he worried that incomes had become so unequal that the nation no longer had a capitalist system but was instead becoming a feudal state. "The oldest and most important conflict in human societies is the battle over the concentration of wealth and power," according to Hanauer. "The folks like us at the top have always told those at the bottom that our respective positions are righteous and good for all. . . . What nonsense this is. We should never forget that . . . the United States of America and its middle class made us, rather than the other way around." Ultimately, extreme inequality should be addressed not just because it is morally right to do so but also because it will renew the legitimacy of America's commonly held social contract. One must wonder, however, whether this can be accomplished without either restoring access to cheap hydrocarbons or developing the alternative energy sources that can both provide energy security and fight the climate crisis.[68]

Educational Inequity
Energy Insecurity + Extreme Conservative Backlash
= Income Inequality = Educational Inequity

Quality education is one of the key pillars of modern economic development. The provision of mass education to advance human capital has been one of the main macroeconomic policies separating advanced economies from struggling economies. According to Robert Allen, "Widespread literacy and numeracy have been necessary (if not sufficient) conditions for economic success since the 17th century. These mental skills help trade to flourish and science and technology to develop. Literacy and numeracy are spread by mass education, which has become a universal strategy for economic development." High literacy levels were very important at the time of America's founding. Average citizens voraciously consumed newspapers, and public participation in the political process was high. The New England states had particularly high rates of both literacy and numeracy, which was related to its reliance on trade to support the economy. As the South became more reliant on extractive institutions like slavery and apartheid, a profound gap in educational attainment formed. Regardless, over the coming two centuries America embarked on a unique journey as it sought to implement universal-education programs in a nation with a large and diverse population. From the beginning, the education system was highly inequitable. Where it worked, it helped drive the country's economic progress; where it failed, it relegated some regions and populations to continued drudgery and deprivation. Despite its clear weaknesses, by the twentieth century the education system had developed into the envy of the world, especially its exceptional colleges and universities.[69]

One of the most distinctive aspects of American educational policy has long been the degree to which it is decentralized. "For most of our nation's history," writes Patrick McGuinn, "control of public education has been left almost entirely in the hands of state and local governments." The first major effort to implement formal schooling came in the 1830s, when Horace Mann began the common school movement in Massachusetts. By the time of the Civil War, this endeavor had resulted in the adoption of common schools in most of the Northeast and the Midwest; it spread more slowly into the South and West. Over time the number of years of schooling mandated increased, but high schools did not begin to appear in most states until the period between 1910 and 1940. Although the federal government had provided some limited support for higher education, most importantly through the creation of land-grant schools, serious debate about more concerted national involvement did not take place until the Third Political Age. Initially, national policy focused on higher education, with the GI Bill of 1944 and the National Defense Education Act of 1958 leading to a rapid increase in the numbers attending college, and especially in the numbers majoring in science and engineering.[70]

The first challenge to the state and local monopoly on K–12 education came from the US Supreme Court, particularly with its decision in *Brown v. Board of Education of Topeka* (1954). The issue in that case was whether segregation of children in public schools solely on the basis of race, even though the physical facilities and other tangible factors were equal, deprived those children of equal educational opportunities. The Supreme Court found in the affirmative, reversing its previous stance that the "separate but equal" doctrine was valid. Chief Justice Earl Warren wrote for a unanimous court that education was not only the foundation of good citizenship but also critical to success in modern society. He argued that segregation robbed children of the chance to engage in discussions and exchange views with other students from different backgrounds, generating a feeling of inferiority that negatively impacted the motivation to learn. He concluded that "in the field of public education the doctrine of 'separate but equal' has no place. Separate educational facilities are inherently unequal." The slow process of truly implementing this decision still continues today, but it did persuade many that the federal government needed to play a larger role in the educational sphere.[71]

When Lyndon Johnson became president, he was determined to enlarge the federal role in education. In 1960, only 2 percent of all school funding had come from the federal government, mostly for special grants, school lunches, and vocational programs. A former teacher, Johnson believed strongly in the importance of properly educating all Americans. In a special message to Congress, LBJ wrote, "Nothing matters more to the future of our country; not our military preparedness, for armed might is worthless if we lack the brainpower to build a world of peace; not our productive economy, for we cannot sustain growth without trained manpower; not our democratic system of government, for freedom is fragile if citizens are ignorant." President Johnson pushed through the Elementary and Secondary Education Act of 1965, which provided more than $1 billion for federal programs aimed at intervening in the education crisis to help poor and minority students. Although the act had serious implementation problems, it became quite popular, and funding more than doubled within a decade. In 1979, with the national government continuing to increase its role in education policy, President Carter created the immediately controversial Department of Education to manage federal programs.[72]

By the time Reagan took office, Republicans had already identified the Department of Education as a danger. The new administration immediately set about reducing regulatory mandates by 85 percent, although without Democratic support they couldn't reduce overall spending. At the same time the National Commission on Excellence in Education was formed, with the intention of showing that there was no need for federal intervention because American schools remained the best in the world. In 1983 the commission released *A Nation at Risk*, which detailed its finding that US schools were actually in terrible shape. The report painted a dark picture, suggesting that the nation's poor education system put national security and economic competitiveness in jeopardy. It utterly changed the public's perception of education policy, placing it in the upper tier of issues concerning most voters. The commission made three recommendations for solving the problem: increase teacher pay, strengthen curricula, and develop standards to measure progress. Unfortunately, the political dysfunction that had taken hold of Washington hindered any efforts to take bold action to adopt and implement these ideas, and the entire issue languished for the remainder of Reagan's tenure. President George H. W. Bush was much more engaged in the issue, but he wasn't interested in addressing the foundational problem. Instead he successfully reoriented the entire debate around issues that were electorally favorable for Republicans: a federal role in standards and testing, school choice, and increased federal funding.[73]

President Bill Clinton moved the Democratic Party to the center on educational issues. His most significant contribution to national policy was the Improving America's Schools Act of 1994, which removed the Elementary and Secondary Education Act's focus on civil rights and argued that schools needed to be made better for all children. This broadened the federal education mandate, although since its focus was almost entirely on standards, it did nothing to address the root causes of the education crisis. Despite winning two presidential elections in which education was a key issue in his favor, the political dysfunction that had infected the government meant that Clinton was unable to push any bold new initiatives, instead working in the margins. When George W. Bush reached the White House, he took advantage of support from Senator Ted Kennedy to gain passage of the No Child Left Behind Act of 2001. The conservative victory saw major provisions requiring "adequate yearly progress" on tests and graduation rates, "highly qualified teacher" certifications, and sanctions that would strip funding to schools if standards were not met. The legislation was a near total failure. Out of the gate, Bush backed away from his promise to Kennedy to fully fund the initiative, as he sought to increase funding for the Global War on Terrorism. More importantly, the entire framework was inherently flawed, particularly because the standardized testing that it required so transformed K–12 curricula that it destroyed the best parts of the American education system, most notably critical thinking, creativity, and promotion of a deep understanding within key fields.[74]

One of the most interesting aspects of American education policy is that the system does not actually suffer from a lack of funding. Education funding has more than doubled in the past forty-five years, far outstripping population growth during that period. State and local governments in particular have increased support significantly. As the funding has increased, however, the ranking of American students has remained static (or fallen) in comparison with students in the best education systems in the world. To be fair, most of those nations have much smaller populations and are considerably more homogenous than the United States. Still, the United States spends more per pupil than all but a few of the world's richest countries. Yet its overall outcomes are not nearly as good. Deep political dysfunction makes it impossible to make the big reform decisions that are necessary to turn the ship. Combined with the financialization of the economy and the rise of the plutocracy, this has resulted in greater inequity within the education system.

As Thomas Friedman and Michael Mandelbaum have pointed out, America actually faces two major education gaps. The first is between rich and poor students. The second is between American students and those in the rest of the world. Although the United States certainly needs to fundamentally reform the education system, the main reason why the nation has fallen so far behind is that it actually operates two education systems, one for rich kids in wealthy cities and states, another for poor kids in impoverished areas. Another way of saying this is that the central reason for poor outcomes is federalism, in this case manifested in an outdated model of funding education primarily at the state and local levels. Any rational political system would have done something about this long ago, but energy insecurity and an extreme conservative backlash destroyed the country's ability to make big decisions on issues like education. This is a huge problem, because as we have seen, increasing educational attainment is correlated with economic advancement. At their best, schools provide students with the skills, knowledge, socialization, and critical-thinking capabilities that are vital for personal and national growth. But many American schools don't adequately deliver these benefits. This is concerning because those with access to high-quality education make far more money over the course of their work lives, leading to more income and wealth inequality.[75]

Federalism is quite inequitable. Because more than 90 percent of all funding comes from state and local governments, students from affluent neighborhoods and states get a world-class education, while everyone else falls further and further behind. This multiplies several intrinsic advantages well-to-do students already enjoy. Wealthier parents are able to provide stable home environments, devote time to reading to their children, access higher-quality childcare, pay for individual tutoring, and spend money on cognitively stimulating experiences. Sean Reardon writes that "high-income families are increasingly focusing their resources—their money, time and knowledge of what it takes to be successful in school—on their children's cognitive development and educational success. They are doing this because educational success is much more important than it used to be, even for the rich." Combined with highly unequal schools, this has resulted in America's bifurcated educational approach. One system sees wealthier school districts performing at levels that make them compare favorably with those receiving the highest global rankings. For instance, if low-poverty states like Massachusetts and Minnesota were independent nations, they would achieve parity with other advanced nations. The other system sees poor school districts performing at embarrassingly low levels that pull down the entire national ranking and, more importantly, threaten America's economic future. Children in high-poverty states like Alabama and

Mississippi are enrolled in an education system that is at best in the middle of the pack internationally. According to Rebecca Strauss, "America's average standing in global education rankings has tumbled not because everyone is falling, but because of the country's deep, still-widening achievement gap between socioeconomic groups. And while America does spend plenty on education, it funnels a disproportionate share into educating wealthier students, worsening that gap. The majority of other advanced countries do things differently, at least at the K–12 level, tilting resources in favor of poorer students." This partially explains the superior academic performance of these countries, which provides them with significant benefits in building a twenty-first-century economy. As Strauss notes, "Historically, broad educational gains have been the biggest driver of . . . economic success; hence the economist's rule of thumb that an increase of one year in a country's average schooling level corresponds to an increase of 3 to 4 percent in long-term economic growth."[76]

Conservatives have long argued that there is nothing wrong with educational federalism and decentralization. In *Conscience of a Conservative*, Barry Goldwater argued that standards were more important than funding, although he didn't explain how to achieve better performance without more money. He contended that states should remain in control even if this led to differing outcomes in rich and poor jurisdictions. This was because he believed federal intervention in education was unconstitutional and that regardless, there was no clear national role. Clearly enhancing macroeconomic security or simply providing equity was not a sufficient rationale. In fact, Goldwater was more concerned with an egalitarian system leading to the neglect of the nation's most talented students.[77] Modern Republicans still largely hold these shortsighted views; if anything, they are even more vehemently opposed to a national role in education. This is ironic, given that it is the people in the states where they win most of the elections who are most disadvantaged by the current arrangement. Ensuring American economic competitiveness in a postcarbon age will require marginalizing these beliefs, followed by dramatic reforms not only in how the government funds education but also in how it is delivered.

Infrastructure Decay
Energy Insecurity + Political Dysfunction
= Fiscal Constraints = Infrastructural Decay

As we have seen, infrastructure development has long been viewed as a key building block for creating a thriving modern society. Nations gain significant benefits not only from expanding their infrastructure but also from properly maintaining it to ensure the benefits are maximized. Infrastructure impacts not only the economic sector but also the overall quality of life for a country's citizens. As the National Council on Public Works Improvement wrote in 1988, "The quality of a nation's infrastructure is a critical index of its economic vitality. Reliable transportation, clean water, and safe deposit of wastes are basic elements of a civilized society and a productive economy. Their absence or failure introduces a major obstacle to growth and competitiveness." The economic benefits are varied, but most fundamentally infrastructure enhances the collective productivity of a nation. Numerous economists have modeled the positive impacts public services play in growing the economy, providing scholarly evidence to support long-held beliefs about the importance of these activities. It has also become clear that capital receives an important boost from public infrastructure, reinforcing the correctness of President Obama's controversial statement during the 2012 election cycle: "There are a lot of wealthy, successful Americans who agree with me. . . . They know . . . if you've been successful, you didn't get there on your own. . . . If you were successful, somebody along the line gave you some help. There was a great teacher somewhere in your life. Somebody helped to create this unbelievable American system that we have that allowed you to thrive. Somebody invested in roads and bridges. If you've got a business—you didn't build that. Somebody else made that happen." While clearly inartful, his statement that businesses and the wealthy benefit greatly from public investments is undeniable. There are also several quality-of-life advantages that flow from high-quality infrastructure. Sewage, waste-management, and water-management systems have dramatically improved human health and public safety and increased agricultural yields. America's vast transportation network has tremendously increased the number of economic opportunities for individuals, farmers, and manufacturers. At their best, transport systems also promote urban density and increase available leisure time and options. Parks provide important public spaces for both social connection and recreation. Finally, energy networks provide the power that is the backbone of our entire modern lifestyle. Although these are but a few examples of the many positive impacts presented by state-of-the-art infrastructure, it is clear that the impacts are immense. David Alan Aschauer, a former senior economist at the Chicago Federal Reserve Bank, argues that past infrastructure investments have been responsible for significant quality-of-life improvements with regard to health, safety, economic opportunity, and leisure time and activities.[78]

Public infrastructure projects were central to the American Rise. As Feliz

Rohatyn writes, "The federal government has traditionally been the indispensible investor in our nation. . . . Activist government, led by bold leaders with vision and perseverance, made far-seeing investments that helped shape America." Some argue that the most important ventures were in transportation. Over the course of two hundred years, there were recurring waves of new developments in transportation. The first wave began in the 1820s with the development of both canals and inland waterways plied by steamboats. This was followed by two waves of railroad development, first from the 1820s to the 1850s and then dramatically picking up speed from the 1860s to the second decade of the twentieth century. By the time these two building phases were complete, America had by far the most impressive rail infrastructure in the world. A fourth wave saw the digging of the Panama Canal from 1904 to 1914, which, while outside America's sovereign territory, tremendously enhanced its economic competitiveness and geopolitical relevance. A final wave saw two parallel developments from the 1930s to the 1960s. One witnessed the construction of modern roads and highways, as well as the erection of thousands upon thousands of bridges. The other beheld the composition of the world's most advanced network of airports and navigation aids. By 1970 the United States had the best transportation system on the globe. Transportation was indeed critical, but a stronger argument could be made for the importance of the complex energy network that powers the nation. Among the many components in this web of interconnections, arguably the three most important are pipelines, power plants, and the electricity grid. Finally, there is little doubt that the creation of vast communications systems has transformed modern life.[79]

For its first two centuries, America invested trillions of dollars to develop the infrastructural components that have helped make it the world's greatest power. Yet in recent decades its dominance has faltered in all the areas described above. Once again, this is the result of the energy insecurity and extreme conservative backlash that have increasingly jeopardized US preeminence. While the county's infrastructure was regularly renewed and expanded through the 1970s, the fiscal crisis that emerged from these converging challenges to American democracy has led to widespread decay and the failure to keep pace with not only other advanced economies but also some rapidly emerging ones. Capital investments were reduced significantly under President Reagan, who chose to focus his attention on large tax cuts and increases in defense spending rather than on modernizing critical infrastructure, particularly in energy and transportation. Little was done to reverse course by the next three presidents, who largely followed his lead. This began to change with

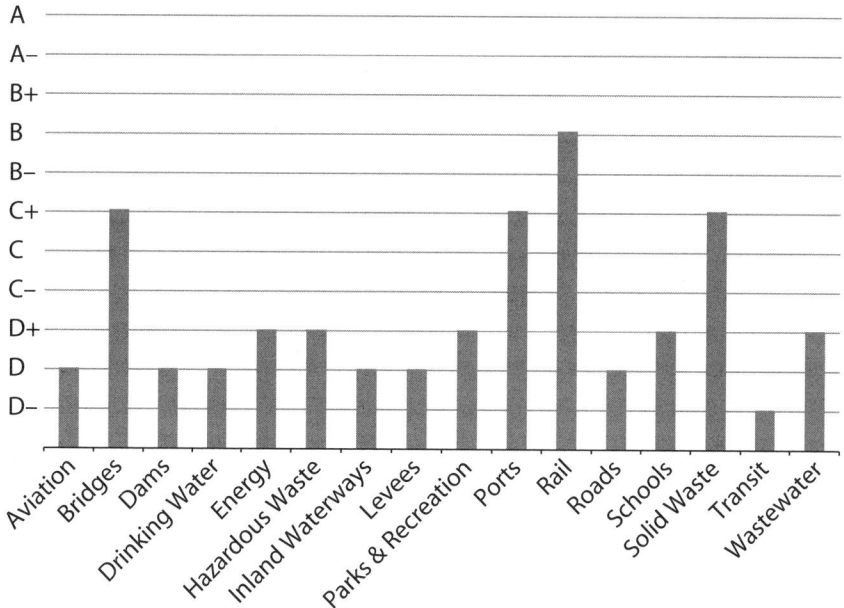

Figure 4.9. ASCE US Infrastructure Report Card, 2017. American Society of Civil Engineers, *Infrastructure Report Card, 2017*, accessed 28 August 2017, tinyurl.com/yc55jxlh.

President Obama's Great Recession stimulus package, but this bump in spending proved to be short-lived as Republicans failed to provide the resources necessary to maintain needed momentum. Consequently, the national infrastructure is in terrible shape.

Every four years the American Society of Civil Engineers(ASCE) publishes a report card on the nation's infrastructure, giving grades in sixteen categories. The most recent assessment, released in 2017, found that the country's cumulative grade had risen to a D+, up from a D in 2009 (fig. 4.9). While improvements had been made in areas targeted by the Obama stimulus, overall things remained in dire shape. American roads and bridges are clearly in bad shape, but the condition of rail and transit networks will be of more concern moving forward. If the country is to transition to a posthydrocarbon age, these systems will play an increasingly significant role. There is some reason for hope in the rail arena, which has experienced a limited renaissance in transporting freight in the past couple of decades, which accounts for its overall B grade. Rail is one of the few industries that began to get ready for increased energy prices early on, so it was comparatively well prepared when companies began to search for more energy efficient long-distance transportation options. According to the

ASCE, "After a period of underinvestment, freight railroads nearly doubled their capital investment from 1990 to 2010 to maximize productivity by replacing aging and inefficient infrastructure as well as shedding lines that were underutilized." Still, there is evidence that as demand increases, congestion costs the economy as much as $200 billion per year (1.6% of GDP). America's passenger rail is in far worse shape everywhere except in the Northeast Corridor, where ridership has doubled since 2000. This is striking given the poor condition of these lines compared with those for state-of-the-art high-speed trains operating in other nations and offering an alternative to intercity air and car travel. Amtrak's overhyped Acela compares poorly with Japanese, Chinese, and European trains because of both poor tracks and inferior engine technology; it averages only about 68 miles per hour compared with 220 miles per hour for Japanese trains. While Amtrak has cautious strategic plans to fix this problem in the Northeast, there is no grand plan for America to create a national network that truly rivals its economic competitors.[80]

Public transit also has been making slow advances in some parts of the country, but overall it is still in incredibly poor condition compared with public transit in other rich nations. Nearly half of all Americans have no access to public transportation (69% in cities, only 14% in rural areas), and there are major regional differences in the comprehensiveness and condition of networks. The Northeast and West have the largest systems, although some are in poor repair. The Midwest and South generally have either poor or nonexistent systems. Across the nation there aren't enough buses, light rail networks, or subways to meet demands. The economic impacts are grim. As the ASCE writes, "Although investment in transit has also increased, deficient and deteriorating transit systems cost the U.S. economy $90 billion in 2010, as many transit agencies are struggling to maintain aging and obsolete fleets and facilities amid an economic downturn that has reduced their funding, forcing service cuts and fare increases." America continues to fall behind European and East Asian economies in creating twenty-first-century networks capable of meeting the needs of a posthydrocarbon nation.[81]

Given how important dams, levees, and wastewater systems will likely be as climate change demands new strategies from both urban planners and farmers, the D and D+ grades in these three categories is particularly worrying. Weather changes related to global warming will result in severe droughts in some areas and heavy flooding in others. They will also result in sea level increases, though there is still hope that these might be relatively modest. While dams are important for producing hydroelectricity, in the future they might be even more important for irrigating cropland as temperatures rise and desertification spreads. Yet tens of thousands of aging dams are in poor repair, and many need to be either removed or replaced. There are more than a hundred thousand miles of levees in the United States, which prevent billions in damages every year ($141 billion in 2011 alone). Nevertheless, a large backlog in repairs endangers large numbers of farms and communities. Interestingly, the cost of repairing both dams and levees is relatively modest, only $125 billion combined. This number could increase, however, as it becomes apparent where new dams and levees might be required in a changing climate. Wastewater and stormwater systems are also in poor condition, requiring nearly $300 billion to bring them up to date. Stormwater systems account for only a small proportion of what needs updating right now, but if the number of extreme weather events like Hurricane Sandy increases, this will likely change. Across the board, much work will be required to create water-management systems capable of ensuring that America can adapt to new climate conditions.[82]

Decades of public- and private-sector underinvestment has led to the decaying functionality of both the electricity grid and pipelines. Aging electricity-distribution networks have led to increasingly frequent brownouts and blackouts, which cost the economy billions of dollars. Another problem with the grid is congestion at key choke points. A far larger problem is the inability of the grid to transport large amounts of clean energy over long distances without major losses in efficiency, or to integrate a significantly greater level of electrons from alternative energy sources into the network, although the Obama stimulus did provide funding to begin the transition toward smart-grid technologies. Pipelines for petroleum (150,000 miles' worth) and gas (1.5 million miles) also have suffered increasing numbers of failures, such as the 2010 Marshal, Michigan, oil pipeline rupture, causing 840,000 gallons of petroleum to spill into the Kalamazoo River, and the 2010 San Bruno, California, gas explosion, which killed eight people and leveled thirty-five houses. Perhaps the energy sector's greatest challenge is the lack of a posthydrocarbon infrastructure (discussed in the final chapter). It is clear from even this short account of infrastructural decay that energy insecurity and an extreme conservative backlash have resulted in serious fiscal constraints on government and undermined this key underpinning of American success. Infrastructure is yet another casualty of the political dysfunction that increasingly threatens national economic competitiveness. Rohatyn concludes: "The aging of our nation's infrastructure has lessened our productivity, undermined our ability to compete in the global economy, shaken our perception about our own safety and health, and damaged the quality of American life. . . . Either we learn from our history and support a program that encourages the federal government to make significant investments to improve the way we live and work; or we complacently let time continue to take its toll on a great, yet aging nation."[83]

Deteriorating International Prestige
Energy Insecurity + Political Dysfunction
= Deteriorating International Prestige

The United States emerged from the Second World War with a superlative international image: the nation had saved the world from fascism. In the coming decades its image remained positive around the globe as it faced off against the Soviet Union in the Cold War. The first significant stain on America's reputation was the poorly conceived and poorly executed Vietnam War, in which America was defeated not only because it faced an enemy protecting its homeland but also because its steadfastness in waging the conflict waxed and waned. The calamity came at the same time as the peaking of domestic oil production, which inexorably led to a further slow decline in popular attitudes as the country more fully engaged in the Middle East. Opinions improved somewhat during the late 1980s, as the standoff with the Soviets entered its final phase. The fall of the Berlin Wall, the disintegration of the Warsaw Pact, and the dissolution of the Soviet Union saw a resurgence in America's prestige as it emerged as the only remaining superpower. Underlying the generally positive opinions about US global leadership, however, were the portents of its rapid fall from grace in the twenty-first century. This was almost entirely the result of the continued high reliance of the American economy on petroleum and its evolving entanglement in Middle East politics in an effort to protect the global flow of that precious resource.

"Petroleum is unique among the world's resources," writes Michael Klare, in that "it has more potential than any of the others to provoke major crises and conflicts." Although British traders had long had connections in the Middle East, in the late nineteenth century the empire began forming protectorates in the region. With the Germans also working to gain a foothold in Mesopotamia, the region ultimately became a critical theater in the First World War. During the war, the British moved north from their protectorate in Kuwait and took control of Iraq, installing Faisal Ibn Husayn as king. The Middle East again became a strategic target during the Second World War, when in 1941 Germans and Italians threatened it from occupied Egypt. A pro-German insurgency attempted to take advantage of Axis air support to dislodge the British from Iraq. There were major battles in Basra, Habbaniya, Fallujah, and Baghdad. Relying on Indian army and Arab Legion troops, the Brits were able to ensure victory in this small but important theater of the larger conflict. Because of the success of this campaign, the supplies of oil needed to keep Britain fighting the Germans long enough for the Americans to enter the war were maintained. As the British Empire's status as the key world power declined after the war, America became the preeminent outside power in the region.[84]

Starting with Franklin Roosevelt, each president during the Third Political Age committed to ensuring that oil flowed out of the Middle East. They understood that, as Paul Roberts writes, "oil was now the linchpin for postwar prosperity, the true currency of geopolitical power. Coal might still produce more total energy, but oil fueled the ships and aircraft, the freight trains and automobiles on which military and commercial dominance was increasingly based. . . . Oil was, for all intents and purposes, *the* fuel of the twentieth century." For this reason, American leaders determined to form a close relationship with Saudi Arabia. This was not surprising given its huge petroleum reserves. By the beginning of the Second World War, two American companies, Socal and Texaco, had already made large investments in the country. During the war, in an effort to protect these interests and to edge out the British, the United States began to provide direct aid to the Saudis. The policy of "solidification" led President Roosevelt to approve Lend-Lease funding for the kingdom, which helped build a relationship between the two nations. In February 1945, after the Yalta Conference, Roosevelt met with King Ibn Saud in Egypt. The former was hoping to secure American oil interests, while the latter was intent on promoting American leadership in the Middle East to offset the threat he believed British influence represented. No notes were taken, so we have little sense of what the two spoke about during their five-hour conversation. But within a few years of their meeting a coalition of companies under the auspices of the Arabian American Oil Company (Aramco) had solidified a concession to drill for Saudi oil. This growing partnership moved forward in 1950, when, under intense pressure from Ibn Saud, Aramco agreed to a 50-50 profit-sharing agreement with the kingdom that would provide long-term stability through the mid-1980s. That same year, President Truman essentially guaranteed Saudi Arabia territorial protection. According to Priest, "Shortly after the proclamation the U.S. Sixth Fleet moved into the Mediterranean Sea, and the air force activated a ring of bases surrounding the Persian Gulf in Greece, Turkey, Libya, and

Saudi Arabia." Forty years later the Truman Doctrine would be the basis for an American invasion of nearby Kuwait to expel Saddam Hussein's forces and prevent an attack against the Arabian Peninsula.[85]

As discussed above, another important relationship in the region was with Iran. After its quarter century as an unofficial protectorate of the United States, growing internal discontent led to the Iranian Revolution and the nationalization of oil production. This in turn resulted in the promulgation of the Carter Doctrine and the Reagan Corollary, which heightened tensions between America and Iran. American support for Saddam Hussein in his war against Iran saw yet another escalation in hostility, which resulted in increased enmity toward the United States within the broader Middle East. By the late 1980s, however, friction with the Iranians had faded somewhat, particularly with the end of the Iran-Iraq War. Iraq steadily became the bigger concern as Hussein began to threaten his southern neighbor Kuwait. This was troubling by itself, but more alarming because if Iraq succeeded in Kuwait, its next target might very well be Saudi Arabia. When Hussein invaded Kuwait in 1990, President George H. W. Bush reacted swiftly to create an international coalition to remove Iraqi forces and protect the Saudis. The stationing of American troops in Saudi Arabia both during and after the war, however, was highly unpopular among radical Islamists. One in particular, Osama bin Laden, decided that his terror organization would begin a jihad to target the trespassing infidels. Al Qaeda launched several attacks aimed at the United States, including the 1993 bombing of the World Trade Center and the 2000 attack on the USS *Cole*. It also attempted to assassinate President Bill Clinton on a trip to the Philippines in 1996, but the plot was uncovered minutes before his motorcade was scheduled to roll over a bridge where a bomb had been planted. Regardless of these attacks, however, America's worldwide hegemony filled both its leaders and its people with an inflated sense of security. As Daniel Yergin wrote after the 9/11 attacks and the beginning of the Iraq War, "It is clear now that, with the end of the Cold War and the resolution of the 1990–91 crisis, and then with lower prices, the world passed into a decade of overconfidence about our energy security—and, indeed, security overall. But turmoil in the Middle East—accentuated by demographic pressures, generational change, and the rise of political Islam; by the threat to political order and infrastructure posed by terrorist organizations; by regional conflict; and by rising demand, market pressure, and price spikes—has brought the issue into sharp focus again."[86]

In 2001, anger over decades of American meddling in Saudi Arabia and other Middle Eastern nations led to the single most devastating terrorist attack in world history. After 9/11, positive public sentiment toward the United States throughout the world soared to heights not seen since the end of the Second World War. Around the globe people showed their solidarity with the American people. In Beijing, tens of thousands visited the US embassy to leave flowers and funeral wreaths. In Berlin, two hundred thousand marched to the Brandenburg Gate. In London, the Queen's Guard at Buckingham Palace played "The Star-Spangled Banner." In France, *Le Monde* ran a headline reading, "We Are All Americans." In Palestine, Yasser Arafat donated blood. And in Tehran, the capital of America's mortal enemy in the region, ten thousand people gathered in Madar Square in a spontaneous candlelight vigil, and an entire stadium observed a moment of silence. By the time President George W. Bush ordered the invasion of Iraq less than eighteen months later, all of this goodwill had been transformed into anger and resentment. Senator Ted Kennedy wrote that "after 9/11 we had the sympathy and concern of the world. . . . Other nations recognized the value of America's leadership. . . . But we squandered this reservoir of admiration and moral authority. Our misguided policies reduced rather than enhanced our influence in the world, created battalions of terrorists, and made it far more difficult to protect our nation and its interests. . . . It inflamed anti-Americanism in much of the Middle East and elsewhere in the world."[87]

The progression of American engagement with the Middle East paralleled the nation's rising dependence on imported oil. The United States began forming relationships in the region during the energy crunch that hit midway through the Second World War, built upon them in the 1950s as imports began to slowly rise, and intensified its efforts spectacularly after the peak in domestic production. Oil was the only reason for US involvement in the region. Some might say that close connections with Israel were a motivation, but there is scant evidence for this. America was somewhat agnostic about the Jewish state through the 1950s. This attitude didn't shift until strategic planners became worried about potential Soviet gains in the region, but once again this was directly related to the fact that there were large oil fields that America's Cold War competitor might snatch up. Over the years the alliance with the Israelis has deepened, largely because it is a counterbalance to Iran, preventing the creation of a legitimate Islamic caliphate in the Middle East. Because many Americans feel a strong kinship with the Israeli people, it is good politics for leaders to support Israel. Yet it is hard to imagine that the United States would have become nearly as entrapped as it is in the region's ever-shifting crises if it weren't for several of the countries' huge stores of hydrocarbons. Still, those fossil fuels did indeed exist and led America to focus much of its geopolitical attention on the Middle East for too many decades. Not only was this hugely expensive—it almost certainly cost more than the benefits received—but in the long run it severely damaged the global reputation of the United States.

For many years, the Pew Research Center has conducted an annual poll that seeks to determine global attitudes toward the United States. As President Clinton was leaving office, the country was viewed favorably by many nations. These figures plummeted even among America's closest allies as President George W. Bush led the nation into an oil-related foreign policy calamity. From 2000 to 2008, American favorability moved from 62 percent to 42percent in France; from 78 percent to 31 percent in Germany; from 83 percent to 53 percent in Britain; from 52 percent to 12 percent in Turkey; from 77 percent to 50 percent in Japan; and from 68 percent to 47 percent in Mexico. The United States was increasingly reviled not because of some innate hatred of Americans or the American way of life and not because of a dislike for democratic or capitalist systems. Instead, it was because of misguided government policies, specifically in the arena of petroleum geopolitics. America had involved itself in a part of the world that would traditionally have been far beyond its sphere of influence, and the result was to anger many in the region, including a small subset of extremists who have successfully sought to hurt the United States. This led to a groundswell of hatred against the nation in the Middle East, but more importantly, it undermined America's role as the world's sole superpower.[88]

None of this was destined to happen. If the citizenry had simply listened to Jimmy Carter in 1979, America wouldn't have turned its back on the alternative energy sources that would have eliminated the nation's reliance on oil. The United States would not have needed to become enmeshed in Middle Eastern politics and very likely would have maintained the moral leadership it had enjoyed for much of the post–Second World War era. Unfortunately, it did not take this path. As Bill Richardson writes, "The United States has failed to invest adequately in the technologies that will help it reduce its dependency on imported oil, free the great powers of the twenty-first century from a similar path of dependency, and address the environmental consequences of U.S. consumption." Thus international confidence in America's moral leadership eroded in yet another arena. By the end of the Third Political Age, the nation's standing was very low. Many at the time believed that America needed to take dramatic actions to change course. Leon Fuerth concluded that "America must display less arrogance, not by hiding it, but by dispensing with it. The country needs to stop glorifying the power and independence of its actions, and instead make it clear that it aims to work with others to build global consciousness in hopes of securing global action. Nothing could more powerfully signify American support for such values than would a resumption of U.S. leadership for purposeful management of the global environment."[89]

Climate Crisis
Conservative Backlash + Political Dysfunction
= Continued Hydrocarbon Use = Climate Crisis

Earth's geological past is a story of major climatic shifts. During the Precambrian Era, from 4.6 billion to 540 million years ago, the planet saw fluctuations between greenhouse and icehouse periods. The Paleozoic Era, from 540 million to 250 million years ago, experienced more cool weather, although warmer weather predominated in the equatorial zones, where fish, arthropods, amphibians, and reptiles evolved. The Mesozoic Era, from 250 million to 65 million years ago, when dinosaurs roamed the planet, was a time of high concentrations of atmospheric carbon dioxide and warm temperatures. The current Cenozoic Era has seen several shifts between greenhouse and icehouse conditions, as mammalian species came to dominate most ecosystems. The Paleogene Period, from 65 million to 23 million years ago, saw a slow transition from greenhouse to icehouse conditions, before temperatures warmed during the Miocene Epoch, from 23 million to 5 million years ago. This period of pronounced mammalian diversity ended with another swing toward cooler temperatures during the Pliocene and Pleistocene Epochs, from 5 million years ago to the beginning of the Neolithic Revolution, which witnessed massive ice sheets forming at the poles. For most of the past million years, there have been recurring ice ages, when those ice sheets reached far south before retreating again to the poles.

The Neolithic Revolution began as the last major ice age came to a close and gave way to the Holocene Epoch, an interglacial period between cyclical ice ages. As warmer and wetter weather predominated, humans became farmers and engaged in animal husbandry. "By 5000 BC, the major climatic shifts that affected humanity were largely over," writes Brian Fagan. "Sea levels stabilized at near-modern levels, the great ice sheets were almost gone, and global vegetation was effectively that of today except when modified by human activity. . . . But this does not necessarily mean that the climate was always benign." Over the next two millennia, during the Holocene Optimum, temperatures reached maximum levels. This was a period when great civilizations such as ancient Egypt thrived because there was far more water available for agriculture. It saw the first cities being born in the Fertile Crescent. It also witnessed the Linear-bandkeramik move quickly across Europe displacing hunter-gatherers. From 3000 BCE to 150 BCE the climate fluctuated between relatively cold and relatively warm periods before trending cooler for a millennium that experienced the rise and fall of the Roman Empire and encompassed the Dark Ages. During the Medieval Warm Period, from 950 to 1250, the Vikings took advantage of warmer temperatures to dramatically extend their explorations, reaching as far as Iceland, Greenland, and Vinland on the North American coast. The Little Ice Age, from 1300 to 1850, brought the coldest weather since the dawn of the Holocene. This period resulted in an early agricultural revolution in Europe, which saw larger farm enclosures working to feed the population. It also witnessed crops like turnips and clover taking on a newly critical role in providing feed for livestock during long, cold winters. More productive agriculture and better infrastructure resulted in the growth of European cities, which in turn paved the way for industrialization in the most advanced countries. Over time the emissions from coal-fired manufacturing began having an impact on the global climate, ultimately ushering in a new era of global warming. Some have labeled this epoch the Anthropocene, because they believe human activities have begun to fundamentally change the planet's climate.[90]

For thousands upon thousands of years, one of the defining characteristics of humanity was its ability to successfully adapt to the shifting climate. Some have suggested that the perils wrought by current global warming will simply provide another example of our species' ability to overcome hardships. While there is no question that human ingenuity is already being brought to bear to adapt to rapidly changing contemporary conditions, the current situation differs from what humans dealt with in the past in two respects. First, there are far more people living on the planet right now. The global population has grown from about five million at the beginning of the Neolithic Revolution to more than seven billion today. Before the arrival of agriculture, when the climate shifted, hunter-gatherers could simply relocate. This is far more difficult now. We have become gradually more wedded to existing farms and ever-larger cities. It isn't a simple proposition to rapidly move global food production and tens of millions of people to new areas. Second, and perhaps more importantly, the potential for warmer temperatures far beyond anything the species has ever experienced calls into question whether our past adaptability is relevant to our present challenges.

* * *

Climate science first began to emerge a little more than a century ago, with significant efforts to understand global warming arising only about fifty years ago. In 1896 the Swedish chemist Svante Arrhenius, a winner of the Nobel Prize in Chemistry, published the first major paper on the greenhouse effect. In developing a model to explain the ice ages, he estimated the levels of atmospheric carbon dioxide needed to change global temperatures. Although Arrhenius's work was somewhat rudimentary by today's standards, his paper made it clear that increased concentrations of carbon dioxide from the burning of hydrocarbons could produce a greenhouse effect that would alter the global climate. This idea was revolutionary at a time when it was far from conventional wisdom that atmospheric compositions had anything to do with climate or weather. Of course, unlike current observers, Arrhenius believed warming would be a beneficial development:

> We often hear lamentations that the coal stored up in the earth is wasted by the present generation without any thought of the future. . . . We may find a kind of consolation in the consideration that here, as in every other case, there is good mixed with the evil. By the influence of the increasing percentage of carbonic acid [carbon dioxide] in the atmosphere, we may hope to enjoy ages with more equable and better climates, especially as regards the colder regions of the earth, ages when the earth will bring forth much more abundant crops than at present, for the benefit of rapidly propagating mankind.

Three years later, the American geologist Thomas Chrowder Chamberlin, a former president of the Geological Society of America, confirmed Arrhenius's basic calculations. There was almost no interest in either discipline for their work, however, so both soon moved on to other areas of study.[91]

Atmospheric study languished for decades as scientists investigated "larger" processes that could explain major climatic shifts, most importantly the Milankovitch cycles (variations in the eccentricity, axial tilt, and precession of Earth's orbit). There was also a belief that even if carbon dioxide could have a large impact on global climate, the oceans balanced emissions by absorbing them before they could cause warming. These beliefs began to shift in the 1940s, when new measurement technologies became available. Furthermore, a Canadian physicist, Gilbert Plass, made critical observations that proved that high atmospheric concentrations did trap solar radiation and create global warming. In the 1950s, an American oceanographer, Roger Revelle, proved that the oceans could never absorb the amounts of carbon dioxide being spewed into the atmosphere. Revelle, who was the director of the Scripps Institute of Oceanography, determined that even though oceans did absorb carbon dioxide, they eventually returned most of it back into the atmosphere. Later that decade, an American chemist named Charles David Keeling, who worked for Revelle, began his now famous measurements of concentrations of atmospheric carbon dioxide as part of the International Geophysical Year. Based at Mauna Loa, he collected data for nearly a half century and produced the Keeling Curve. This simple chart clearly showed that the share of carbon dioxide in the atmosphere had increased from 318 parts per million in 1958 to 380 parts per million in 2005, the year Keeling died.[92]

There are many theories about why it has taken so long to take action on climate change, when the essential facts were established by the 1960s. First, one early setback was the result of the period of global cooling from 1940 to 1970 (discussed below), which critics latched upon to question the global warming hypothesis. Second, the natural caution of the scientific community, combined with the public's general lack of scientific and mathematical knowledge, almost certainly delayed action. The early assessment reports from the United Nations Intergovernmental Panel on Climate Change did not proclaim unequivocally that climate change was caused by humans and would have devastating impacts. While the organization certainly argued that most data pointed in that direction, it also contended that it might take a decade or more to make those beliefs irrefutable. Although various national academies of science where far more assured, it wasn't until 2007 that the IPCC proclaimed that "most of the observed increase in global average temperatures since the mid-20th century is very likely [90 percent confidence] due to the observed increase in anthropogenic greenhouse gas concentrations." Even then the organization had less confidence that the warming had been responsible for more extreme weather events and water insecurity, and its predictions for sea level rises could be spun as relatively insignificant. While strong arguments can be made that even with these uncertainties the American government should have taken action, it is somewhat disingenuous to argue that the world's most important climate science organization has long been signaling that the debate was settled. Third, many environmental leaders have correctly blamed the failure of the media to play the role of referee in the global warming debates. Television networks, newspapers, and magazines largely failed to examine the weight of the arguments on each side, which allowed well-organized climate deniers to manufacture doubt. At the same time that academics were providing steadily more compelling evidence that climate change was caused by humans and would have devastating impacts, the media coverage suggested that legitimate questions persisted. Fourth, some contend that a global environmental movement was needed to make the problem more visible. It could just as easily be argued, however, that activist overreach and a public backlash were partially responsible for political inaction. Finally, a false dichotomy was created between economic growth and sustainability. Climate deniers, as well as a few respectable economists, suggested that any effort to fight the climate crisis would crash the economy. This argument was made despite mounting evidence that environmental regulations have little negative impact on growth and that government investments in key technologies can have a large positive impact.

Several of the above theories have merit, but they ignore two more important reasons why it took so many decades to take the climate crisis seriously. First, while it is clear that global warming will have significant worldwide consequences, Americans have an intuitive sense that their nation is in a far better position than most other countries to adapt to a changed environment. The United States is quite wealthy, with not only the world's largest economy but also a high per-capita GDP, which would provide it with the resources to deal with new conditions. This might include building sea walls around major cities or constructing flood-prevention systems. It is also an expansive nation, with a relatively small population given the amount of territory it encompasses. Although its population of more than 320 million, which is slightly more than 4 percent of the global population, makes it the world's third most populous nation, after China and India, it ranks 179th in population density, with only 83 people per square mile. This means that if climate change has particularly devastating impacts in certain regions of the nation, it would be possible to relocate the population accordingly. This would most likely be necessary as a result of drought and desertification, but the United States is in an enviable position with regard to water security. Although large sections of the American West are drying up because of less rainfall and the draining of underground aquifers, there are still huge water supplies available east of the Mississippi. The country also has northern rivers, like the Columbia or even the Yukon, which could be partially diverted to irrigate thirsty cities and parched farmland. Even more promising is the fact that Americans are extremely inefficient in their water use, which means there is a lot of room for improvement. Finally, the nation has more cultivated land than any other country. It currently has 1.7 million square kilometers growing annual and permanent crops, about 18 percent of the global total. America has more cultivated land than both China and India, which are trying to feed populations that are four to five times larger (currently China is accomplishing this goal). While America has more farmland than anyone else, there is still a significant portion of arable land that it doesn't even choose to farm. As cities have expanded, they have swallowed up perfectly good agricultural lands. In addition, agriculturalists have simply stopped farming in certain areas of the nation. In Vermont, for instance, 75 percent of the land was once cleared of trees and used for both farming and animal husbandry. In the past century, however, that figure has flipped; now 75 percent of the land is forested. Similarly, if the United States stopped growing corn for environmentally dubious ethanol production, it would have even more prime agricultural land available, much of it in the wetter eastern half of the country. Finally, the American diet relies heavily on animal proteins, but if it shifted toward a healthier, more sustainable omnivorous diet, even more cropland would be available. This is all by way of saying that even with fairly significant climate impacts, the United States might be able to successfully adapt to the new conditions, assuming that it would be willing to invest a large portion of its national wealth in the endeavor. Many think that while poor people in places like the Middle East and Africa might suffer, the United States can simply return to its traditional isolationism and weather the storm.

Second, I would argue that a false sense of energy security, combined with generational dynamics, made it a foregone conclusion that America would initially fail to take the necessary steps to address climate change. The result was to create a climate *crisis*. The Silent and Boomer Generations came of age believing that energy security was a longstanding historical fact. More importantly, they had never known a time when hydrocarbons had not endowed the nation with great wealth and a previously unimaginable degree of comfort. Consequently, these two generations were not predisposed to accept arguments that they should sacrifice their cozy lifestyles because of the far-off threat posed by global warming. But climate science didn't evolve in the best of worlds. Instead, it was advanced while the nation was struggling with burgeoning, if unacknowledged, energy insecurity and during the deepest conservative backlash in the nation's history. So despite the growing consensus within the scientific community regarding humanity's role in the calamity, it was easy for the hydrocarbon and financial-led plutocracy to manufacture uncertainty and thereby thoroughly confuse the American people. It was easy for conservative leaders to persuade voters in the electorally dominant Silent and Boomer demographics to ignore the problem, particularly because most of them wanted to anyway. It was easy for large numbers of Americans to focus on the mythical individualism they believed was their birthright rather than on inconvenient societal crises. The result was decades of political and technological inertia.

* * *

Although for some time it could be plausibly argued that climate science wasn't mature enough to mandate aggressive government actions to address global warming, that level of certainty was achieved long ago. Using a variety of techniques, from drilling ice cores to studying tree rings to examining corals, scientists have been able to assemble a comprehensive data set regarding the planet's climate history. To this they have added massive amounts of data gathered from Earth- and space-based instruments, some of the former dating back more than a century ago. They have confirmed that natural cycles do indeed change the climate. Volcanic eruptions can cause both warming and cooling. Milankovitch cycles do seem to have a limited impact on climate variations. The carbon cycle and changing ocean currents can also shift climatic conditions. While it has become evident that natural cycles do affect the climate, these fluctuations cannot explain current global warming. Instead it is predominantly the result of an increasing greenhouse effect caused by the emission of specific gases, mostly associated with burning hydrocarbons, into the atmosphere. When these gases are released, they trap solar radiation within the atmosphere, which leads to rising planetary temperatures.[93]

The rise in greenhouse gas emissions is intimately connected with the INNATE revolutions discussed at length in this book—the Industrial Revolution (IR), the Transportation Revolution (TR), the Electrification Revolution (ER), and the Agricultural Revolution (AR). Carbon dioxide is a product of burning hydrocarbons (IR/TR/ER/AR) and trees (AR) and of making cement for urbanization. Under normal conditions, it can remain in the atmosphere for up to two hundred years. Methane is an product of the mining and transport of hydrocarbons (IR/TR/ER/AR), as well as of raising livestock (AR) and decaying organic material in landfills. It is particularly troubling because it is twenty-one times more efficient than carbon dioxide at trapping heat. The only good news is that it only stays in the atmosphere for about a dozen years before breaking down. Nitrous oxide is emitted by manufacturing plants (IR) and farms (AR) and by hydrocarbon-fired power plants (IR/ER). It can remain in the atmosphere for more than a century. Carbon dioxide, methane, and nitrous oxide represent 97 percent of annual emissions. The other 3 percent are from fluorinated gases like hydrofluorocarbons, perfluorocarbons, and sulfur hexafluoride, all by-products of various industrial processes (IR). They are considered to be particularly potent and can remain in the atmosphere for very long periods (e.g., perfluorocarbons have an expected lifetime of more than fifty thousand years). Finally, it is important to remember that water vapor is by far the most abundant greenhouse gas; there is one hundred times as much water vapor in the atmosphere as there is carbon dioxide. While water vapor is not directly produced by human activities, additional planetary warming caused by humans can lead to increased evaporation and thus added water vapor.

Another way of looking at the relationship between the INNATE revolutions and climate change is shown in figures 4.10 and 4.11. Although figure 4.10 only shows a moment in time, looking at emissions in 2010, it provides a sense of the sources, sectors, and greenhouse gases that cause global warming. The left side of the chart reveals that about two-thirds of all emissions are directly related to the burning of hydrocarbons (IR/TR/ER/AR). The other third are direct emissions from land-use changes (AR), livestock (AR), energy-industry releases, and expenditures and losses from mining, refining, and processing hydrocarbons. We can see that globally the Industrial Revolution contributed the largest share of emissions, trailed by the Agricultural Revolution and the Transportation Revolution. It is slightly more difficult to determine an exact number for the Electrification Revolution, because it encompasses several sectors. Electricity is the dominant source of emissions from residential, commercial, and public buildings, but it also played a part in the Industrial and Transportation Revolutions. Electricity is probably at least on par with transportation, perhaps representing slightly more emissions. Regardless, what clearly emerges from this chart is that almost all global emissions are products of the INNATE revolutions, which is why understanding how these sectors developed is so important if we are to appreciate energy's role in creating modernity. The right side of figure 4.10 shows that social growth has resulted in the emission of greenhouse gases that now threaten humanity.

Figure 4.11 provides a picture of how American contributions to climate change differ from the global mean. What is most fascinating is that the Industrial and Agricultural Revolutions are so much smaller, while the Transportation and Electrification Revolutions are significantly larger. There are many reasons for these differing ratios. On the industrial front, America's manufacturing sector has relatively low energy intensity as measured by delivered energy per dollar of industrial-sector shipments. The United States also

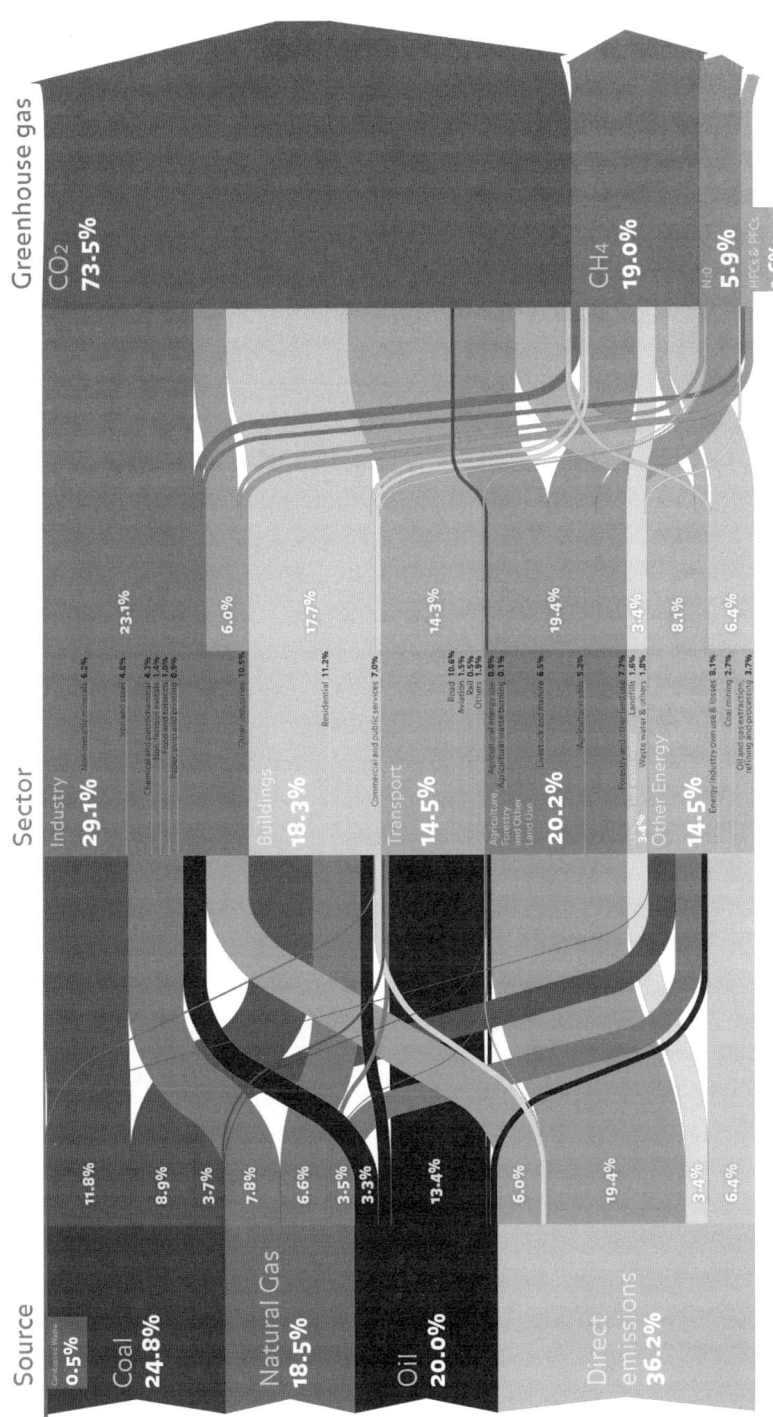

Figure 4.10. World Greenhouse Gas Emissions, 2012. Ecofys, "World GHG Emissions Flow Chart 2012," accessed 26 September 2017, tinyurl .com/y9nnjy7t.

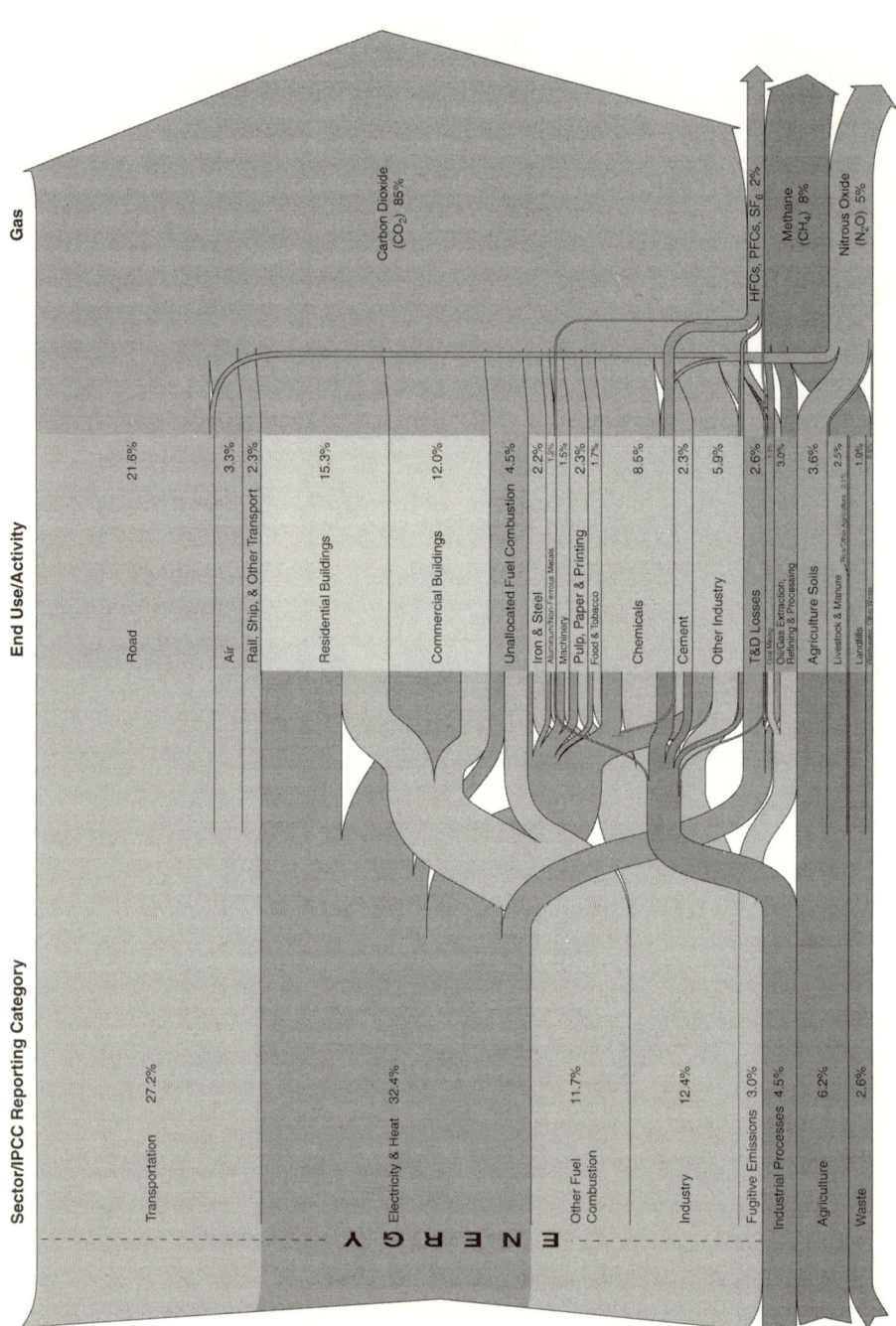

Figure 4.11. US Greenhouse Gas Emissions, 1990–2003. World Resources Institute, "U.S. Greenhouse Gas Emissions Flow Chart," accessed 24 March 2015, tinyurl.com/nrpr763.

has a much bigger service sector than most other countries, which helps explain why commercial buildings use so much more energy. With regard to the Agricultural Revolution, while America uses a significant amount of oil and petrochemicals, it doesn't have large amounts of emissions related to land-use changes, because trees were mostly cleared from agricultural areas long ago (chiefly before the beginning of the Second Agricultural Revolution). The Transportation Revolution has had a much bigger impact in the United States, because low population densities resulting from suburbanization have resulted in many more miles being driven in personal cars. This has been matched by a longstanding failure of most American cities (with some exceptions generally along the coasts) to invest in comprehensive public transit. The nation also largely turned its back on efficient passenger rail in favor of cars and airliners. Finally, the Electrification Revolution has had a significant impact partially as a result of the large service sector but far more because American homes are so much larger and energy intensive than those even in other advanced economies, a result, once again, of low population densities. While it is troubling that with only 4 percent of the world's population the nation contributes 15 percent of emissions, there is reason to hope because the manufacturing and agricultural sectors are relatively energy efficient (and becoming more so every year). There is no doubt that charting a new course in both the transportation and building sectors will require a major government effort, but the nation has a good sense of how to bring energy efficiency to both areas.

Climate Central, an independent organization made up of leading scientists and journalists, has written that "once we invented the steam engine, climate change was pretty much inevitable."[94] Once the industrial revolution began, the stage was set for additional revolutions in transportation, electrification, and agriculture. These INNATE revolutions helped push the world's population from seven hundred million to more than seven billion. As a result, so much carbon dioxide has been added to the atmosphere that the planet's natural processes can no longer remove it or absorb it. Oceans are already heavily acidic because they have soaked up so much carbon, which has led to ecosystem breakdowns. Deforestation is a double whammy, because it not only discharges massive amounts of carbon dioxide into the atmosphere but removes one of the primary sources of natural sequestration. And we are still adding huge amounts of the gas to the air every year, so the problem gets worse and worse. Carbon dioxide would normally remain in the atmosphere for only a couple hundred years, but there is evidence to suggest that because of the existing imbalance, it could take millennia for it to be removed—and that is assuming that we stop adding more. This means that the high existing concentrations have already baked a certain level of heating into the system. At this point it is just a question of whether we keep intensifying the problem. One of the main worries in this regard is whether we might trigger one of several potentially devastating positive feedback loops. For example, warming could destabilize ocean methane hydrates or high-latitude permafrost, adding nearly unimaginable amounts of methane into the atmosphere.

Even if we are able to avoid such catastrophes, the business-as-usual trend will lead to significant negative impacts. Now that concentrations of atmospheric carbon dioxide have passed 400 parts per million, levels that haven't been seen in the past 650,000 years, the repercussions of continued inaction will be cataclysmic. The most serious consequence of climate change for human populations will be water scarcity, so that it won't be possible to get freshwater to the cities or to agricultural lands that depend on irrigation. Reduced rainfall and disappearing glaciers are already creating water insecurity, and we can expect expanded drought conditions and desertification to make things worse in the future. Given that the globe is expected to have more than nine billion people to feed by midcentury, this is the gravest of concerns, particularly in vulnerable poor countries. Another major impact of climate change will be rising sea levels. Already there is a scientific consensus that we have locked ourselves into several feet of increases (seas have already risen eight inches since 1900), which will consume low-lying land and increase the potential for destructive storm-related flood tides. If heating continues unabated, we will see the world's foremost land-based glaciers melt away into the ocean. The loss of either the Greenland or the West Antarctic ice sheet would raise global sea levels by up to twenty feet, while the demise of East Antarctica would raise levels by 160 feet. Any of these outcomes would lead to the forced relocation of hundreds of millions of refugees, while also threatening hundreds of coastal cities, many of which are the engines of the global economy. Extreme weather events are already happening more frequently, and these severe storms (with massive storm surges, as seen with Katrina and Sandy) and heat waves will only get worse in the future. There is significant concern that climate change will result in negative health consequences, ranging from malnutrition to dangerous infectious-disease vectors. Finally, the climate crisis has already led to significant biodiversity loss and habitat destruction—which some have labeled the sixth mass extinction. Not only does this raise serious ethical questions about our responsibility to protect the larger biosphere, it calls into question whether we can survive without the species that are being destroyed. In the coming years we must answer one fundamental question: Do we wish to continue to grow and thrive as a species? If the answer is yes, we must find a way to stop emitting greenhouse gases into the atmosphere and to adapt to the global warming that is already destined to take place. It will require Americans to change the way their political systems work. It will require them to reassess how best to provide national security. It will require them to consider how best to ensure both macroeconomic and microeconomic security. And finally, if Americans hope to lead the charge to save our species, it will require them to undertake a sustainability revolution capable of providing the nation and the world with climate security in the Fourth Political Age.[95]

CHAPTER FIVE

Gas and National Renewal in the Fourth Political Age

For the past quarter century, scores of prominent commentators have made comparisons between the Roman and American Empires. Neoconservatives saw America's Cold War triumph as evidence that it had become a New Rome. They gloried in the idea that the United States had become the most powerful country in history and that the only limits to its reach were psychological. These voices dominated for many years, particularly as America's armed forces were deployed to impose its imperial will in the Middle East and Central Asia. These notions showed strain as the Iraqi occupation began to go sideways, but they fully crumpled during the Great Recession. A new wave of pundits proclaimed that instead of being the New Rome, the American Empire was beginning to unravel just as the Roman Empire had collapsed in the late fifth century. These critics point to various comparisons between the two imperial powers, suggesting that these shared weaknesses signal a common fate. America had become decadent and lazy. America lacked the will to meet its superpower obligations. America had become far too insular. American foreign policy was overextended. America had a fundamentally broken political system. While each of these assertions contains a sliver of truth, the idea that the loose similarities between the two empires have doomed the latter to permanent decline is dubious.[1]

Among the literally hundreds of causes that have been proffered for the fall of Rome, two are especially compelling. At a basic level, the entire Roman Empire was limited by the productivity of first-millennium agricultural techniques, which never performed much above subsistence levels. Peter Heather writes that although Roman agriculture generally flourished and many rural areas saw prosperity, "there is nothing in the archaeological or written evidence to gainsay the general picture of a late Roman countryside at or near maximum levels of population, production, and output." Because farming was central to the entire economy, any disruption to an already fragile system would be disastrous. Just such a crisis occurred in 375, when the Huns emerged from the Great Eurasian Steppe and began a chain reaction that resulted in a series of invasions of Roman territory. With their 130-centimeter bows and ability to fire accurately from horseback, the Huns swept westward pushing various barbarian tribes before them. According to Heather, "The Huns fell upon the Alans, the Alans upon the Goths and Taifali, the Goths and Taifali upon the Romans." An initial invasion in the Balkans was eventually halted when Theodosius was forced to make peace with the Goths, but a generation later the movement of the Huns onto the Great Hungarian Plain spurred three Germanic invasions of the Western Roman Empire and created a new emergency. In 405 the Goths under Radagaisus launched a failed incursion over the Alps into Italy. The following year, the Vandals, the Alans, and the Suevi swept across Gaul to Iberia. Two years after that, the Burgundians moved into the Rhine valley. These shocks in turn led to the Gothic sacking of Rome, a mutiny in Britain, and the rise of the usurper Constantine III in Gaul. A decade later, Flavius Constantius engineered a return to stability by making peace with the Goths and removing Constantine III, but Britain had essentially been lost and Iberia was in chaos.[2]

After Constantius's death, his successor, Flavius Aetius, built upon his successes. "All in all," writes Heather, "Aetius' achievement during the 430s was prodigious. Franks and Alammanni had been pushed back into their cantons beyond the Rhine, the Burgundians and Bagaudae had been thoroughly subdued, the Visigoths' pretensions had been reined in, and much of Spain returned to imperial control. Not for nothing did Constantinopolitan opinion consider Aetius the last true Roman of the west." In the meantime, however, the Vandals and the Alans had jumped across the Mediterranean and begun their conquest of Roman Africa. The loss of these rich agricultural lands led to a huge reduction in tax income and put more pressure on landed classes elsewhere to support the government. Before Aetius could return these rich provinces to Roman control, however, the empire faced a new threat. In 440, Attila the Hun marched across the Danube and laid waste to the Balkans, forcing Constantinople's capitulation. Ten years later he invaded Gaul, taking Trier and laying siege to Orleans, before Aetius took the field and defeated Attila's forces at the Battle of Catalunya Fields. The following year, Attila invaded Italy, but after initial success his disease-ridden and hungry troops were forced to retreat; their leader died shortly thereafter following a stretch of heavy drinking. Although Attila ultimately failed, beating him back was so costly that the Western Empire was unable to take back Roman Africa, return Britain to the fold, or prevent the Suevi takeover of Iberia. After the assassination of Aetius by Valentinian III, Roman control in the west faded away over the next two decades as many landowners began to shift their allegiance to new rulers who allowed them to maintain control of their holdings. With farmland that could produce only a limited surplus, the heavy taxes that had been required to hold the empire together in the face of recurring invasions had sapped their resources. Since the empire was no longer able to provide even basic security, they had little choice but to transfer their loyalty. As a result, by 476 the Western Empire had ceased to exist. Its economic system had been far too precarious to address the recurring crises of the fifth century, leading to its final demise.[3]

Does this sound at all like the situation that America faces in the twenty-first century? It is a nation that has the most productive agricultural system ever created. Rather than operating at subsistence levels, it has seen productivity more than double since the mid-twentieth century, with inputs ticking up only marginally. Whereas the vast majority of Romans were farmers, only 1 percent of Americans work in agriculture. They are able to generate such large food surpluses that the government has allowed urban sprawl to swallow up productive land and created subsidies to limit production or redirect crops into such boondoggles as ethanol production. And new technologies like genetically modified crops promise additional production gains in the future. So the differences between the Romans and the Americans in this key area are huge; there is limited risk that any but the most massive disruption would prevent US farmers from feeding the citizenry. While there is a nontrivial possibility that climate change could create such dislocation, it seems unlikely that the nation is on the verge of being invaded. Since independence, the mainland has been invaded only once, and that was more than two hundred years ago. The United States has certainly faced a limited number of domestic attacks, but although temporarily destabilizing, these never threatened the nation's very existence. Some have argued that decades of immigration across the southern border is the equivalent of the barbarian invasions sixteen centuries ago, but these hyperbolic and xenophobic beliefs run contrary to almost all the evidence regarding the positive impacts of continued settlement of new peoples in the United States. These realities on both the agricultural and national security fronts make it hard to see a parallel between the fall of Rome and the fate of America in the twenty-first century.[4]

Other commentators, such as Kevin Phillips and Niall Ferguson, have argued that the decline of modern imperial powers like the Spanish, the Dutch, and the British is a closer fit. Phillips has argued that both the Dutch and the British dominated world affairs because of their superiority in using first wind power and then coal-fired power and that their declines were linked to a loss of energy hegemony. As discussed earlier, Phillips believes that the United States rose to the top because of its supremacy in producing and utilizing petroleum. Although he is correct that America's ability to leverage oil played a large role in its rise to world leadership, his contention that this global position is threatened because of a growing scarcity of hydrocarbons simply doesn't jibe with reality. Hydrocarbons will certainly continue to grow more expensive in the future, and one hopes that we will choose to stop using them in an effort to maintain planetary habitability, but the role played by American companies in advancing the recovery of unconventional sources of petroleum and gas (discussed below) makes it seem unlikely that any other country is going to surpass America in the ability to provide energy security.[5]

Ferguson contends that the British Empire, the best precedent for understanding the American Empire, was dismantled because of the overwhelming costs of defeating more oppressive great powers in both world wars. The British Empire, he writes, "did the right thing, regardless of the cost. And that was why the ultimate, if reluctant, heir of Britain's global power was not one of the evil empires of the East, but Britain's most successful former colony." Although this is an interesting argument, Ferguson seems determined to overlook the most important weakness of the British Empire, namely, that it was a small island nation with relatively few natural resources. Despite having the ships and the financial system necessary to create the largest empire in history, it was never destined to maintain domination over far-flung colonies forever. Ferguson somewhat persuasively argues that the British weren't as cruel as many of the other colonial-era powers, but the reality of spreading modernity was that it was increasingly difficult for a small number of bureaucrats to oppress large native populations who wanted to share in the world's growing prosperity. America's position in the early twenty-first century is quite different from the British Empire's in many ways. To begin with, it is not a small island nation. By area, it is the world's fourth largest nation, and it has the third largest population. Although it only controls about 6 percent of the global landmass (the United Kingdom has only 0.04 percent), its great advantage is that this landmass is mostly contiguous and almost entirely populated with American citizens. Unlike the British at the height of their empire, the United States doesn't face a huge population actively seeking to boot it out of their territory. At the same time, the nation faces no threat from its northern or southern neighbor, and because of nineteenth-century territorial expansion it has large oceans protecting it on the east and west. The British also didn't have a protective shield of nuclear bombs. While America does have hundreds of overseas military facilities, with nearly a quarter million troops stationed abroad, these bases primarily project national power rather than seeking to enslave native populations. So once again, faced with this reality regarding the geographic extent of the "home countries," internal security, and national security, it is hard to see a clear parallel between the fall of the British Empire and the fate of America in the twenty-first century.[6]

* * *

Unlike the Romans in the fifth century, the United States will not fade away over the coming decades. In many important ways it is far better positioned than the British Empire was in the early twentieth century to maintain its role as the world's preeminent geopolitical and economic power. That said, however, there is no doubt that during the Third Political Age's conservative cycle the nation began a slow relative decline, the result of the increasing costs of energy insecurity, the renewal of former great powers, and the rise of new state actors on the international stage. There is some evidence that things are looking up for America as the Fourth Political Age begins, but the country almost certainly won't return to the outsized position it enjoyed during the previous political age. America never sought to be a global empire, and it will be a stronger nation if it slowly sheds some of the responsibilities of the global law-enforcement officer. It would be a mistake for the United States to return to the isolationism of its past, as the Trump administration is trying to do, but by discarding its mantle as an imperial power it may be able to return to its republican ideals in the coming decades. This would free up funding needed to provide both economic and climate security while actually increasing national security by allowing the government to focus more attention on geographic regions actually critical to America's future development.[7]

It was inevitable that America's relative power would ultimately decline. US supremacy was artificially high after the Second World War because the homeland had suffered very limited destruction and the industrial buildup needed to fight the war had created an imposing economy. The reemergence of both Western Europe and Japan, aided significantly by the Marshall Plan, began to cut into the United States' authoritative standing. The Vietnam War further eroded its influence. China's split with the Soviets and its eventual turn toward a socialist market economy positioned it for a return to great-power status. Meanwhile, America's policy innovation began to lag behind that of both established and rising powers. As Paul Kennedy writes, "The United States was not helped by certain other secular trends which were occurring in its economy: fiscal and taxation policies encouraged high consumption, but a low personal savings rate; investment in R&D, except for military purposes, was slowly sinking compared with other countries; and defense expenditures themselves, as a proportion of national product, were larger than anywhere else in the western bloc of nations. In addition, an increasing proportion of the American population was moving from industry to services, that is, into low-productivity fields." As detailed at length in the previous chapter, the loss of energy security and the inability to adjust to it contributed to this problem.[8]

In the past several decades, as American-sponsored globalization has taken hold, a truly world order was birthed. Perhaps the most important result is that the planet is more peaceful than ever before in recorded history. "It feels like a very dangerous world," writes Fareed Zakaria. "But it isn't. Your chances of dying as a consequence of organized violence of any kind are low and getting lower. The data reveal a broad trend away from wars among major countries, the kind of conflict that produces massive casualties." Many national leaders have determined that it makes more sense to grow their economies than to engage in armed conflicts. While larger economies have undeniably had a bad impact on the natural environment, they have also raised a couple billion people out of abject poverty. What will America's role be in this newly globalized world? Zakaria suggests that it may be entering an era of uni-multipolarity, with the United States playing a key leadership role but sharing hegemony with many other powers. He believes this could be the case because not only does America still have by far the largest national military but it is still ranked as the most economically competitive great power (trailing only the much smaller Switzerland and Singapore in the Global Competitiveness Index). Despite the obvious problems with educational equity, it also still has the world's best higher-education system. Perhaps its greatest strength is its continued demographic vibrancy. Almost every other current great power, with the exception of India, has an aging population, which puts great pressure on a government to provide broad-based security to its citizens. America has the great advantage that people still want to immigrate there. Even as the nation continues to engage in a damaging immigration debate, there is no denying that it is still among the world's leaders in assimilating new people. This is an area where its major Eurasian competitors significantly lag behind.[9]

Zakaria may be correct that we are entering a period of uni-multipolarity, but there is reason to hope that in the coming political age America will help create a truly multipolar world. Paul Starobin writes that to accomplish this goal, we will need to see a return of nationalism. But rather than a nationalism focused on destroying other states, we need a nationalism based on national pride. The rise of Donald Trump seems to have been propelled by a desire for this type of redirection, although, strangely, the candidate did this by bashing recent progress. America remains a great country. While its economy is imperfect, it is the wealthiest nation in history. While its government is imperfect, it has maintained stability within a democratic form for nearly two and a half centuries. While its global leadership is imperfect, it has led the way in making the world a safer place for all people. These are incredible accomplishments. There is no doubt that none of this would have occurred without America's foundational hydrocarbons and the INNATE revolutions. But these fossil fuels have come to endanger the global environment. Playing a leadership role in a successful transition to cleaner forms of energy that safeguard modernity, however, could help bring about a return of national pride. To accomplish this goal, America will likely have to first fix its broken political system. This will involve destroying the existing plutocracy and returning the nation to its democratic and republican roots. Political reform will provide a political system able to renew and expand upon the commitment to providing economic security. This must be accompanied by a significant geopolitical repositioning of the country. The United States must significantly scale back its international security role and demand more from other great powers. At the same time, all Americans must strive to become better global citizens, something that is mostly only true of elites today. As Zakaria writes, "At the end of the day, openness is America's greatest strength . . . historically, America has succeeded . . . because of the vigor of its society. It has thrived because it has kept itself open to the world—to goods and services, to ideas and inventions, and, above all, to people and cultures." Once America restores order by reforming its political system, reestablishing economic security, and reintroducing itself as an international leader, it will be ready to truly commence a sustainability revolution capable of ensuring climate security.[10]

Fourth Political Age
Progressive Cycle, 2009 to Present?

The Fourth Political Age began during the worst economic crisis since the Great Depression. This Great Recession was the result of both energy insecurity and the risky behavior of plutocratic financial elites. It coincided with two momentous demographic changes within the electorate: a growing number of nonwhite voters and an expanding generational divide. In Paul Taylor's view, "The fact that both are unfolding simultaneously . . . will put stress on our politics, families, pocketbooks, entitlement programs and social cohesion." The 2008 election saw the true arrival of minority voting blocs on the national stage. America is rapidly moving toward a majority nonwhite population. In 1960, whites made up 85 percent of the population. By 2010 they made up only 64 percent, and by 2060 their share will fall to 43 percent. African Americans will continue to make up 14 percent of the population, but Asian Americans' share will jump from 5 percent to 8 percent, and Hispanic Americans' share, which was 16 percent in 2010, will soar to 31 percent, while the share of other nonwhites of native or mixed-race background will double to 6 percent. Much of this growth is the result of higher fertility rates among nonwhite groups, but it is also the product of ongoing Hispanic and Asian immigration. Since the Democratic Party supported civil rights legislation in the 1960s, white voters, particularly white men, have steadily shifted their allegiance to an increasingly conservative Republican Party. At the presidential level, the shift in white voting patterns translated into an uphill battle for Democratic nominees in general elections. This is one reason why the party's only successful presidential candidates during the Third Political Age's conservative cycle were from the South. By the twenty-first century, however, things had begun to change. The growing proportion of African Americans, Hispanic Americans, and Asian Americans within the electorate meant that if they turned out to vote in large numbers, they could overwhelm white voters, who were divided between the two major political parties (although leaning heavily Republican). George W. Bush was able to win two presidential elections by capturing a relatively high percentage of Hispanic votes, but the overall picture shifted radically in the next two contests, in which the electorates were the most diverse in American history. In 2008, Barack Obama received 95 percent of African American votes, 67 percent of Hispanic American votes, 62 percent of Asian American votes, and 66 percent of the votes of other nonwhites—with high turnout. In 2012, President Obama built upon these advantages as the overall minority population increased from 26 percent to 28 percent of the overall electorate. In his reelection bid, he won 93 percent of African American votes, 71 percent of Hispanic American votes, 73 percent of Asian American votes, and 58 percent of the votes of other nonwhites. This support provided the political capital needed to pursue a progressive agenda.[11]

While there is no doubt that the growth in the nonwhite population is already having a critical impact on American politics, the emergence of Generation X (my own generation) and the Millennial Generation as major electoral actors could be more important. For the past two decades, Republican candidates have received a small majority of the votes in national elections from older members of Generation X (who turned 18 during the Reagan Revolution). Democratic candidates have received a somewhat larger majority of votes from younger members of that same generation (who gained voting eligibility during the Clinton presidency). In the early 1990s, many liberal activists in my generation suggested that the nineties would make the sixties look like the fifties. This was wishful thinking. It didn't come about because my age cohort was largely split on many of the controversial issues of the day. The older members of the generation were mostly economic libertarians, while the younger members were largely economic progressives—the entire generation is generally liberal on social issues. Over the years, however, the generation as a whole has shifted left on a wide range of issues, although this hasn't consistently translated into more support for Democrats. Regardless, Generation X'ers are now emerging as the progressive and pragmatic leaders that America's history of recurring political cycles would have predicted. Their time has come.[12]

During Generation X's formative years, the term most often used to describe its members was *slackers*. They were seen as lacking ambition, avoiding work, socially disaffected, politically apathetic, and highly cynical about everything. According to Jeff Gordiner, "Since Xers grew up in the leviathan shadow of the boomers, a sense of apartness played a role in forming our identity from the start." While Silents and Boomers viewed X'er slackerhood in a negative light, the generation itself didn't share that assessment. Richard Linklater, who directed the 1991 movie *Slacker*, defined a slacker as "someone who's being responsible to themselves. It's not avoiding responsibility; it's finding your own path through [society's] maze of programming and pressures." As X'ers followed this path, the generation changed in many ways. When they came to understand that they needed to make money to enjoy the leisure time that was so treasured, they rapidly entered the workforce. X'ers made this shift just as the Third Industrial Revolution was becoming fully ascendant. Gordiner writes that "if a scientist had wanted to introduce avarice into the X bloodstream, he couldn't have come up with a better mode of mass infection than the dot-com boom. . . . The dot-com boom reached out a gloved hand and said: *Luke, I am your father.* That's because the people building these new companies were *not* corporate tools—not yet, at least." This ideal was spoiled somewhat by the rush of business-school graduates to Silicon Valley, which eventually pushed the boom over the edge and created a bust. Still, X'ers had not only been exposed to a workplace ideal worth pursuing; they had also been imbued with a sense that they were a generation of change agents. "We were fluent in corporate jargon," argues Gordiner. "We had skill sets; we had 401(k)s; we had 529s. Nobody tarred and feathered us with the word slacker anymore. In fact, because we were Xers, because we were people who were said to have both a loafer's love of freedom and a lizard-eyed respect for commerce, it was believed that we were uniquely trained for twenty-first century combat."[13]

As X'ers have arrived on the political scene, they have begun to play the critical role of negotiating the future while dealing with the widening generational divide between aging Boomers and rising Millennials. Increasingly, they are the actors creating balance in the middle of the recent political storms, although this role has been significantly hampered by the nations fundamentally broken political system. One reason they can play this role is that they have received far less government support as rising adults compared with other recent generations, so their expectations for their elderhood are much more modest. This will make it easier for them to accept the sacrifices required to prepare for a needed renaissance in America. Neil Howe and Richard Strauss predict that X'ers "will favor investment over consumption, endowments over entitlements, and the needs of the very young over the needs of the very old. Whether by raising taxes, by freezing the money supply, by discouraging debt, or by shifting public budgets toward education, public works, and child welfare, elder 13ers [X'ers] will tilt the economy back toward the future." This generation is already seeking change in the world of government institutions and economic norms. X'ers have become increasingly distrustful of government, but rather than creating apathy, these misgivings have begun to manifest themselves in a desire to reform the system. Although older X'ers roughly split their presidential votes in the 2008 election and gave a slight edge to Mitt Romney in 2012, younger members of the generation provided approximately a 10 percent edge to Barack Obama in 2008 and increased their support in 2012. Given the liberal lean of younger X'ers, this bodes well for a future progressive national agenda.[14]

What is significantly better for potential progressive reforms is the political arrival of the Millennial Generation. This age cohort has voted for the Democratic Party in every national election since the first of its number became eligible to vote. In 2008 it made up fully 18 percent of the electorate and gave Barack Obama an enormous thirty-four-point edge (the largest generational margin since 1972). Although this figure dipped slightly in 2012, Millennials still favored the incumbent president by twenty-three points, as their share of the electorate ticked up by at least 1 percent. Given that they overtook the Boomers as the largest living generation in 2015, Millennials will be the foundation of the current Fourth Political Age's progressive cycle. Thus far, their political participation in presidential elections has been far more consistent than was true for previous generations at the same life stage. Unfortunately, they have not maintained that enthusiasm in midterm elections, which is undoubtedly the reason for major Republican Party wins in both 2010 and 2014. Like all generations, however, as they get older and enter the workforce and have children, they will begin to show up to vote in most elections. The only question then is how progressives can maximize their support for needed reforms in the coming years and decades.

Who are these Millennials? What do they believe? This generation was born between approximately 1981 and 2000, which means that they were raised during a period of tremendous economic growth. Although they came of age during a major economic crisis, their earlier experiences have led them to remain optimistic about the nation's future. According to the Pew Research Center, this optimism has shaped their generational personality, which it describes as "confident, self-expressive, liberal, upbeat and open to change." This is the most ethnically and racially diverse generation in American history. They are far more secular than Silents, Boomers, and even X'ers, which likely contributes significantly to their progressive policy preferences. If current trends continue for the next few years, they will likely become the best-educated generation the United States has ever seen. Their life goals differ from those of other recent generations, with being a good parent and having a successful marriage placed above economic goals like having a high-paying job or owning their own house. That said, they don't seem to be in a hurry to do either of those things, marrying and giving birth later than their own parents. Finally, Millennials are the most technologically connected generation the nation has ever seen. All of its members are digital natives, having grown up in a world in which computers, the Internet, and smart phones were the norm. Although their tech savvy might not mean they will have the necessary skills to excel in academia or the workforce, they have the foundational digital abilities to quickly learn how to advance in the new knowledge economy. Consequently, Millennials' technological exceptionalism will help define their generational personality.[15]

From a political perspective, what is most distinctive and promising about Millennials is how they view the government. They are much more supportive of an activist government, believing it should do more to solve societal problems. They have far less confidence in the market's ability to play such a role. Interestingly, the decline in their support for President Obama in 2012 was likely not the result of shifting beliefs but instead a reaction to the belief that he hadn't done enough to change the way the government operates and to gain approval for progressive legislation. Millennials are much more liberal on social policy, "where they stand out for their acceptance of homosexuality, interracial dating, and expanded opportunities for women and immigrants." They aren't as supportive of an assertive American national security policy, favoring withdrawal from both Iraq and Afghanistan at higher levels than older generations. They don't view the military as key to creating a safer world. Finally, nearly three-quarters of Millennials believe not only that climate change is happening but that government should take action to limit greenhouse gas emissions; the proportion fell slightly to about two-thirds for X'ers and Boomers and plummeted to only half for Silents. These figures take into account the fact that climate regulation will generate new costs. It is also interesting to note that even Millennials who self-identify as Republicans are far more likely than older conservatives to favor climate action. One of the goals for progressive leaders will be to translate these beliefs into more political engagement, particularly in local and midterm elections. If this can be accomplished over the next several decades, there is a better than average chance that progressives can turn the ship of state in a more positive direction, one that will guarantee America's long-term national strength.[16]

*　*　*

The Fourth Political Age began with a generational war to gain control of the White House. In 2008 there were three key battles in this conflict, which pitted different peer groups within the two major political parties against one another. The first battle was for the Republican nomination. Of the dozen major GOP candidates, only three seriously challenged to be the party's selection: Senator John McCain of Arizona, Governor Mike Huckabee of Arkansas, and Governor Mitt Romney of Massachusetts. Not surprisingly given the party's reliance on older voters, the campaign boiled down to whether a Silent (McCain) could hold off the challenges from a couple of Boomers (Huckabee and Romney). Things started off favorably for Huckabee when he won the Iowa caucuses; then McCain began to turn things around with wins in New Hampshire, South Carolina, and Florida. The previously moderate Arizonan, who had been written off by the national media, had swung far to the right to pick up these wins. McCain followed up his early triumphs by earning a big victory on Super Tuesday, winning 574 delegates to Romney's 231 and Huckabee's 218. This result precipitated Romney's exit from the race, leaving only Huckabee to mount a serious fight. But McCain maintained his momentum, winning most of the remaining contests and ultimately forcing Huckabee from the race.

Thus, the Republican Party had selected a seventy-two-year-old white man who had been required to run as a far-right conservative as its standard-bearer in what promised to be a change election.

The second battle of the election cycle was for the Democratic nomination, which turned out to be far more interesting and revealing in many ways. Although ten major Democrats vied to be the party's nominee, the race quickly boiled down to a contest between Senator Hillary Clinton of New York and Senator Barack Obama of Illinois. Once again this was a generational clash, with Clinton trying to hold on to the presidency for the Boomers and Obama attempting to bring the X'ers into office. Although much of the commentary on this duel focused on gender and race, it was far more about differing generational visions. For Clinton the domestic and foreign policy battles of the previous several decades loomed large, while Obama wanted to turn a new page in both areas. As had been the case with her husband, Senator Clinton was seen as a smart and savvy moderate who would be able to get things done. She was a foreign policy hawk who had supported the ill-fated invasion of Iraq. She was a hard-working campaigner but rarely wowed the voters. More than anything, however, she was seen as a known quantity that wouldn't offer voters much novelty—despite being the first woman to have a genuine shot at the presidency. Obama, on the other hand, was young and charismatic. George Packer writes that on "rare occasions . . . a leader can become the object of an intensely personal, almost spiritual desire for cleansing, community, renewal." In 2008 Obama was that candidate. In almost every major speech, he inspired voters to believe in "hope and change." Although he was a centrist, a more radically liberal domestic policy identity was thrust on him by critics and supporters alike. As Chuck Todd writes, Obama was "someone whose very skin color promised a freshness, a rejuvenation, a different kind of politics." Far more internationalist than Clinton, he had opposed the American intervention in Mesopotamia, which became perhaps the defining difference between the two candidates for primary voters. In sum, he was everything a candidate is expected to be in a major realignment election. He was an X'er who could bring younger people to the polls in support of progressive policies, while also representing the change that so many nonwhite voters sought. It was a generational battle for the ages, and it had a largely predictable outcome.[17]

When the primary season began, Clinton held nearly a twenty-point lead in national polls. But nomination fights are not contested nationally; they are a state-by-state slog in which candidates try to capture more delegates than their opponents. The Obama team, led by David Axelrod and David Plouffe, had fashioned a brilliant electoral strategy aimed at building upon their candidate's natural advantages with young and nonwhite voters to beat Clinton in the delegate race. A key focus was on caucus states, where field organizing and electioneering provided the Obama forces with an opportunity to get more of their supporters to the schools, churches, libraries, and houses where votes are cast. This was a main reason for Obama's stunning win in the Iowa caucus, where, because of overwhelming support from younger voters, he won by nine points over Senator John Edwards of North Carolina (Clinton finished third). This triumph focused significant public attention on his candidacy, which he capitalized on with a captivating victory speech in which he told his Iowa supporters: "On this January night, at this defining moment in history, you have done what the cynics said we couldn't do. . . . We are choosing hope over fear, we are choosing unity over division, and sending a powerful message that change is coming to America." Obama failed to clinch the nomination with a win in the New Hampshire primary, however, as Clinton stormed back to gain a narrow victory over her younger rival. Even so, in a dynamic that would prove to be Clinton's undoing, Obama tied Clinton in awarded delegates. Although Clinton won the popular vote in the Nevada caucus eleven days later, Obama won more delegates because of his superior organization in rural areas. A week later he pummeled Clinton in the South Carolina primary, capturing more than twice as many votes as she did. This all set the stage for Super Tuesday, when twenty-three states and territories would cast votes to determine the disposition of 1,681 delegates. Although Clinton won the big electoral prizes of California and Massachusetts, Obama won every caucus state and several primary contests and ended the night with a slight delegate advantage. This set the Illinois senator up to sweep the remaining eleven contests in February, which provided him with a small but meaningful lead in the delegate count, which he would never relinquish. Early March saw Clinton earn a win in Ohio and a split decision in Texas, but Obama still emerged from the month with more delegates after victories in smaller states.[18]

March also saw race raise its ugly head with allegations about controversial sermons delivered by Obama's former pastor, but the candidate turned the tables on his critics by delivering a speech on race that proved he was willing and able to effectively counterattack to regain the high ground. He told the American people:

> I chose to run for the presidency at this moment in history because I believe
> deeply that we cannot solve the challenges of our time unless we solve them

together—unless we perfect our union by understanding that we may have different stories, but we hold common hopes; that we may not look the same and we may not have come from the same place, but we all want to move in the same direction—towards a better future for our children and our grandchildren. This belief comes from my unyielding faith in the decency and generosity of the American people. But it also comes from my own American story. I am the son of a black man from Kenya and a white woman from Kansas. I was raised with the help of a white grandfather who survived a Depression to serve in Patton's Army during World War II and a white grandmother who worked on a bomber assembly line at Fort Leavenworth while he was overseas. I've gone to some of the best schools in America and lived in one of the world's poorest nations. I am married to a black American who carries within her the blood of slaves and slaveowners—an inheritance we pass on to our two precious daughters. I have brothers, sisters, nieces, nephews, uncles and cousins, of every race and every hue, scattered across three continents, and for as long as I live, I will never forget that in no other country on Earth is my story even possible.

In this speech, one of the finest in American history, Obama cemented his following among young and nonwhite voters and reminded many Boomers that the issues they had fought for in their own youth still called for national attention. By this time it was a near certainty that Obama would be the nominee because of his marginal but persistent delegate lead. Still, Clinton continued to fight, and for two months the candidates traded wins in ten primaries. Although the New York senator gained ground, it wasn't enough to retake the lead. With Obama ahead when the primary season ended in early June, Democratic superdelegates rushed to endorse him to ensure his selection at the Denver convention in August. Thus ended a party clash between X'ers and Boomers, with the younger generation coming out on top.[19]

While we will never know whether Hillary Clinton could have won in the general election against John McCain, Obama's nomination nearly ensured a Democratic win in this final generational battle of the 2008 electoral cycle. Specific policy issues were certainly important in this contest, but the central dynamic was change versus experience. From the outset it seemed clear that the American electorate favored the former. Barack Obama campaigned on progressive issues, including universal health care, full employment, environmental sustainability, and a foreign policy that would include ending the war in Iraq and closing the detention center at Guantanamo Bay. As Clinton had done, McCain focused less on the issues and instead tried to persuade the public that it was too risky to turn over the reigns of government to such an inexperienced candidate. Once again, this approach did not resonate with voters, particularly after McCain's selection of Alaska governor Sarah Palin as his running mate backfired as it became clear that she was completely unprepared to serve in such a high-level position. Although McCain, a former naval aviator and prisoner of war, would have preferred a campaign focused on national security policy, the electorate was far more concerned about economic policy as the financial system began to implode. The Arizona senator looked out of touch on this issue not only because of his own personal wealth but also because of his inconsistent and sometimes incoherent discussions of the key issues facing the nation. Obama, on the other hand, seemed to fully grasp the problems and the needed solutions, while at the same time displaying a calm that instilled confidence about his ability to address the crisis. Largely on the basis of his effective arguments regarding the restoration of both macroeconomic and microeconomic security, Obama won clear victories in all three presidential debates. By the time Election Day rolled around, an Obama victory was a certainty. He won nearly 53 percent of the popular vote and beat McCain in an Electoral College landslide, 365 to 173. Obama also had coattails, helping his party enlarge its majorities in both the House of Representative and the Senate. With a coalition dominated by young and nonwhite voters, President-elect Obama and the Democratic Party were set to commence the progressive era of a new political age. The question now was whether they could gain support for the policies needed to tackle several converging crises resulting from the damage energy insecurity had wrought in the past several decades.

* * *

Barack Obama entered the White House during the worst economic crisis since the Great Depression; at the same time, the country was still waging two overseas wars. The American people were scared and also angry about the weakening of the middle class in recent decades. This was partially the result of reduced per capita energy use (fig. 5.1). After the domestic oil supply peaked, it began to decline, and real wages stagnated for the bottom 90 percent of income earners, falling by $17,867 from 1979 to 2007.[20]

Combined with the extent to which America's political institutions had deteriorated since 1981, the challenges Obama faced were nearly as serious as those confronting Franklin Roosevelt seventy-six years earlier. In his inaugural address, the young president laid out his agenda for returning the country to the progressive path that had characterized the long American Rise.

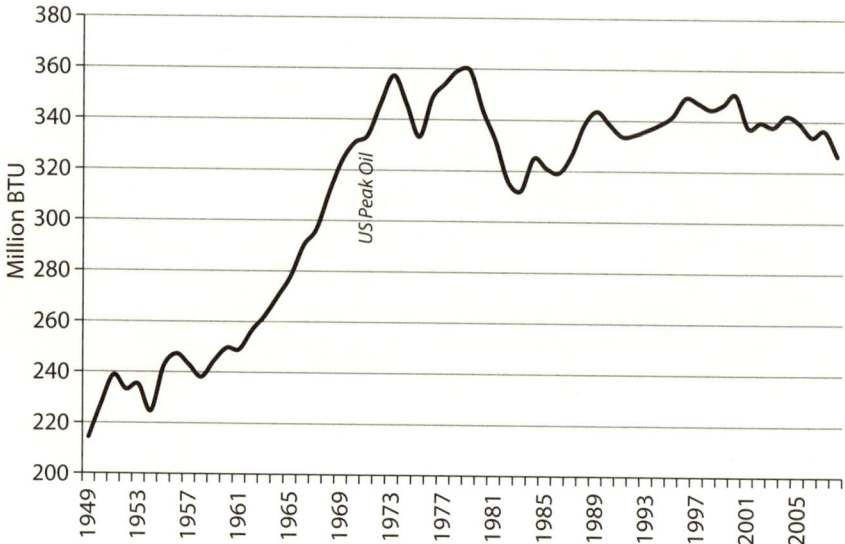

Figure 5.1. US Energy Use Per Capita, 1949–2008. US Energy Information Administration, *Annual Energy Review, 2009.*

This would include a renewal of national and economic security, as well as the adoption of climate security as a formal government responsibility. In the national security sphere, Obama argued that it was time to discard the notion that the country couldn't maintain its ideals while also remaining safe. "Recall that earlier generations . . . understood that our power alone cannot protect us," he said, "nor does it entitle us to do as we please. Instead they knew that our power grows through its prudent use; our security emanates from the justness of our cause, the force of our example, the tempering qualities of humility and restraint." With this statement he made clear that his administration would renounce the policy of preemptive attack and the use of torture, which had been championed by his predecessor. He promised an American withdrawal from Iraq and a just peace in Afghanistan. He vowed to fight to make the world safe from nuclear weapons, an issue he had championed in the Senate. Turning to domestic policy, he echoed Jimmy Carter's "Crisis of Confidence" speech, while laying out his key objectives:

> Our economy is badly weakened, a consequence of greed and irresponsibility
> on the part of some, but also our collective failure to make hard choices and
> prepare the nation for a new age. Homes have been lost, jobs shed, businesses
> shuttered. Our health care is too costly, our schools fail too many—and each day

brings further evidence that the ways we use energy strengthen our adversaries and threaten our planet.

These are the indicators of crisis, subject to data and statistics. Less measurable, but no less profound, is a sapping of confidence across our land; a nagging fear that America's decline is inevitable, that the next generation must lower its sights.

Today I say to you that the challenges we face are real. They are serious and they are many. They will not be met easily or in a short span of time. But know this America: They will be met.

In this key section of the speech he laid out his agenda for restoring economic security and introducing climate security. As President Polk had done so successfully in the nineteenth century, Obama laid out a four-point agenda for his presidency that included economic recovery, health care reform, education reform, and a response to global warming. Even with large majorities in Congress, this was an incredibly optimistic program. He concluded by asking the American people to support these momentous efforts: "America, in the face of our common dangers, in this winter of our hardship . . . let us brave once more the icy currents, and endure what storms may come. Let it be said by our children's children that when we were tested we refused to let this journey end, that we did not turn back nor did we falter; and with eyes fixed on the horizon and God's grace upon us, we carried forth that great gift of freedom and delivered it safely to future generations." In the coming years, as the Fourth Political Age began in earnest, Obama began the difficult task of matching the rhetoric with action. Although the road would be filled with obstacles and he would face constant attacks from his political enemies along the way, he accomplished all his major domestic and foreign policy goals, although perhaps not to the extent needed.[21]

At the beginning of the Fourth Political Age one of the key differences between older and younger generations was how they defined liberty. This debate had once again arisen within the electorate, with libertarian conservatives becoming defenders of negative liberty. This ideological positioning formed the backbone of the Tea Party movement, which is overwhelmingly composed of older, white Americans. President Obama became the proponent of a renewal of positive liberty, with government playing its critical role in ensuring that all citizens could discover happiness. Young and nonwhite voters supported him because they intuitively understood the need for an activist government not only capable of solving the nation's domestic problems but also dedicated to safeguarding its international competitiveness. Unfortunately, the institutional damage done during the previous conservative cycle made it far easier for his political opponents to weaken Obama's efforts to prevent an economic collapse and return the country to a more virtuous path.[22]

The new president's key task upon taking office was to stabilize the national economy and set the groundwork for an eventual recovery, which most experts at the time believed could take many years. While still campaigning for the presidency in October 2008, Obama had strongly advocated for President Bush's bailout package, which provided $700 billion to prop up endangered financial institutions. In his first weeks in office, his administration worked with Congress to gain passage for a Keynesian stimulus package. This legislation temporarily increased public spending to replace lagging private spending, while cutting taxes to put even more money into the economy. It was clear that Obama wasn't going to follow the Hoover model; he was going to act to save the economy. Although some economists were arguing at the time for a $1 trillion to $2 trillion multiyear stimulus, the pragmatic young president determined that a slightly more modest figure of $787 billion over two years was more politically feasible given the potential for a filibuster by Senate Republicans. Obama worked hard to make sure that relief programs would target needed national investments in infrastructure, renewable energy, health, and education. The stimulus also included an expansion of unemployment insurance, whose benefits were later extended to cover the jobless for a longer period of time. It also included payroll tax credits to provide a short-term jolt to the economy. There is near universal agreement among economists that the stimulus package had a huge positive economic impact, holding the unemployment rate to only 10 percent, far lower than what had been seen in the Great Depression or what was being seen in other nations during the Great Recession. The month after the stimulus passed, Obama made the controversial decision to bail out the American auto industry. The government provided loans to both General Motors and Chrysler, taking a majority ownership stake in the former, and forced both companies to reorganize. Later in the year he also supported the Cash for Clunkers program, which provided a boost to the domestic car market. The result was that this critical economic sector emerged from the crisis far stronger. There is little doubt that these collective macroeconomic security policies pulled the nation back from the brink and set it back on a positive course. As Paul Krugman, a frequent Obama critic, grudgingly wrote, "The bottom line on Obama's economic policy should be that what he did helped the economy, and that while enormous economic and human damage has taken place on his watch, the United States coped with the financial crisis better than most countries facing comparable crises have managed." By 2014, just five years after he took office, the contrast with the situations in economic powers ranging from Europe to Japan to China was evident. They were all still struggling, whereas America's unemployment rate had fallen to 5.5 percent and its annual growth rate was a respectable 2.4 percent.[23]

After beginning the long process of stabilizing the economy, Obama's next agenda item was to guarantee that this type of crisis did not occur again in the future. It is important to remember that the economic upheaval had been the result of the erosion over three decades of the macroeconomic policies that had been largely responsible for America's economic stability after World War II. Obama hoped to return the nation to that earlier path, although he faced major challenges from an entrenched financial elite. While many commentators were calling for the deep restructuring of the financial sector, Senator Chris Dodd and Representative Barney Frank elected to push a more modest set of reforms. President Obama ultimately decided to support this approach, even though it didn't restore Glass-Steagall's prohibition of merging investment and banking activities in a single company. Still, the legislation provided important new regulatory power to the national government that, although it might not prevent future crises, would at least endow officials with the tools to act quickly in an emergency.

By far the most controversial legislative effort made by President Obama and his congressional supporters was in the area of health care reform. This microeconomic security policy had essentially been on the national agenda since the New Deal, but a succession of presidents either had chosen not to push for universal (or near universal) coverage or had failed in their attempt to gain approval of such a program. After each abortive effort, the resulting backlash was so powerful that it relegated health reform to the fringes for a generation. Truman tried and failed. Nixon tried and failed. Clinton tried and failed so disastrously that it led to his party's losing Congress for the first time in decades. Many liberals questioned Obama's decision to take on the issue, worried that it would prevent action on other agenda items. But he believed providing health insurance was critically important for individual economic security. During the contentious year when it was drafted, he faced criticism from both the Left and the Right as he tried to move the debate forward. Rather than attempting to gain approval for a single-payer system or a public option, as was desired by his base, Obama once again fell back on the pragmatic progressivism that had come to define his political approach. Having learned from the mistakes of his Democratic predecessor, he rightly concluded that gaining passage would require a more moderate approach that would secure the support of conservative members of his own party. Although the centerpiece of the resulting bill had emerged the previous decade from the Heritage Foundation, Republican leaders determined to maintain party discipline and prevent any members of their congressional caucus from either participating in the legislative process or voting for the bill. Despite the continued power of the Senate filibuster, Democrats were able to maneuver the statute to final approval and signature by March 2010.

The objective of this revolutionary bill was to increase access to quality and affordable health insurance. A variety of mechanisms were employed. At its foundation the statute promulgated an individual mandate, which required all nonexempted Americans to purchase health insurance or pay a fine. This provision was intended to broaden the risk pool to include younger and healthier people, driving down overall costs. State and federal exchanges were created to provide a marketplace where insurers could compete for the newly insured; these companies were required to meet certain minimum care standards. Other measures prohibited insurance companies from denying coverage based on preexisting conditions, guaranteed coverage for children under their parents' insurance until they reached age twenty-six, and various Medicare payment reforms. Perhaps most contentiously, the act included both an expansion of Medicaid and subsidies for low-income Americans to pay for coverage. In the years after the passage of the legislation and as it was slowly implemented— sometimes not very successfully, as in the rollout of HealthCare.gov—the act came under continuous legislative and legal attack. Obamacare, the name given to it by conservative opponents and then embraced by supporters, emerged mostly unscathed when the Supreme Court upheld its constitutionality. Despite these assaults, Obamacare has been a clear policy success. One year after all its provisions had gone into effect, the law was succeeding according to several metrics: average premiums were lower than projected, and cost growth had moderated; approximately fifteen million Americans had gained coverage; the overall program costs were far below estimates; and the newly insured were happier with their coverage than those who had employer-provided insurance. As Steven Ratner writes, "The verdict is indisputable: Its disastrous 2013 rollout notwithstanding, the Affordable Care Act has achieved nearly all of its ambitious goals. Most important, just three key provisions—creation of exchanges with subsidies for those who qualify, expansion of Medicaid and minimum standards for insurance plans—have benefited at least 31 million Americans.

Millions more have taken advantage of other features, such as the inability of insurance companies to deny coverage based on pre-existing conditions and the ability to include children up to age 26 in a parent's plan." President Obama deserves a great deal of credit for not only gaining approval for this critical microeconomic security policy but also overseeing its implementation. It was proof positive that the nation had entered a new progressive political cycle based on support from young and nonwhite voters.[24]

Since the National Commission on Excellence in Education's 1983 release of *A Nation at Risk*, education had been a central feature of American politics. By the time Obama took office, however, there was an increasing belief within the county that President George W. Bush's No Child Left Behind reforms had largely failed. High-stakes testing had created an environment in which educators no longer focused on what had long made the US system great, namely, teaching skills, imparting knowledge, instilling social norms, and most important, fostering critical thinking. Instead, far too much emphasis was placed on successful test taking. The punitive nature of the "adequate yearly progress" provision took funding away from the schools that needed it most. Once in the White House, Obama moved to address these issues, because he understood the macroeconomic importance of solid educational performance for America's international competitiveness and future growth. Although his actions in this area were not as bold as in health care, his Race to the Top initiative has been generally well received. Working with Education Secretary Arne Duncan, Obama created a $4.35 billion grant program in which states competed for funds by putting forth innovative reforms. Perhaps the most favorable outcome of the policy was the adoption of the Common Core State Standards, which had been developed by the National Governors Association and the Council of Chief State School Officers to create consistent educational standards across the country, thus better preparing students for postsecondary success. While the effectiveness of these standards continues to be debated, early results in Kentucky, the first state to adopt them, have been considerably positive. It is clear, however, that not nearly enough has been done in this important sphere.

One major criticism of the Obama administration from both the Left and the Right is that it lacked a clearly articulated national security doctrine. However, Robert Worley writes that "commentators are perfectly justified in advancing their own preferred strategy. But those who claim there is no strategy aren't qualified participants in the debate." Obama's pragmatism led him to adopt a flexible approach that allowed him to make realistic decisions about when and where America should engage. Still, neoconservatives criticized Obama because of his unwillingness to maintain America's longstanding role as the global cop. He had indeed begun moving away from this role, however, because he understood that continuing to carry most of the international security burden seriously limited the nation's ability to make needed changes at home. Thus, he charted a future course according to which the globe's other great powers would be forced to become increasingly involved in providing international security. Progressives had seemingly become modern-day isolationists, criticizing Obama because he had not completely removed the nation from all foreign policy engagements. This would have been a disastrous direction for the country to take, especially in light of how economically interconnected the world has become, which was why he instead developed a strategy of selective engagement. Rather than becoming involved everywhere at once, Obama determined to employ American military power only where there was a vital national interest. He also sought to ensure that engagement took place only where the costs were limited and the probability of success was high. In sum, although ideologues on both sides of the political aisle might have disliked this approach, there is no doubt that a coherent Obama doctrine existed.[25]

Ironically given all the criticism he received, Obama was highly successful in the national security arena. As Bill Scher writes, "Under President Obama, the Democratic Party has successfully swiped the national security mantle from the Republicans, precisely because he has repeatedly met politically risky foreign policy challenges instead of ducking them." His reversal of America's hugely damaging torture policy shortly after taking office began the process of healing relationships with the international community. He signed the New START (Strategic Arms Reduction Treaty) agreement with the Russians, capping the number of strategic warheads each country could deploy at 1,550 (a two-thirds reduction from START I). He pulled troops out of the costly engagements in both Iraq and Afghanistan, but in 2011 he provided air support for rebels in Libya trying to overthrow Muammar Gaddafi. He engaged in a multinational program that used drones to find and kill leaders of major international terror organizations, which, while extraordinarily successful at achieving its primary objective, resulted in far too many civilian casualties. Obama coordinated the establishment of significant economic sanctions against Iran aimed at crippling its economy. This led Iran to the negotiating table and resulted in a deal to postpone its nuclear-weapons development in order to give moderate leaders time to solidify their political power base. Obama also began the process of normalizing relations with Cuba after decades of tensions between the two nations. Perhaps his only widely acclaimed act was ordering SEAL Team Six to undertake a highly risky mission aimed at killing Osama bin Laden in a compound inside Pakistan. Defense Secretary Robert Gates, who had previously served under three Republican presidents, later wrote: "I was very proud to work for a president who made one of the most courageous decisions I had ever witnessed in the White House." Although it has not received nearly the same level of media coverage, Obama also began a strategically critical pivot away from the Middle East toward East and South Asia. This move took into account not only the high cost of America's continued involvement in the Middle East but also the historic economic developments that are fundamentally reshaping East and South Asia. When he won the Nobel Peace Prize during his first year in office, many argued that President Obama hadn't done enough to earn the award, but the cumulative effect of his many policy initiatives intended to make the world more stable and prosperous built a case that the prize was deserved, although awarded somewhat prematurely.[26]

* * *

America's changing demographics were only part of the story of the new political age. The central thesis of this book is that energy security provided the United States with the resources necessary to become an economic and geopolitical power and that the relative loss of this energy security starting in 1970 resulted in a far more damaging conservative cycle in the second half of the Third Political Age. But just as the country was falling into the Great Recession, a technological revolution in hydrocarbon extraction was reshaping American energy security. Understanding how this came about requires some understanding of the "Resource Triangle," which suggests that at the top of the hydrocarbon pyramid sits sweet crude and methane-rich gas. But as society descends into broader portions of the structure, the quality of the supply decreases significantly; this is where one finds oil and gas trapped within shale formations. There is a lot more of it, but it is less pure and more expensive to extract. Two decades ago, many believed that it was pointless to even consider expending capital on trying to reach these unconventional reserves. But over just a few short years everything changed, as three major innovations totally altered the future of hydrocarbons.[27]

For at least two millennia, humans have used easily accessible gas. About 500 BCE, the Chinese gathered surface outflows, channeled them through bamboo tubes, and burned the gas to distill seawater into potable water. Greek cults and Zoroastrians used such seepages in religious ceremonies. But it wasn't until the eighteenth century that modern gas exploitation began. The British produced

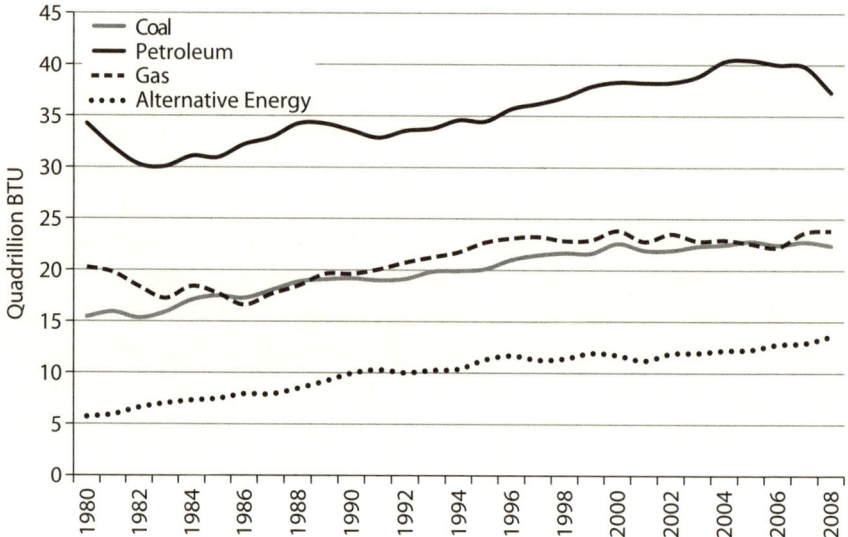

Figure 5.2. US Energy Consumption by Source, 1980–2008. US Energy Information Administration, *Annual Energy Review, 2009.*

coal gas (or town gas) and utilized it to illuminate homes and fire streetlamps. Naturally occurring gas was difficult to use, however, without extensive pipeline networks to transport it from wells to consumers. The process of building such an infrastructure began in America during the 1890s, when a 120-mile pipeline was constructed between Indiana gas fields and Chicago glass factories. The experiment only lasted eight years, however, having completely depleted the gas reserves. Three decades later, the United States began building the far more extensive pipeline network that would crisscross the entire nation; by the twenty-first century this web covered more than three hundred thousand miles, not counting local distribution systems. Gas had become a key component of the hydrocarbon mix, as can be seen in figure 5.2. Yet, as the turn of the millennium approached, proved North American gas reserves had begun to plummet, falling nearly 50 percent in the United States alone. In 2005, however, annual production began to skyrocket. The shale-gas boom had arrived.[28]

The story of this transformation began with the birth of the American oil industry in the 1860s. Edward A. L. Roberts, a largely overlooked figure, was the early pioneer. Roberts invented a tool called the petroleum torpedo, which was designed to fracture rock within an oil well to create artificial seams and gain access to previously inaccessible reserves. As one contemporary observer noted, "The explosion caused the oil and water to shoot out of the well some thirty feet into the air, and made the ground groan like a great monster in the agonies of death." The flow of petroleum would accelerate considerably after the detonation. After obtaining a patent, Roberts enjoyed a monopoly on this fracking technology that made him a very rich man. The technology faded after oil production in Pennsylvania peaked, but the idea resurfaced again in the 1930s. Dow Chemical, experimenting with ways to make oil wells more productive, investigated whether hydrochloric acid might be used to dissolve rock and create artificial fractures. The results were astounding, particularly in limestone formations, where extraction companies would see production numbers increase eightyfold. Unfortunately, the treatment did not work in the far more prevalent sandstone formations. The next breakthrough came a few years later, when Floyd Farris, a researcher at Stanolind Oil & Gas, came up with the idea of forcing pressurized water into wells to create artificial fractures. After the Second World War, he was joined in testing this notion by his colleague Bob Fast, who conducted a trial in the southwestern Kansas gas fields. The general idea seemed feasible, but they couldn't keep the fractures open. Moving their trial to East Texas, they tested whether including sand in the mixture would keep the channels open. It worked. "Oil companies adopted fracking rapidly, writes Russell Gold. "By 1955, less than a decade after the first experiments, more than one hundred thousand wells had been fracked. In the coming decades other geological engineers, most notably Al Yost at the Morgantown Energy Technology Center, would advance the various technologies needed for a full-blown revolution."[29]

At this point, George Mitchell, the Father of Fracking, entered the picture. Born to Greek immigrants in Galveston, Texas, he earned a geology degree from Texas A&M and then pursued a career as a wildcatter. He founded Mitchell Energy, which struck oil using hydraulic-fracturing techniques in Wise County above the Barnett Shale. In the early 1970s, after reading *The Limits to Growth*, a book about the effects of interactions between planetary and human systems, Mitchell became a committed proponent of sustainable growth. Certain that society needed to stop using dirty coal and oil but equally convinced that renewables were not technologically ready to replace them, he began advocating for gas as a bridge fuel to a new energy economy. And he believed his leases northwest of Fort Worth might prove valuable in achieving this outcome. Mitchell Energy geologists had "noticed that every time their wells passed through shale rock in search of conventional pockets of oil and gas, instruments registered a significant gas presence." It had long been believed that it was useless to try to extract this gas, because the rock where it settled was too impermeable. But Mitchell wanted to figure out how to do it. Starting in 1982, his company began fracking the Barnett Shale. For more than fifteen years, it experimented with many approaches, including explosive charges, hydrochloric acid, and a gel that was pumped under high pressure into the ground to create artificial fractures, with very little success, but on 11 June 1998 everything changed. One of Mitchell's field engineers, Nick Steinsberger, "suggested a new and revolutionary idea. He wanted to use water instead of gel. And not just a little water, but a massive amount of water—four or five times as much water by volume as the typical slug of gel." He added a chemical mixture, which made the water slippery and reduced friction (as well as killing any subsurface bacteria). His final brainstorm was to wait until fractures had been opened in the shale before he pumped in sand to keep them ajar. On that fateful day in the summer of 1998 his ideas worked brilliantly. That first well produced roughly 50 percent more gas in ninety days than the best conventional wells at the time. In the aftermath of Steinsberger's leap forward, Mitchell Energy adopted the technique in all of its existing wells, finding that it even worked in wells in which the far less effective gel had been tried.[30]

The breakthrough wasn't immediately made known within the oil and gas industry, but when George Mitchell decided to sell the company, it was revealed in order to demonstrate to potential investors that a gas company could still be very profitable. Devon Energy eventually stepped in to purchase the company. This proved to be important to the industry as a whole because Devon engineers added a critical innovation to the fracking process, horizontal drilling. At the time, this approach was being employed in fewer than 10 percent of all wells. Directional drilling had been around since the 1940s. The idea was that an initial vertical shaft would be drilled to a suitable depth, and then, using slightly bent pipes (approximately 2.5 degrees), one could slowly adjust the orientation to fully horizontal. Applied to gas fields being fracked, it allowed the well to access a much larger area of the gas-rich layer of shale, ultimately delivering seven times more gas to the surface than traditional vertical wells. Horizontal drilling thus fundamentally changed the economics of the fracking industry.[31]

Early in the first decade of the twenty-first century, then, things were looking up for the gas industry. One thing was missing, however. The sector didn't have nearly enough money to drill wells to extract gas from the many shale formations found in the Northeast, the Upper Midwest, the South Central and Mountain Western regions, as well as the Monterey Shale in California. Then the final actor in the story of the gas boom, Chesapeake Energy CEO Aubrey McClendon, appeared. The Oklahoma native had been in the oil and gas business for more than a decade when the revolution in fracking technology changed the dynamics of the industry. Unlike many of his counterparts, he had an astoundingly high risk tolerance. This served him well as the competition to obtain shale-drilling rights engulfed the sector in the first decade of this century. McClendon excelled in this battle, largely because he was willing to leverage his company far beyond what most executives believed was advisable. Although this risk-taking would ultimately come back to bite him and the company during the Great Recession, until that time it helped him transform the hydrocarbon landscape. Russell Gold writes that "Chesapeake didn't pioneer either of the two technologies that brought about the shale revolution. . . . What did Chesapeake do? McClendon figured out how to finance the revolution, tapping into Wall Street's deep reservoir of capital. He sold the revolution to the world's bankers." In the process, McClendon became an unlikely spokesperson for the environmental movement. He formed a short-lived partnership with the Sierra Club in arguing that society needed to stop using coal and oil and that gas could support the crucial bridge to an energy economy dominated by solar and wind power. Although he would die tragically in 2016, shortly after being indicted on antitrust charges, McClendon was a critical actor in the still unfolding story of the gas boom.[32]

The results have been astounding. Fracking and horizontal drilling removed fears of peaking hydrocarbon supplies, particularly after they became commonplace in the oil business as well. In 2005 the United States imported 60 percent of its petroleum. By 2014 this figure had fallen to 26 percent, and the Energy Information Administration forecasts that the nation could become a net exporter within a decade, assisted by only modest growth in American consumption. If anything, the prediction for gas is even more positive, with the United States expected to be a net exporter in just a few years and its position likely to improve even more in coming decades. As Tyler Priest writes, "Just when the decline of the American oil empire appeared irreversible, technological breakthroughs in extracting hydrocarbons from new geological frontiers . . . surprisingly regenerated North American oil and gas resources. Combined with steady gains in energy conservation, these resources possibly herald a new era of [energy] abundance for the United States." While environmentalists should be worried about this hydrocarbon renewal, there is no doubt that it has largely eliminated the form of energy insecurity that had dominated since the 1970s, although it remains unclear whether this fact has yet dawned on the American electorate or its leaders.[33]

* * *

In 2016 America experienced one of the strangest presidential election cycles in its history. Yet its outcome wasn't all that surprising given whom the two major political parties nominated. The race saw one major political party taken over by an unprincipled insurgent, while the other major party selected a nominee who was disliked by many within its own coalition. In the Republican contest, a reality television star with a checkered business record won the party's nomination, defeating what many had argued was the deepest field the conservatives had assembled in years. How did a man who had no political experience, who seemed incapable of telling the truth, accomplish this goal? In many respects, it was simply the end result of decades of cynical politics. Since the 1980s, the Republican Party had been arguing that government was broken and had done their best to actually break it. They had also suggested that the entire political class was corrupt. At the same time, the party had waged a war against facts, believing that experts were inherently dangerous. In areas ranging from economic policy to geopolitics to climate science, conservatives demonized the very notion that evidence should have any bearing on politics or policy. Then along came Donald Trump, the ultimate outsider. A man who had built his entire business career on lies, who had made bilking millions from common people an art form. A man with a talent for self-promotion who had no compunction about lying to voters. A man with a chip on his shoulder who was eager to embrace an alarming brand of both white nationalism and isolationism that proved popular with a large segment of the population. The mainstream field of established politicians never had a chance. Over many years they had cultivated a coalition of primary voters who could no longer distinguish right from wrong; as a result Trump became the party's nominee by taking a wrecking ball to his opponents.

For decades now, leaders from the Silent and Baby Boom generations have ruled the Democratic Party. Even at their best they governed within the constraints of the conservative political cycle that began during the 1980s. At their worst, leaders like Bill Clinton actively embraced conservative ideas in the areas of welfare reform, law enforcement, and financial deregulation. Even President Barack Obama, who succeeded in gaining passage of progressive health care legislation and took executive action to advance climate security, failed to take strong enough action to challenge a financial system that benefits a small share of American citizens. In 2016, the top two contenders for the party's nomination were from these generations. Senator Bernie Sanders was a member of the Silent Generation, and former secretary of state Hillary Clinton was a Baby Boomer. Although the race wound up being far closer than many had expected, Clinton always had the inside track to the nomination. Yet the reasons for her win are revealing. Her strongest asset was an impeccable resumé: she was arguably the best-prepared presidential candidate in the past century. There was also a great deal of genuine and warranted excitement about the possibility of electing the first woman president, although that excitement was felt mostly by older members of the party. As had been the case when she ran eight years earlier, however, Clinton also had glaring weaknesses. She was not a naturally gifted campaigner. Most Republicans and many Democrats disliked her (even if the reasons were manifestly unfair). And her politics were far too moderate for the moment. Millennials and many nonwhite voters viewed her policy prescriptions as insufficiently progressive. Instead, they were attracted to the ideas of a seventy-four-year-old white man from a small rural state. While almost any other member of the Silent Generation might have been seen as wholly out of touch with these voters, Bernie Sanders's brazen support for socialist policies endeared him to voters who believed this was the direction the party and the nation needed to move. Tens of thousands showed up at his rallies. They donated millions of dollars to his campaign. They protested at the Democratic National Convention, angered at a party establishment that had failed to understand that it would be difficult to win without a candidate who could energize them—even if the other party had a loathsome nominee.

The general election campaign can be summed up fairly easily. Hillary Clinton dominated in all the areas that traditionally would have determined the winner. She raised more money. She developed a better field organization. And she crushed Donald Trump in all three televised debates. Meanwhile, Trump relied almost solely on large rallies and free media to get his anti-immigrant, antitrade, and anti-Muslim message out to the American people. His racist, misogynistic, and xenophobic rhetoric was matched with bold lies about how he would seek to make the nation great again—despite the reality that the United States was already in an enviable position both economically and geopolitically. Rather than conducting an evidence-based campaign that would counter Trump, Clinton focused on how dangerous it would be to elect a man with such an unstable temperament. This probably would have been enough to win under ordinary circumstances, but when Russian hackers and FBI Director James Comey intervened on Trump's behalf, the approach ultimately failed.

Although Clinton won a popular majority of nearly three million votes, Trump was able to win narrow victories in three traditionally Democratic states— Pennsylvania, Michigan, and Wisconsin—to prevail in the Electoral College. White voters who felt abandoned in a globalized economy had flocked to the polls, particularly in rural counties, as they hadn't done in recent memory. The Silent and Boomer generations surged to the polls to vote for a candidate who they believed could protect them from a more diverse world, while enough Millennials and X'ers either stayed home or voted for a third-party candidate. And traditional conservatives and plutocrats held their collective noses and voted for a candidate whom they abhorred but who frightened them less than the alternative. Given the depth of the political dysfunction that developed during the final cycle of the Third Political Age, it is not entirely surprising that the establishment of a sustained progressive cycle in a new political age would face major challenges. Although the result is still quite alarming, it doesn't undermine the overarching thesis of this book. Quite the opposite. Instead it reinforces the notion that the peaking of domestic oil production in 1970 contributed to considerable political dysfunction, while also revealing major weaknesses in the US Constitution.

It is too early to tell whether the outcome of the 2016 election will be the full derailment of the progressive era that began in 2009. There is no doubt that President Trump has surrounded himself with advisers and appointees who want to take drastic executive actions to turn the clock back by eliminating regulations established by President Obama, particularly in areas such as trade, immigration, and climate. While the executive has a great deal of leeway to take these actions, he will likely face concerted legal challenges from a variety of quarters. This could be seen within days, when the new administration encountered stiff resistance in the courts after it attempted to implement an anti-Muslim travel ban. There is also no doubt that most members of the Republican-controlled House of Representatives and Senate would like to return the nation to an imagined past, but it isn't clear that they have the right plans or sufficient votes, particularly in a Senate where the filibuster can still be used to stop passage of new bills and the repeal of existing legislation. This was evidenced in the opening months of the new administration when attempts to repeal and replace Obamacare met with a series of humiliating defeats for President Trump, Speaker Paul Ryan, and Majority Leader Mitch McConnell. As I write this, a concerted resistance movement has begun to form. These efforts have resonated with the Millennials, the X'ers, and nonwhite demographic groups that will form a growing majority of the electorate in coming elections. While President Trump and the Republican Congress will likely try to pursue an agenda that could do great damage to the nation, this could also simply be a Jeffersonian interlude within a generally progressive cycle.

The remainder of this chapter focuses on key policy directions the government should take to enhance national and economic security, and the final chapter focuses on what should be done to ensure climate security. Unfortunately, however, none of the needed reforms will be carried forward unless progressives regain control of the White House and retake Congress. American voters face a real choice between parties. One party has been taken over by extremists who want to return the United States to a past that no longer exists. The other party wants to prepare the country for continued growth and leadership in this new century. The choice should be an easy one, but the forces of backwardness and inequality have the financial and quasi-constitutional backing that makes them a continued danger to democratic republican ideals. My great hope is that the electorate will soon deal the plutocrats such a dramatic series of defeats that the nation will return to a state of affairs in which it has two political parties both fundamentally interested in making the future better for *all* people. Americans are never stronger as a people than when they have two political parties articulating different policy mechanisms to accomplish a shared vision, with members able and willing to find the compromises that seek the best from both viewpoints. The nation now has the energy security to make this happen, but to realize this restoration of American politics the electorate must select leaders who can make needed critical changes to the structure of government.

Restoration of Democratic Republicanism
The Need for Constitutional Reform

America's federal government and most of its state and municipal governments have become exceedingly dysfunctional in recent decades, largely because of cyclical trends and the impacts of energy insecurity. But at a root level these developments simply revealed deep flaws in the US Constitution. This governing charter is a brilliant document in many ways, particularly with regard to the rights protected in the first ten amendments. Yet many of its principal articles are riddled with governing mechanisms that have long outlived their usefulness. As Sanford Levinson writes, "We must recognize that a substantial responsibility for the defects of our polity lies in the Constitution itself." In numerous books, he differentiates between the "Constitution of Conversation" and the "Constitution of Settlement." The former is characterized by the majestic generalities found in the Commerce Clause, the Necessary and Proper Clause, and the Bill of Rights. This is the territory where the federal judiciary lives, the arena where much of the contemporary constitutional debate lives. Levinson argues, however, that many modern problems are the result of foundational problems in the latter constitution, the portions of the document that receive very little scrutiny today. These permanent rules might have been necessary two centuries ago, particularly to stitch together a united nation from states with different interests, but they should have been abandoned long ago. They are among the primary reasons for the US government's inability to function properly and make the smart decisions necessary for the nation to stay ahead in an ever-changing world. In the Fourth Political Age, progressives should be far more focused on fashioning significant revisions to the Constitution of Settlement as well as making one tweak to the Constitution of Conversation. While this notion has barely registered in current policy debates, young and nonwhite voters must ensure that these reforms not only reach the national agenda but are acted upon quickly. The philosophical objectives of these political changes should be twofold. First, progressives must endeavor to institutionalize as much *democracy* as possible in the designation of elected representatives. Second, they must restore as much *republicanism* as possible to the functioning of government. Only these changes will create the type of government needed for the nation not only to remain competitive but to address the challenges posed by the climate crisis in the coming century and beyond.[34]

Given the mismatch between constitutional structures and contemporary needs, it is time to seek constructive change. Article 5 provides two methods for amending the Constitution. The first, which has been used twenty-seven times since 1789, requires adoption of an amendment by a two-thirds majority in both congressional chambers and ratification by three-fourths of the states. In the best of circumstances, this has proven to be a difficult process. Americans plainly don't enjoy these optimal conditions today. "If one must choose the worst single part of the Constitution," writes Levinson, "it is surely Article V, which has made our Constitution among the most difficult to amend of any in the world. . . . The near impossibility of amending the national Constitution not only prevents needed reforms; it also makes discussion seem futile and generates a complacent denial that there is anything to be concerned about." Still, there is another method for amending the Constitution, one that has never been used. It requires two-thirds of the states to call for a convention for the specific purpose of adopting new amendments to be ratified by three-fourths of the states. This method has the advantage of bypassing Congress, although it would still be extraordinarily hard to cobble together support for such a convention. But this doesn't mean that progressives shouldn't be working toward this goal. It would be an objective worthy not only of the movement but also of the newly empowered young and nonwhite voters who will come to dominate the electorate. Organizing around a call for a constitutional convention and pressing for amendments that would restore democracy and republicanism could help unite the vast majority of Americans who would like to return power to We the People. If this were to happen, it would help ensure that the national government was structured properly to create the prosperous common future for which most citizens yearn.[35]

* * *

In order to institutionalize as much democracy as possible in the election of representatives, a vital objective for the current political age must be an amendment that offers broad-based campaign reform—call it the Democracy Amendment. The United States can no longer accept the consequences of congressionally mandated regulations in this area being ruled unconstitutional, as happened in *Citizens United v. Federal Election Commission* (2010). It is time to incorporate these foundational democratic principles into the nation's governing document. In taking on this difficult task, progressives must embrace the best practices developed by other nations to democratize their election processes. The first priority should be a section that provides clarity regarding the authority to regulate campaign finance. This is not the time or place for half measures, given how fundamentally important it is to remove the influence of money from the political process. While proponents of reform ranging from Senator Tom Udall (D-NM) to retired Supreme Court justice John Paul Stevens have called for a relatively modest amendment that simply confers upon Congress the authority to regulate contributions and expenditures, the electorate need much stronger language that guarantees their constitutional role in selecting America's leaders. This is crucial if we are not only to restore balance between democracy and capitalism but also to incentivize the adoption of the progressive policies necessary to ensure comprehensive security for all Americans. A proposed amendment introduced by Representative Rick Nolan in April 2015 provides language that would be an appropriate starting point for a constitutional convention:

> SECTION 1. The rights protected by the Constitution of the United States are the rights of natural persons only. Artificial entities established by the laws of

any State, the United States, or any foreign state shall have no rights under this Constitution and are subject to regulation by the People, through Federal, State, or local law. The privileges of artificial entities shall be determined by the People, through Federal, State, or local law, and shall not be construed to be inherent or inalienable.

SECTION 2. Federal, State, and local government shall regulate, limit, or prohibit contributions and expenditures, including a candidate's own contributions and expenditures, to ensure that all citizens, regardless of their economic status, have access to the political process, and that no person gains, as a result of their money, substantially more access or ability to influence in any way the election of any candidate for public office or any ballot measure. Federal, State, and local government shall require that any permissible contributions and expenditures be publicly disclosed. The judiciary shall not construe the spending of money to influence elections to be speech under the First Amendment.[36]

This strongly worded amendment makes clear that We the People includes only natural persons and gives the government the authority to regulate artificial entities. It also gives the government the distinct right to set contribution and expenditure limits, removing any ambiguity that has previously existed about congressional powers. Adding these provisions to the US Constitution would be an appropriate democratic outcome.

America's naïve faith in the ability of politically motivated state legislatures to control the redistricting process is unusual in modern times. While this approach is still used by a few advanced nations, including France and Italy and South Korea, far more have tasked either existing electoral commissions or specially created nonpartisan redistricting commissions to carry out this critical duty; some state governments have already taken this action. The Bipartisan Policy Center's Commission on Political Reform, cochaired by former Republican leader Trent Lott and former Democratic leader Tom Daschle, has endorsed the creation of nonpartisan redistricting commissions by state governments. These commissions are quite clearly needed, but an approach that relies on states to take such an action is plainly inadequate. This is another area in which new constitutional direction is needed. Justice Stevens proposed specific language for such a provision, which would be an excellent third section for the Democracy Amendment:

SECTION 3. Districts represented by members of Congress, or by members of any state legislative body, shall be compact and composed of contiguous territory. The state shall have the burden of justifying any departures from this requirement by reference to neutral criteria such as natural, political, or historic boundaries or demographic changes. The interest in enhancing or preserving the political power of the party in control of the state government is not such a neutral criterion.[37]

This obligation would begin the process of returning the House of Representatives to the entire electorate, not solely to the politically active members of one political party. The likely outcome would be a diminished disparity between the national popular vote in these elections and the number of seats won by each political party. Adding this provision to the US Constitution would be an appropriate democratic outcome.

After more than two centuries, it is finally time to get rid of the antidemocratic Electoral College. Incredibly, there are still those who argue that this constitutional provision is fair, which is why it hasn't been abolished. They seem to ignore the truth, which is that political elites and small states gain significant influence under the Electoral College and thus have a vested interest in its maintenance. They argue that it is in the national interest to protect geographic diversity, even if that results in a violation of the judicially protected principle of one person, one vote. The Electoral College was never intended to be democratic. It was given preference at the Constitutional Convention almost solely because a popular vote was undesirable for southern slaveholders. Many more northerners were eligible to vote, particularly because African Americans were quite obviously not part of the electorate. The Electoral College was simply a means of placing more power in the slaveholding states by providing a mechanism for counting slaves without providing them any of the rights of citizens. Fifteen decades after the end of the Civil War and five decades after the end of Jim Crow, it is amazing that America maintains an institution born of its original sin. Beyond this moral objection, there are significant historical and contemporary problems with the system that combine to make the need for a change obvious. As discussed above, two of the last five elections ended with the selection of a president who had not won the popular vote. While these outcomes are bad, far more troubling is that the system leads presidential candidates to essentially ignore large parts of the country. Large states like California and Texas receive almost no attention, because everyone knows that the former is going to be won by a Democrat and the latter is going to by won by a Republican. Thus, the vast majority of campaign resources and events are concentrated in "battleground" states. Interestingly, of the thirteen smallest states, which are supposedly protected by the Electoral College, only New Hampshire is competitive; the others are ignored during the general election. Because so many states are "safe" for one party, turnout in those jurisdictions is depressed because potential voters know they won't have any impact on the outcome; this harms the democratic process for congressional, state, and local elections. Critics have long suggested that the Electoral College protects rural voters, but it is hard to understand why this would be considered an appropriate rationale in the twenty-first century for a country in which 80 percent of the population lives in urban areas. To protect rural voters, a system is being used that systematically disenfranchises the vast majority of the electorate. Additionally, there are almost no data suggesting that candidates would actually ignore rural areas; it certainly doesn't happen in statewide elections. Finally, abolishing the Electoral College has extraordinary support among the American people. For five decades, Gallup polls have found that strong majorities of voters would like to move to a direct popular vote to select the president.[38]

The time is overdue for eliminating a system that was only selected to protect the power of slaveholders and has long violated the cherished ideals of equality and democracy. As George C. Edwards III writes, "A central theme of American history is . . . the democratization of the Constitution. What began as a document characterized by numerous restrictions on direct voter participation has slowly become much more democratic."[39] As nonwhite populations, who have long been the targets of efforts to suppress their votes, continue to grow, it is past time to abolish one of the last vestiges of the Constitution's original undemocratic compromises. Since 1789 there have been more than seven hundred proposed amendments intended to modify or to do away with the Electoral College—more than for any other constitutional reform. These resolutions provide a wealth of potential language for such a constitutional improvement. The two primary goals for such a provision should be the termination of the Electoral College and the creation of a system for the direct election of both the president and the vice president by a majority of the people, with instant-runoff voting. Modeled partially on a resolution cosponsored by Representatives Jesse Jackson Jr., Lloyd Doggett, Barney Frank, and Fortney Pete Stark, the following would be an excellent fourth section of the Democracy Amendment:

> SECTION 4. The President and Vice President shall be elected jointly by the direct vote of the citizens of the United States. The ticket receiving a majority of the votes cast for President and Vice President shall be elected, with a ranked instant run-off system guaranteeing that result. Congress shall enforce and implement

this section by appropriate legislation, including, but not limited to, procedures for carrying out an instant run-off, rules for inclusion on the general election ballot, and a process for selecting replacements in the case of the death of the President-elect or Vice President-elect before that person is inaugurated.

This constitutional language would eliminate the aristocratic and plutocratic nature of the Electoral College. It would solidify the principle of one person, one vote. By requiring a majority of votes from the American electorate, it would guarantee that a newly elected president had the clear legitimacy to push her or his policy agenda. Including an instant-runoff system would not only assure a majority outcome but also potentially encourage additional parties to participate in presidential elections. These would all be positive democratic outcomes.

One final section is needed in the Democracy Amendment to help ensure fairness in the election of national representatives. While state governments certainly still play an important role in the federal system, their retention of a role in managing national elections is a relic of the past that should be discarded. This will require language that overrides that found in article 1, section 4, clause 1 and article 2, section 1, clause 4. At the same time, given historical and contemporary attempts to disenfranchise citizens, there is a need for a guaranteed right to vote. Finally, it is time to make Election Day a national holiday with strict limitations on the amount of time employees can be asked to work so that they have adequate time to get to the polls. Based partially on a resolution cosponsored by Representatives Mark Pocan and Keith Ellison and legislation sponsored by Senator Bernie Sanders, the following would be an appropriate fifth section of the Democracy Amendment:

> SECTION 5. The times, places, and manner of holding elections for President, Vice President, Senators, and Representatives shall be prescribed by the Congress. Federal Election Day, the Tuesday next after the first Monday in November in each even-numbered year, shall be a paid public holiday. Every citizen of the United States who is of legal voting age shall have the fundamental right to vote in any public election held in the jurisdiction in which the citizen resides. Congress shall enforce and implement this article by appropriate legislation.

This section could and should be used to adopt national standards for voting registration and compulsory voting. The former is needed to protect voting rights, the latter to increase the percentage of the electorate that actually participates in national elections. President Obama endorsed this idea, specifically citing the Australian model, which uses modest fines to boost turnout well above 90 percent. These would all be positive democratic outcomes.

* * *

To both renew republicanism and ensure that elected leaders have the ability to make the smart decisions necessary for continued American leadership, a vital objective for this political age must be an amendment that offers broad-based institutional reform—call it the Republican Amendment. Since at least the time of the founding, many prominent Americans have argued that the country should have adopted a parliamentary rather than a presidential government. In the former, the same political party controls the executive and legislative branches because members of Parliament select ministers from among their number. Elections have clear consequences because the party in power can take action to adopt and implement the initiatives they campaigned on, along with other agenda items that arise between elections. Another advantage of this approach is that all elected members of the government are selected at the same time. Finally, in a parliamentary system there is often an incentive to avoid extremism that might offend the electorate and lead to the downfall of the government. Quite obviously, most of the framers of the US Constitution wanted to steer clear of such a structure. They wanted to avoid British-style government. Yet others, most importantly James Madison and Alexander Hamilton, arrived at the Philadelphia Convention hoping to gain approval for an American parliamentary system. This was not fated to be, however, and since that time the United States has muddled through with a configuration with far too many veto points. Over time it has become ever more clear that this scheme was not adequate to meet the challenges of an increasingly complex world. Yet out of a misplaced sense of tradition, Americans have maintained these various checks and balances for far too long. One result has been that only with leaders of great capacity or during periods of significant crisis could the government make the smart choices necessary to push the nation forward. Another result has been that the concentration of power in a plutocracy has enabled them to stymie all action within the legislative branch, creating the vetocracy discussed earlier. This brought about the growing mismatch between public and private services that so worried John Kenneth Galbraith. It is likely too much to hope that the United States could transition from a presidential to a parliamentary system during this political age. But progressives should be able to take actions to begin moving the nation in this direction and thus make government more responsive and efficient. In taking on this battle, reformers must seek to return authority to elected representatives and empower them to react nimbly and enact big changes when necessary. It will be necessary to eliminate the institutions at the foundation of the American Empire, which has never suited the country or the people, and seek national renewal in a restored American Republic. In so doing, progressives can help guarantee the country's position on the world stage for the remainder of this century.

Removing or reducing the influence of various veto points institutional-ized in the Constitution of Settlement should be one of the most important goals of political reformers. This is not to suggest that all veto points should be done away with immediately, but it is to say that the separation of powers as institutionalized by the Constitution have become a hindrance to national competitiveness, prosperity, and happiness. An obvious place to start is the amendment process itself. Protecting the states from the actions of the federal government might have passed muster two centuries ago, but Americans can no longer afford to base their policy decisions on this aristocratic framework. As Patrick Henry warned regarding the difficulty of the amendment process during the Virginia ratifying convention, "To suppose that so large a number as three-fourths of the States will concur, is to suppose that they will possess genius, intelligence, and integrity, approaching to miraculous. It would indeed be miraculous that they should concur in the same amendments, or, even in such as would bear some likeness to one another. For four of the smallest States, that do not collectively contain one-tenth part of the population of the United States, may obstruct the most salutary and necessary amendments."[40] Today a single legislative chamber in the thirteen smallest states, which represent less than 5 *percent* of the national population, can prevent an amendment from being adopted or ratified. Is this really acceptable? I think not. To seek change, progressives need to look no further than the constitutions of the fifty state governments. Nearly every state constitution requires only a simple-majority vote in each legislative chamber to adopt a new amendment, followed by a simple-majority vote of the people for ratification. Americans should accept nothing less at the national level. Therefore, the following language, modeled on the original article 5, would be an appropriate first section of the Republican Amendment:

> SECTION 1. The Congress, whenever a majority of both chambers shall deem it necessary, shall propose amendments to this Constitution, or, on the application of the legislatures of a majority of the several states, shall call a convention for

proposing amendments, which, in either case, shall be valid, to all intents and purposes, as part of this Constitution, when ratified by a direct majority vote of the citizens of the United States.

A renewed republic requires this more democratic amendment process, which depends upon the collective wisdom of elected representatives but also guarantees that We the People have a voice in the process.[41]

The US Senate should be abolished. As the constitutional scholar Daniel Lazare writes, "The U.S. Senate is by now the most unrepresentative major legislature in the 'democratic world.'" As discussed in the previous chapter, this is the result of equal representation for each state regardless of population. Wyoming's 584,000 people have the same voting weight as California's 38.8 million, so that a voter in the former has sixty-five times the influence of a voter in the latter. A majority of Americans live in only nine states, yet they have fewer than 20 percent of the seats in a legislative body with enormous political power. As discussed, the Senate filibuster makes things even worse, providing members representing about 10 percent of the national population with the ability to obstruct the entire legislative process. In *Federalist 10*, James Madison fretted over the potential for the takeover of government by a majority faction. He should have spent more time thinking about the probability that the constitutional system would lead to de facto minority rule. Although many other advanced nations, such as Great Britain, Germany, and the Netherlands, have either eliminated or severely limited the power of their upper legislative chamber, the US Senate's authority has never been more overwhelming.[42]

One might ask what makes the Senate worth keeping. For small states, it is clear that they receive outsized influence in the governing of the nation. The plutocrats benefit from a system in which they only need to influence a small subset of senators to maintain their perquisites. But the common people, including, ironically, those in small states, suffer greatly from a governing situation that prevents Congress from working effectively to pass the smart policies supported by large majorities of the citizenry. And things will only get worse, because as big states continue to get bigger and small states remain stagnant, the representative inequalities will be even more magnified. Some argue that eliminating the unconstitutional filibuster is sufficient, but this doesn't do nearly enough to overcome the wholly undemocratic nature of the legislative branch's upper chamber. Although the people may not seem ready for such a radical departure from political traditions, the progressive movement should put this at the heart of its political agenda and seek to convince the electorate of the merit of the idea. It should promote the following language, which would be an appropriate second section of the Republican Amendment:

> SECTION 2. Article 1, Section 3 of the Constitution of the United States is hereby repealed. All powers previously reserved to the Senate shall be transferred to the House of Representatives. The House of Representatives may not by its rules require supermajority votes, except where provided for elsewhere in the Constitution of the United States.

In addition to making Congress more democratic, this amendment would have the equally important outcome of removing a damaging veto point from the constitutional system. To meet twenty-first-century challenges, Americans must have a government that can react much more quickly then was necessary in the eighteenth century. The world moves faster, yet the government has failed to keep pace with a more streamlined republican system. Abolishing the Senate would protect the traditional separation of powers between the three branches of government, while providing a unicameral legislature with the ability to more rapidly adopt needed policies to address national problems. The above language would also prevent a newly empowered House of Representatives from adopting procedural rules such as the filibuster. A renewed republic requires this more democratic and efficient approach.

A variety of other reforms aimed at promoting democracy and more effective government should be included in several additional sections of the Republican Amendment. One such change would involve increasing the number of representatives in the House. When the Constitution was first adopted, House members each represented 30,000 people; today that figure is closer to 725,000. If single-member districts are to be maintained, at the very least the total number of representatives should be raised so that they can be on familiar terms with their constituents and more closely adhere to their wishes. James Madison was so concerned about the potential for the development of overly large districts that he introduced an amendment that would have created a process for increasing the size of the House of Representatives. Unfortunately it was never ratified. In 1929, Congress passed legislation that permanently fixed the size of the chamber at 435 members and established a formula for automatic reapportionment after each decennial census. This decision was made despite the fact that the nation was experiencing explosive population growth. Like many other such structural decisions, the adoption of the apportionment plan was related to the continuation of southern apartheid. Most other advanced democracies have constituencies that are closer to 100,000 people, but they also have smaller populations overall. The congressional scholar Larry Sabato has suggested slightly more than doubling the number of House members, which would result in one seat for every 350,000 people. Madison favored something closer to 50,000, but he lived at a time when the country had a quite small population. It seems that the sweet spot might be closer to Sabato's number, but Madison's concerns should be taken into account. Thus, progressives should seek to approximately triple the size of the House by having one representative for every quarter million people. This is reasonable. It is a number that shrinks constituencies and brings representatives closer to the people, while also ensuring that the chamber is not too big to function effectively.[43] As the population continues to grow, the size of Congress would grow. Some fear that such a large body would be somewhat unwieldy. But the House no longer operates as some mythologized nineteenth-century deliberative body. Most of the real work is done in the dozens upon dozens of committees and subcommittees, which would not be significantly harmed by having more members. Therefore, the following language, modeled on Madison's original amendment, would be an appropriate third section of the Republican Amendment:

> SECTION 3. There shall be one representative for every 250,000 persons, until the number shall amount to 1,600, after which the proportion shall be so regulated by Congress that there shall be not fewer than 1,600 representatives, nor less than one representative for every 300,000 persons, until the number of representatives shall amount to 2,400, after which the proportion shall be so regulated by Congress that there shall not be fewer than 2,400 representatives, nor more than one representative for every 350,000 persons.

Another excellent change would be to extend House terms to four years. This would have multiple benefits. While many early Americans worried that even a two-year term was too long, over time it has become apparent that it is in fact too short. New representatives cannot learn the governing process adequately to play a meaningful role in such a short period. And because of the frequent elections, representatives spend far too much time campaigning. Moreover, four-year terms would mean that presidential and congressional elections could take place at the same time. This would result in far greater voter interest in the campaign, which would translate into greater turnout. (In recent presidential elections turnout has been approximately 20 percent higher than in midterm elections.) Quadrennial elections for all national officeholders would also mean that the same electorate would be responsible for the makeup of the executive and legislative branches.[44] The following language, modeled on article 1, section 2, clause 1, would be an appropriate fourth section of the Republican Amendment:

SECTION 4. The House of Representatives shall be composed of Members chosen every fourth year, in concert with the Presidential election, by the People of the several States.

In recent years, a major criticism of Congress, partially linked to the current two-year election cycle, has been how little time it actually spends in session. Perhaps just as important is how little time members spend in Washington during a session. It has been common for members to only show up a few days a week, spending the rest of the time in their districts raising money and conducting constituent services. This means they don't have a lot of time to govern. It also means that members don't really know one another, because there is little incentive to socialize. Before the arrival of the airliner, members knew one another outside work, which made it harder to openly attack one another personally. Since that time, things have changed for the worse, and Congress is a far nastier institution. The Bipartisan Policy Center's Commission on Political Reform has recommended that Congress schedule five-day workweeks in Washington, with three-week sessions followed by one-week recesses. The following language would be an appropriate fifth section of the Republican Amendment:

SECTION 5. Congress shall schedule three-week sessions composed of at least five-day workweeks, followed by a one-week recess, in at least ten months of each calendar year.

One casualty of the dysfunctionality of the federal government is the annual budget process, which has been so wrought with problems in recent years that Congress rarely gets around to actually passing a budget. The reasons for this are partially political and ideological, but it is at least equally the result of government's seeking to pass a new budget every year. A number of groups, including the Bipartisan Policy Center's Commission on Political Reform, have advocated for multiyear budgets. With a change to a four-year election cycle, it would make sense to have a four-year budget that can be amended as needed. The following language, partially drawn from the Commission on Political Reform, would be an appropriate sixth section of the Republican Amendment, which assumes the adoption of both sections 2 and 4:

SECTION 6. Congress shall adopt quadrennial budget resolutions and appropriations bills before the beginning of the fiscal year most directly after the commencement of its term, and shall have the authority to amend these resolutions and bills through the regular legislative process as deemed necessary.

A final reform that would dramatically improve the efficiency of the federal government is related to the approval of presidential nominations. It has become common practice in recent decades to delay action in this area through various procedural rules, most importantly, but not limited to, the Senate filibuster. This has been done almost entirely to pressure the president to nominate individuals who are more ideologically acceptable to a *minority* of senators. This undemocratic tactic has had the effect not only of eroding a critical executive power but of causing the executive and judicial branches to operate inefficiently because administrators and judges have not been approved. President Obama endorsed a reform that would require an up-or-down vote for all presidential nominations within ninety days. This would not remove congressional powers to provide advice and consent; it would simply guarantee that the process would move forward expeditiously. The following language, partially drawn from the nonpartisan organization No Labels, would be an appropriate final section of the Republican Amendment:

SECTION 7. Congress shall confirm or reject all presidential nominations within 90 days of the nomination being received. If a nominee is not confirmed or rejected within 90 days, the nominee shall be confirmed by default.

There is no doubt that the passage of both a Democracy Amendment and a Republican Amendment would be extremely difficult politically. Nevertheless, they should be at the center of the progressive movement's agenda. The electoral and governance processes are no longer working for the vast majority of the American people. Measures short of amending the Constitution cannot be relied upon and therefore are not acceptable. Given what we have learned about the cyclical nature of national political history, it is important to ask whether the electorate is willing to wait for another progressive cycle before these fundamental problems are fixed. If they don't act within the next couple of decades, reformers might not get another chance until the end of the twenty-first century. Because of the undemocratic character of article 5, there is probably no option but to seek state-by-state support for a constitutional convention to debate and adopt these amendments, followed by ratification elections in those same states. A campaign to gain approval for a constitutional convention in the requisite number of states could be a beneficial organizing framework for the entire progressive movement, first at the state level and then nationally. Americans can certainly move forward as a people without these structural changes, but it will be far harder to gain support for a sustainability revolution, which is so badly needed to provide climate security. This is a worthy goal for a new generation of American leaders. The time is now.

National Security Policy
Adoption of the American Geopolitical Cross

As we saw in chapter 4, the costs to America of domestic peak oil and the continued propping up of the hydrocarbon system have easily surpassed $500 billion annually since 1970. Not only has this been a drag on the national economy but it has placed significant fiscal pressure on the federal budget. Priest writes that "the costs of using the U.S. military to protect foreign oil fields, pipelines, and sea lanes, the price of which has never been registered at American gas pumps, increasingly endangers U.S. fiscal health." These costs paved the way for a troubling rise in political dysfunction and a collective failure to make the big decisions needed to thrive in the twenty-first century. Combined with an increasingly interventionist foreign policy aimed at maintaining the free flow of petroleum, the result was a markedly weakened international position. But given the contemporary reality of fading energy insecurity and the rapidly declining costs associated with renewable energy sources, one is forced to ask a very simple question: why is America still so heavily engaged in the Middle East and Central Asia? It is time for progressives to seriously consider, as the Obama administration began to do, whether the nation should almost entirely shift away from a military and diplomatic focus on these regions. Since the United States no longer needs the oil, and the benefits associated with maintaining the free flow of petroleum are far outweighed by the costs, it is past time to take a hard look in the mirror and determine what is best for America's future. "Decision-makers in Washington must face the awkward and enduring fact that the sum total of the United States' global interests and obligations is nowadays far larger than the country's power to defend them all simultaneously," writes Kennedy. It is time to adopt a smarter national security policy, one that reduces American commitments in regions that have limited geopolitical importance, focuses more attention on regions that will be critical in this century, and reinvests freed-up resources to make the country stronger at home.[45]

The US economy is not heavily dependent on imports, which represent only about 16 percent of GDP, far below the percentage in most other G20 nations. Very little of this trade is with the Middle East and Central Asia. These regions account for barely 1 percent of US imports and less than 4 percent of exports (and this latter number could fall precipitously if regional military ties and annual arms deals become a thing of the past). By 2014 only 5 percent of the petroleum consumed domestically came from the Persian Gulf, and that number will almost surely fall in coming years. Thus, it is ever more clear that the nations in these regions cannot endanger American economic interests and pose no direct military threat. Despite these emerging realities, however, the budget for US Central Command, which oversees US operations in these regions, still accounts for a staggeringly large share of noncore military spending.

Neoconservatives might argue that the continued threat of state-sponsored terrorism provides a rationale for remaining engaged in these regions. But this argument misses the point. Middle Eastern or Central Asian nations only sponsor terrorism aimed at American interests because America has inserted itself into their politics. Given that the rationale for military engagement in these regions has disappeared, US withdrawal would ultimately eliminate the need to pursue jihad against the Great Satan. Thus terrorism, at least from that part of the world, would no longer need to be a major concern of the American government. The United States can no longer afford to fight a global war on terrorism when that entire conflict is largely predicated upon the continued involvement in a region where its economic and political interests are quickly waning.

Organizations like the American Israel Public Affairs Committee (AIPAC) would argue that continued U.S. engagement is needed to provide support for Israel, America's closest ally in the region. Given America's large Jewish population, which is nearly as large as Israel's and at least ten times as large as that in any other nation, this close relationship makes sense in many ways. It is also true that American Jews are an important demographic in national politics, which has influenced the national commitment to the Middle East and Central Asia for many decades. That said, an objective observer would be forced to conclude that without the larger goal of guaranteeing the flow of petroleum from the broader region, the practical need to maintain such close ties with Israel would be significantly diminished. It seems highly unlikely that America would completely abandon its longtime partner anytime soon, but there is ample evidence that it can no longer afford to allow a sentimental attachment to influence its overall national security posture. For too long the United States has spent too much to maintain stability in the Middle East and Central Asia in order to protect the flow of petroleum, but now those funds are needed to safeguard America's collective future in a postcarbon world that requires a focus on climate security.

*   *   *

The erosion of American interests in the Middle East and Central Asia is but one of the developments that will influence national security policy in coming decades. Another critical development is what Zakaria calls "the rise of the rest." Over the past four decades, just as the United States was suffering from its post-peak oil malaise, several developing nations experienced rapid economic growth, with more than two billion people shifting into the world of markets and trade. One result is that the United States now faces economic competition from a far larger group of nations, with rising powerhouses ranging from China to India to Brazil now gobbling up a larger share of global GDP. For a long time, this threat wasn't taken seriously. Policymakers saw this phase of globalization as entirely advantageous for Americans, who would be able to invest capital in these emerging markets and enjoy the flood of low-cost consumer goods from abroad while maintaining their leadership in high-end and cutting-edge sectors. As the potential danger has become more apparent, the United States has begun to slowly react to the increased international competition. But it has yet to make the big macroeconomic decisions necessary to ensure the continued centrality of the American economy in the coming century. As a result, the country's geopolitical stature has declined as the reputations of these new powers have generally improved, a trend that dramatically picked up speed after the most recent presidential election.[46]

This situation is particularly troubling as we are just entering a phase when geopolitics is becoming extremely important once again. As Peter Zeihan argues, the United States "turned geopolitics off" during the Cold War by creating a system of global commerce protected by American economic and military power. As a consequence, economics eclipsed traditional geopolitics, giving rise to a far safer global order despite the threat posed by the Soviet Union. However, America's rationale for maintaining this globalized system is diminishing. In the absence of a superpower that represents an immediate threat and without the need to continue to protect the global flow of petroleum, it is becoming ever more difficult to understand why the United States continues to shoulder nearly the entire burden of providing global security, particularly when one considers the demands on resources at home. The logical conclusion, therefore, is that as America positions itself for this new era, a serious rethinking of its national security posture is overdue. "Washington has to move out of the eighth century AD," writes Zakaria, "adjudicating claims between Sunnis and Shias in Baghdad, and move into the twenty-first century—to China, India, Brazil—where the future is being made." This could undermine the global order if other great powers don't renew their commitment to stability, but it is past time for the United States to force them to take on more responsibility instead of allowing their political elites to snipe from the sidelines. America would be unwise to completely return to the pre–Second World War policy of isolation from the rest of the world, but it should significantly shrink its international commitments and focus more attention on climate security. Because other global competitors are often just as dependent on hydrocarbons, there is an incentive to move quickly. As Timothy Mitchell writes, "The leading industrialized countries are also oil states. Without the energy they derive from oil their current forms of political and economic life would not exist. Their citizens have developed ways of eating, travelling, housing themselves and consuming other goods and services that require very large amounts of energy from oil and other fossil fuels. These ways of life are not sustainable." Thus, the United States must move expeditiously to reduce hydrocarbon use and adopt more viable future energy technologies.[47]

As the sole global superpower in an age of uni-multipolarity, America will continue to exert enormous influence. But rising powers such as China, India, and Brazil will become more prominent players on the world stage, giving the United States an opportunity to reduce its international commitments while influencing new developments to preserve global security. Because of its limited and decreasing economic connections with the Middle East and Central Asia, they should be the primary focus of the American geopolitical contraction. Given the reliance of Europe, China, India, and Japan on Persian Gulf oil, it makes far more sense for them to play a role in securing the flow of petroleum through the Straits of Hormuz and beyond. The United States should move to disentangle itself from the region, forcing these other governments to take on more responsibility. China, which has quickly become the globe's second most important power and considers Asia to be its natural sphere of influence, would likely move to fill the void left by the United States. Because it lacks a serious blue-water navy, however, it is unclear whether China could fill the void on its own. This would provide an opening for America to continue to influence events from afar, primarily via its growing partnership with India, a country with which America has much in common, including democratic institutions, vibrant markets, a shared language, and the rule of law. An alliance between the two nations would help provide a counterbalance to the Chinese in the Persian Gulf and the Indian Ocean; if implemented intelligently, such an alliance could benefit everyone. While there is no doubt that the transition away from significant American engagement in the Middle East and Central Asia could be perilous, beginning to develop a system of shared hegemony would help ease the way toward a new global security regime, which would significantly benefit the United States.

As it shifts away from geopolitical commitments in the Middle East and Central Asia, the United States will need to renew its focus on the areas where it has deep economic interests. Picture a map of the world. America's geopolitical strategy should be concentrated on a large, somewhat oddly drawn cross whose vertical line stretches from Canada to Chile and whose horizontal line stretches from the Far East (plus India) to Western Europe, with the continental United States serving as the critical axis. The nations within these areas account for approximately 85 percent of American exports and 90 percent of its imports. Why would America spend a significant portion of its national treasury to provide security in other regions of the world? The only possible reason is misplaced hubris, and this is a poor rationale for continuing to play a central role in areas outside what I call the American Geopolitical Cross (AGC). This does not mean that the United States cannot join coalitions of global powers that are attempting to provide security in the Middle East, Central Asia, South Asia, Eastern Europe, and Africa. This also doesn't mean that it should not be providing significant, well-designed nonmilitary assistance to friendly countries in these regions. But it does mean that they should essentially disappear from America's core strategic planning.

Instead of focusing on regions where it has limited or no national interests, America should concentrate on solidifying relationships with countries in the AGC. The first pillar of this strategic approach must be strengthened relations with the country's two most important trade partners, Canada and Mexico. Mexico in particular ought to be seen by all Americans for what it is: a rapidly emerging nation enjoying vibrant demographic and economic growth whose friendship makes the United States stronger. Maintaining the bond that has been forged between the three nations, specifically via mechanisms like the North American Free Trade Agreement, will help ensure that the continent remains the world's fulcrum point both economically and geopolitically. The second pillar should be reinforcement of the union formed with Great Britain and Western Europe during the Cold War, when America helped guarantee regional security in the face of threats posed by the Soviet Union. The deep affection between Americans and Europeans has weakened somewhat since the fall of the Berlin Wall (while America's relationship with the British has strengthened), particularly as the United States engaged in a reckless foreign policy in the Middle East and Central Asia. With a general withdrawal from that region, however, the time would be ripe for renewing the linkages between the two societies. Given a shared history and traditions, and in some cases a common language, there is little reason to doubt that all involved have much to gain from a recommitment to one another.

The third pillar ought to be a general reconciliation with Latin America. There is a deep distrust toward the United States in this enormous region, for good reasons. For more than a century US leaders treated the countries to the south as children to be shepherded (sometimes through the barrel of a gun) rather than partners to be embraced. Consequently, US moral authority in the region has reached an all-time low, opening the door for China to gain influence and make commercial inroads. While this is perhaps a natural result of an increasingly globalized world, it does not justify America's failure to take the threat to its economic competitiveness more seriously. With a population of six hundred million and national economies with tremendous room for growth, Latin America represents one of the most promising markets for exports and imports in the coming century. Because of its proximity and historical ties to these nations, the United States must do a much better job of forging connections throughout the region.

The final pillar must be the continuance and enhancement of America's commitment to maintaining stability in the Far East (plus India). While this certainly includes maintaining strong connections with important regional allies like Japan and South Korea, more specifically it means paying vigorous attention to the ongoing relationship with the Middle Kingdom. There was a period before the 9/11 attacks when the Pentagon began to wrongly view China as a growing security threat. Although America received little payback from its nearly single-minded focus on terrorism, it did temporarily derail this dangerous development. Thus America was able to work with Chinese leaders to develop a mutually beneficial arrangement whereby armed conflict was avoided and they instead collaborated to maintain global stability. Given the deep economic ties already existing between the two countries, neither has much to gain from a military rivalry. That said, there is little doubt that the United States wants to maintain influence throughout the region, while China is attempting to extend its sphere of influence. America can certainly maintain its influence by projecting military power, but it would be better served by developing smarter strategies aimed at some form of shared hegemony in the region. If armed conflict can be avoided, there is good reason to believe that America's natural economic and geopolitical advantages will keep it competitive in the region. This might mean that China's global influence will increase, particularly in regions where the United States has few (or at least diminishing) national interests; however, there seems little reason to fear such a development. Although China has autocratic institutions that are rightfully viewed with some suspicion, there is no denying that it is increasingly following an economic path that tracks with America's larger world-view. That being the case, the United States should happily welcome the opportunity to shed some of its responsibilities for maintaining global stability to this near-peer power. This would free up funding needed to provide both economic and climate security, both of which will have spillover effects because America is always at its best when it is strong at home. The objective of national security policy moving forward, therefore, should be to implement these four pillars as efficiently as possible.

* * *

During his time in office, President Obama began to shift American national security policy in important ways. At a fundamental level, his administration argued that the United States could no longer afford to play the role of global cop by itself. His rationale was that the domestic economy was suffering because of the need to continually divert resources abroad, too often to fund interventions with questionable benefits. Obama contended that America needed to rebuild its strength at home, partially by identifying new fiscal resources in order to make strategic investments in a more prosperous and sustainable future. Some of these funds can be obtained by redefining America's national security posture with relation to the rest of the world, specifically by shifting the geopolitical focus to the AGC's four pillars. To do this, however, it will be important to adopt strategies that allow for projecting power within the AGC while also playing a role in coalitions intended to maintain stability in the regions where national commitments are fading.

An important first step will be a renewed dedication by the United States to soft power approaches that complement hard power. America has few outright enemies within the AGC, so it should be working to maintain good relations rather than playing the bully. Outside the AGC, soft power will be the primary means for influencing events. This will require a partial or even total dismantling of the current unified combatant command structure, whereby American military leaders to some extent displace diplomats in interactions with other nations. At the very least this would mean abolishing both CENTCOM (responsible for the Middle East and Central Asia) and AFRICOM (responsible for Africa). Part of this process should surely be a significant downsizing in the number of military bases operated in these theaters, if not their complete elimination. Conceivably, the United States could also build more fruitful relations with nations within the AGC by also abolishing the remaining combatant commands and/or reducing the number of bases being operated in other theaters.

Joseph Nye, who introduced the concept of soft power in the 1980s, suggests that it comes in three types—culture, political values, and policies. Although it can be argued that American soft power has declined recently, which is very likely true, predating the Trump administration Nye still ranked the United States third behind only Great Britain and Germany. China, by comparison, ranks dead last out of thirty nations. This ranking combines objective metrics and subjective international polling to compare "the relative strength of countries' soft power resources; assessing the quality of a country's political institutions, the extent of their cultural appeal, the strength of their diplomatic network, the global reputation of their higher education system, the attractiveness of their economic model, and a country's digital engagement with the world." America ranks first in higher education, culture, and digital engagement. While this is of course good news, it does not mean that there isn't room for improvement in each of these areas. The United States cannot rest on its laurels here; rather it must ensure that it continues to improve in each area to maintain its lead. America ranks highly, but its diplomatic network could certainly be strengthened. Focusing on the AGC and dismantling specific combatant commands could provide an opening to commit more funding for, and pay more attention to, the diplomatic efforts of the State Department.[48]

Although there is much that is positive in these the soft power rankings, in the areas where many think the United States is strongest danger lurks. What would surprise most Americans is how poorly their country ranks with regard to both its political institutions (24th) and its economic model (9th). A longstanding, nearly undiminished sense of exceptionalism has led many to conclude that the United States still leads the world in these areas, but the data and global perception tell another story, one that progressives must take seriously if they seek to build upon the country's existing soft power capacity rather than allow it to fall further behind. The world has not failed to notice that political dysfunction is increasingly gripping governmental institutions, making it ever more difficult to portray American political values in a positive light.

Fortunately, the answer to the problem isn't new fiscal resources but simply a willingness on the part of the United States to recognize where it is failing and try to find solutions. The American people have done this in the past, time and again, but today's generations have yet to follow the example of their forebears, particularly with regard to reforming the constitutional system. With regard to their economic model, Americans have strayed too far from the intelligently regulated market economy that flourished during the Great Compression. The result has been an erosion of the strong middle class that was at the heart of the nation's economic success story. Once again, the good news is that policymakers have already mastered most of the approaches needed to turn the economic ship. Over a century and a half, Americans built an economic system that was the envy of the world. But then they lost sight of what We the People were trying to accomplish, instead allowing the interests of the plutocrats to outweigh the common good. To solve this problem, progressives must recommit to the economic approach that served the nation so well for so long and tweak it in important ways to address the unique challenges of the twenty-first century. If the United States takes this path, its soft power capacity will help assure national security.[49]

In the meantime, the United States still has the largest military in the world. The question, then, is how to best employ hard power to achieve strategic objectives moving forward. Nye suggests that beyond securing the homeland, America's international role should be to provide or coordinate the provision of shared public goods. This means the promotion of regional balance (particularly in the AGC) and open markets, while ensuring equal access to global commons. The latter objective should increasingly be the focus of American hard power. "America has an interest in keeping international commons, such as oceans, open to all," writes Nye. "Today, however, the international commons include new issues such as global climate change, preservation of endangered species, and the uses of outer space, as well as the imperfect 'virtual commons' of cyberspace." Although America will certainly deploy soft power in some of these arenas, there is little doubt that to be effective it will have to also exercise its economic and military might to protect these global commons. In some arenas, such as climate security, this will require significant domestic commitments that thus far have not been forthcoming. But once this necessary precondition has been met, the United States will have the leverage to influence the actions of other major nations.[50]

Securing the homeland and playing a part in the provision of global public goods requires that America reconsider how it is currently employing its armed forces. As Nye writes, "A return to traditional prudence must be part of a twenty-first century smart power narrative. Global leadership does not require global interventionism. . . . A stomach for empire or colonial occupation is one of the important ways in which the American political culture differs from that of nineteenth-century Britain. . . . At the same time, universalistic values and a temptation to intervene on the side of 'the good' are also in the nature of the political culture. Prudence rests in understanding both international and domestic limits and adjusting objectives accordingly." While a new focus on the AGC will help the nation return to a policy of prudence, this will still require forces that can both project American power and protect critical global commons. Without doubt the US Navy will be the key military branch for accomplishing both of these objectives. In particular, its aircraft carrier fleet provides the country with an unparalleled capacity to pursue them. Because carriers can launch both aircraft and helicopters, a naval force need not depend on local bases to project air power. This means that America can influence events on land, as well as ensure the freedom of the seas that is so critical to the global trade network. As a nation, the United States has an overwhelming lead in this area; its ten *Nimitz*-class supercarriers by themselves have twice the capacity of the rest of the world's navies combined (and this doesn't include smaller *America*-class amphibious assault ships). Furthermore, if the government maintains support for a strong navy, it will build upon the existing dominance in this sphere as the next generation *Ford*-class supercarriers and additional *America*-class ships enter the fleet. Thus far, no other great power has shown an ability or willingness to match American strength in this arena. Although maintaining these forces is expensive, the benefits provided by the carrier fleets allow the United States to not only protect its own interests but also bolster its international reputation by providing vital global public goods. It is crucial that this strategic capability be maintained in the future.[51] What would returning America's national security structure to a focus on naval forces, the key branch of the armed forces until the mid-twentieth century, mean for land and air forces? The Pentagon needs to change the way it responds to threats. On the one hand, it needs to recognize that in the post-Vietnam era the nation has yet to fight a war against a peer competitor. The globalized economy and nuclear weapons have dramatically reduced the chances of such a great-power conflict, so when the American military goes to war it is against far inferior armed forces. Yet, the services have only slowly been evolving to account for this reality. Assuming that this paradigm doesn't change—and there is scant evidence that it will—it is time to transition land and air forces to far smaller units capable of moving much faster. This means the increased use of Special Forces troops and drones and far less dependence on massed infantry and armor as well as crewed aircraft. One likely result of this shift is a reduction in the overall size of the armed forces, with the subsequent budgetary reductions providing new resources for other national priorities. On the other hand, the United States also needs to take a leadership role in creating a global force capable of stepping into the void after successful combat operations to ensure an equally effective effort to win the peace. This global force, staffed and funded by the entire international community, would seek to avoid the mistakes made by the United States after combat operations were concluded in the Iraq War. The objectives would be to return stability to the affected region, rebuild damaged and critical infrastructure, and work to foster needed connections to the globalized world. Once again, this would allow America to maintain its role as a global leader, while reducing the financial commitments associated with exporting security to every corner of the globe.[52]

* * *

In the coming century and beyond, America will need resources to mitigate and adapt to the climate crisis. Changing its national security posture is one of the most obvious ways to find the needed funding. The nation can save tens of billions of dollars, if not more, by shifting away from its longstanding mission of protecting the free flow of petroleum. It can provide additional savings by concentrating its geopolitical strategy on the AGC, which would allow the country to focus on far more cost-effective soft power strategies to maintain and/or rebuild strong friendships and also develop new global security partnerships with emerging powers. Finally, radically altering its military structure to focus on naval forces, modernized land and air forces, and leadership of an international postconflict reconstruction arrangement would result in reductions in the number of personnel in the armed forces, with the associated fiscal savings. Although much of the money freed up by these various policies would help fund a needed sustainability revolution, it could also be used to build American strength at home by promoting enhanced economic security.

Macroeconomic Security
Restoring Economic Stability

Finding solutions to the climate crisis must occur in tandem with the equally important aim of promoting macroeconomic security. Meeting these objectives must start at home, but because of the outsized influence of the American economy, their effect will be felt globally. While a sustainability revolution, discussed in the next chapter, is a natural meeting point between climate security and macroeconomic security, additional policies in the latter sphere will be just as important, largely because they will create the strong economic foundation necessary for the nation to adequately adapt to changing conditions. As President Barack Obama has argued, "This one trend—climate change—affects all trends. This is an economic . . . imperative that we have to tackle now."[53] Fortuitously, America is well situated to make an economically beneficial transition to a low-emissions economy. As discussed earlier, for the first time in four decades it has the hydrocarbon resources, particularly unconventional oil and gas, to deliver energy security in the near term. This is critical to getting its political and economic house in order, as well as to beginning the process of reducing overall emissions, specifically by eliminating coal usage. Because certain climatic impacts are already baked into the system, the government needs to foster macroeconomic security to ensure that society is resilient enough to adapt to future shocks.

* * *

The appropriate starting point for any return to the macroeconomic security enjoyed in the post–Second World War era will be to reverse the financialization of the economy. President Obama made a good start with the Dodd-Frank financial regulations, which restored some of the rules that had been jettisoned in previous decades. In particular, the new Consumer Financial Protection Bureau has safeguarded prospective homebuyers in the mortgage market and taken on predatory lenders. Additionally, the Treasury Department now has the authority to systematically bail out important financial institutions without rewarding risk-taking bankers and investors. Although these early successes have been important, far more needs to be done. The coming decades will be perilous enough—with the impacts of climate change resulting in increased stress on the economic system—without a financial sector that rewards dangerous risk taking to continue with business as usual. Although the "too big to fail" provisions of Dodd-Frank are a satisfactory stopgap, financial security needs to be further augmented by legislating super-banks out of existence. Nations that prohibit the existence of overly large banks, such as Canada, have weathered financial crises far better than nations with huge financial institutions capable of bringing down an entire economy. Congress must admit that repealing Glass-Steagall was a colossal mistake that has to be corrected. Senators Elizabeth Warren (D-MA), John McCain (R-AZ), Maria Cantwell (D-WA), and Angus King (I-ME) have introduced legislation that would once again create a firewall between traditional banks, which have FDIC-insured savings and checking accounts, and financial institutions engaged in investment banking, insurance, swaps, hedge fund activities, and private-equity activities. These are the two most needed changes in the financial system, but other ideas ranging from limits on executive remuneration to fees on financial transactions are worthy of serious attention. It bears repeating that, as Sheila Bair, the former Republican-appointed chair of the FDIC, has argued about the Great Recession, "During the run-up to the crisis far too much money was directed towards booming, oversupplied property markets. The bust that followed is clear evidence that capital was misallocated and could have been put to more productive use in areas such as energy, infrastructure or the industrial base." In short, it is past time for Americans to discontinue the reckless financial behavior that took hold during the Third Political Age and return to the fundamental principles that made America an economic powerhouse. Not only would this restore financial stability to the economic system but it would give the nation the resiliency needed to face climate-related crises in the future.[54]

* * *

To achieve macroeconomic stability and resilience, beyond reducing the role of financial institutions within the national economy, the government should work toward balanced investments in the future. These investments would renew America's commitment to its own standard model for development (which included a stable banking sector), with funding directed toward growing human capital via education and immigration; repairing and expanding key national infrastructures; and research to foster leadership in the Third Industrial Revolution. Although all these investments would be required even in the absence of the looming climate crisis, the great benefit is that they would contribute significantly to the inauguration and continuation of a sustainability revolution. At the heart of the American ethos is a single passage from the Declaration of Independence: "We hold these truths to be self-evident, that all men are created equal, that they are endowed by their Creator with certain unalienable Rights, that among these are Life, Liberty, and the Pursuit of Happiness." For more than two centuries the nation has been perfecting its definition of this central tenet of the social contract. At least in principle, the clause now includes all citizens—men and women, white and nonwhite, young and old. In practice, however, the most important institutions for ensuring that all persons will have an opportunity to develop their gifts and chase their dreams are failing to accomplish those objectives. As a result, American economic mobility remains near the bottom of the Organization for Economic Co-operation and Development (a group of mostly wealthy nations). While at its root this is a function of the undemocratic nature of American political customs, the inability of the education system to provide equity of opportunity is nearly as disastrous. As discussed in the previous chapter, this is largely the result of the decentralized nature of educational funding, with wealthy suburban children getting a world-class education, while urban and rural poor children are left behind. By the time these different populations are preparing to leave the P–12 system, the gap is so wide that children from the former group gain ready access to the superlative higher-education system, while their less fortunate peers join the permanent underclass. Not only is this morally abhorrent, it is economically stupid: the nation is failing to fully develop all its human capital at a time when its global competitors are doing just that. This would be a perilous approach even at the best of times, but given the challenges that will be faced during the coming decades of climate insecurity, it could prove truly calamitous. Not only does the failure to develop its human capital make the nation less resilient but it means that Americans aren't cultivating the talent needed to solve their collective problems.[55]

While many educational reforms are needed to create a fully equitable system, three approaches should form the core of national interventions. Instead of focusing on the never-ending battles about curriculum, the federal government should focus on what it does well. To start with, it is very good at raising tax revenues and effectively distributing them where they are needed. This is essential, because the first thing the federal government should do to improve equity in public education is take over the funding of that system. Rather than getting bogged down in the debates about what should be taught in local schools, however, it should simply send that money back to school districts, while ensuring that a satisfactory level of funding is available for every pupil, regardless of the wealth of the state or the municipality in which he or she lives. This would nominally increase the size of the federal budget, by approximately $500 billion annually (approximately the amount currently spent at the state and local levels), but the bureaucracy required to transfer those revenues to schools need not be extensive. Local school boards would still be in charge of the curriculum, as well as of hiring and firing principals and teachers. But all schools, regardless of location, would have a base of approximately $10,000 a year to spend on each student. This would provide the resources necessary for universal preschool, paying teachers adequately, funding the arts and athletics, and properly equipping schools for the twenty-first century. This would also begin to right one of the great injustices of modern times, the dramatic disparity existing between what the nation spends on the youngest Americans and what it spends on the oldest. The United States spends nearly three times more on senior citizens than on children, which is terribly shortsighted for a nation that wants to remain internationally competitive. And this doesn't even include the massive costs those children will have to deal with when the bill comes due for the reckless greenhouse gas emissions of those seniors. Properly funding an equitable education system capable of addressing the problems caused by those earlier generations would help right this historic wrong.

A second policy intervention that is badly needed nationally is a widespread upgrade of school buildings. There are more than eighty thousand K–12 schools in the United States, approximately half of which were built at least a half century ago. The American Society of Civil Engineers gives this school infrastructure a D+ grade, arguing that there is a backlog of at least $270 billion for building maintenance and repair, which does not include the wholesale replacement of many outdated facilities. This is terrible news for the nation. Research suggests that there is a direct causal link between the quality of school buildings and academic outcomes. Modern buildings are tied to better student achievement. Features ranging from air quality and acoustics to temperature control and lighting affect not only scholastic success but also social development. Most of the worst buildings are in the poorest neighborhoods and towns, which lack the tax base and/or political will to ensure that students are studying in the palaces they deserve. Since the federal government excels at managing massive, strategic infrastructure projects, it would be appropriate to create a national program to either upgrade or build anew the country's schools. Not only would this stimulate the economy by generating tens of thousands of construction jobs but it would provide long-term benefits in the form of more accomplished and better-adjusted citizens. Additionally, it could include a national architecture and design competition, bringing both beauty and functionality to America's schools, which would go a long way toward showing parents and students in the poorest neighborhoods and towns that their leaders care about their future. And once again, it would not involve the federal government in local decision making about curriculum and personnel. Rather it would be a more effective redistribution policy that helps enforce educational equity.[56]

The single most important factor in student success is the quality of teachers. Study after study has shown this to be true. And the top systems of education in the world have taken direct steps to ensure that excellent teachers make it into the classroom. Finland, whose education system is frequently ranked highest in the world, requires that all teachers have a master's degree—which is paid for by the state upon acceptance into a highly selective program. Singapore, another top performer, focuses heavily on teacher training as well, developing a "comprehensive system for selecting, training, compensating and developing teachers and principals, thereby creating tremendous capacity at the point of education delivery." Prospective teachers are identified in secondary school, drawn from only the top third of academic achievers. All future teachers attend the same highly selective National Institute of Education at Nanyang Technological University. Many other top-performing nations follow a similar model, with the best students being actively recruited into the teaching profession and paid accordingly. In these countries, educators hold a status similar to that of other professionals like lawyers and doctors. This is far from the case in the United States. The most talented students choose career paths that are more lucrative or prestigious. America doesn't pay teachers enough to draw the finest and most suitable students, and educators are held in less esteem today than in the past. This must change. If it is to cultivate the citizenry that will be needed to remain globally competitive and to be truly resilient, the nation must once again revere its teachers. For this to happen, how teachers are educated must be completely reformed.[57]

Becoming a teacher must be made more difficult, requiring attendance at a highly selective, government-sponsored teaching institute, most logically housed at an existing land-grant university. The government can and should set stringent requirements for the curriculum at these colleges, which must be funded at the federal level (although operated by the states). Perhaps most importantly, students who are accepted to study at these institutes should attend free of charge as long as they agree to teach for a set number of years after graduation. This should be supplemented by the passage of a national minimum wage for P–12 educators, starting at approximately forty thousand dollars annually (for a nine-month contract), with raises to keep the profession competitive within the broader economy. This would help guarantee that the ablest college students would seriously consider a career in education. Above all, it would provide the nation with the teachers needed to have a genuinely world-class education system once again. It is unlikely that any macroeconomic policy would prove more beneficial as the United States faces the challenges of the twenty-first century.

Improvements to the education system should be supplemented with a far more enlightened approach to both legal and illegal immigration. The United States is in the rare position of still being a place where millions upon millions of people would like to make a life. And the nation has proven over many centuries to be exceptionally good at assimilating these newcomers and making them Americans. With developed economies around the world struggling with the challenge of an aging population, the United States is fortunate to have a relatively high fertility rate, as well as receiving more immigrants than any other nation (approximately one million annually). These advantages allow it to maintain a workforce that is large enough to pay for the social insurance programs that benefit retired Americans, while delivering the talent necessary to grow the national economy. But this will only continue to be the case if Americans reach consensus on the importance of continuing to be a place where the globe's strivers want to live. This translates, quite simply, into the need for significantly more enlightened immigration policies.[58]

In view of the fact that the flows of undocumented workers in the country have reversed in recent years, with a larger number of people leaving than coming in (largely because there are better economic opportunities in their home countries, particularly Mexico), this is the appropriate time to address the legal status of the workers who are here already. The nation can no longer allow the xenophobia of one political party to stop it from making decisions that are in the best economic interest of the nation. Thus, a pathway to citizenship must be created for the eleven million undocumented immigrants in the United States. These workers often fill jobs that Americans don't want to do, while at the same time providing space for job upgrades among those already in the workforce. Providing a pathway to citizenship would also maintain a pool of low-skill workers who are willing to relocate to areas of the country where many citizens are unwilling to go, at a time when geographic mobility is decreasing overall. Although it is critical to finally put the national angst about illegal immigration behind us, the United States must also streamline the system of legal immigration. Because of existing weaknesses in the education system, there are vital sectors of the economy in which America doesn't have adequate numbers of college graduates. While over the long term the nation must fix the fundamental problems in the education system, in the near term it should be actively recruiting highly skilled workers. As a starting point, it should effectively staple a green card to the diploma of any international student who graduates from an American college or university. Beyond that, the government should be more intentional in filling gaps in important economic sectors with legal immigrants. Combined with a more effective education system, this smarter approach to immigration would allow the United States to continue to benefit from demographic growth while ensuring that future generations are highly entrepreneurial. The country would thus be able to outcompete both advanced economies and rapidly developing nations in the decades to come.

During the most recent conservative cycle in American politics, the country failed to adequately maintain, repair, and revolutionize its national infrastructure. In addition to making the transformative internal improvements that have been essentially ignored for several decades, ranging from power grids to high-speed rail to renewable technologies, the government needs to create institutions that are better able to maintain and enhance basic services. If the United States hopes to keep its edge as a leading economy, it must have the best transportation, water, and sanitation systems in the world. The country has failed to do so not because it lacked the funds but because of a collapse of political will and an unwillingness to ask everyone to pay their fair share. Rather than continuing to take this critical infrastructure for granted, the government needs innovative interventions capable of putting it back on the political agenda, where it belongs. The obvious way to begin is simply to provide departments and agencies responsible for infrastructure development with adequate funding for maintenance and repair; however, a national infrastructure reinvestment bank is also needed to underwrite renewal of existing public works and development of new ventures. In 2007, Senators Christopher Dodd and Chuck Hagel first proposed legislation to create such an institution. The bank would use relatively modest federal funding to incentivize large local or private investments in prioritized projects. If implemented correctly, this new organization would include a planning council that for the first time in American history would adopt a coordinated national infrastructure plan— something that is done in many other countries. This is critical to the country's continued macroeconomic security and has the added benefit of providing necessary support for a sustainability revolution.[59]

Just as the Great Depression occurred during the transition between the First and Second Industrial Revolutions, the Great Recession took place in the shadow of the transition between the Second and Third Industrial Revolutions. Britain dominated the initial industrial revolution; America ruled its successor. Among the big questions facing the world today is who will lead the way in yet another iteration. Arguably no nation is better positioned to take on this role than America, but if it is to do so, it must revive its commitment to federal funding for cutting-edge research and development; in particular, it must fund not only the defense sector but also nondefense programs. Both defense and nondefense research spending have plummeted in recent decades, from more than 1.2 percent of GDP in 1976 to less than 0.8 percent in 2016, with defense spending continuing to outpace nondefense spending. In public- and private-sector research spending combined, the United States falls outside the top ten globally, with key competitors such as Germany and Japan spending billions more annually. Advanced information technologies, agile robotic manufacturing, three-dimensional printing, nanotechnology, biotechnology, clean energy, and groundbreaking new materials define the Third Industrial Revolution. In this new world, jobs will demand creativity rather than simple repetition. These innovations will lead to a redefinition of commerce, transforming the international economy. While manufacturing jobs have recently gravitated to nations that could provide cheap labor, this dynamic has already begun to change as companies move plants closer to their customers, where they can react quickly to new developments in the marketplace. Although financing the best education system in the world and streamlining regulatory processes are essential to American leadership in the Third Industrial Revolution, the government must also provide more consistent public research funding and incentives for private funding. This will be particularly critical in fostering developments in the needed sustainability revolution discussed in the next chapter.[60]

Microeconomic Security
Reducing Income and Wealth Inequality

I argued above that in order to fight the climate crisis, the United States needs macroeconomic policies that would make it resilient enough to adapt to future shocks. The same can also be said for microeconomic security. The coming decades are likely to present significant challenges to the nation both because of global warming and because of emerging economic realities. The nation enjoys many advantages in facing these tests, but comprehensive microeconomic security would build upon them in many ways. As a starting point, social insurance programs would provide a safety net for the unlucky victims of either climate change or the ongoing economic transition into the Third Industrial Revolution. Of equal importance, however, these programs would sanction the type of entrepreneurial and measured risk-taking necessary to find the public and private solutions to the crises the country faces. Without them society would swing too far toward a risk-averse public policy, which would lead to economic stagnation. To better provide microeconomic security, the government must do a much better job of balancing the investments it makes in support of the elderly and the young. This means that it needs to ensure that the high-quality universal health care that is provided to seniors is available to all Americans. While providing income support for the elderly is critical, we must never lose sight of how important it is to also offer support to the poor and the unemployed—not just because it helps them but because it makes the entire fabric of society stronger. It also means that government should provide universal childcare and early-childhood education, as well as nutrition and health programs for growing children. This type of collective risk sharing will make the nation more robust, which will be crucial as it faces up to global warming.

While many social welfare programs need to be improved and others need to be created, in the coming decades much of the focus within the sphere of microeconomic security is likely to be on income and wealth inequality. As we saw in the previous chapter, both income and wealth inequality began an inexorable rise at the outset of the most recent conservative cycle. This has been incredibly damaging to the nation, eroding economic growth and also dividing Americans as never before. With the probable exception of climate change itself, income and wealth inequality have become the biggest challenges facing the nation. There are, however, many policy interventions that can be used to reverse these damaging trends. All that is required is the political will to adopt them. In light of the progressive leanings of both Generation X and the Millennials in this area, there is reason to believe that the public resolve exists to support progressive political leaders willing to take on the existing plutocracy.

A smart package of policies aimed at reversing dangerous levels of income and wealth inequality would start with higher marginal income tax rates, asking the rich to pay more into the system that funds the economic security programs that are important for the whole nation. This should be considered a basic component of the social contract, not only because it is fair but also because it has proven to benefit everyone over the long haul. Higher marginal income tax rates can be supplemented with a nonprogressive value-added tax, or VAT, which is a tax on the amount of value added to a good or service during its life cycle. This is a more sophisticated consumption tax than a simple sales tax. Although it requires better accounting, it is far more transparent. While one might initially want such a tax to be progressive or at least to have exemptions for goods like food and clothes, it is politically smarter to charge everyone the same rate. If designed correctly, a VAT can be used to raise vast revenues to pay for the public services discussed earlier in this chapter. To take the politics out of this approach, it is best to have the tax apply equally to everyone, so the wealthy cannot claim to be paying more than their fair share. While this might somewhat harm the poor at the outset, as the interventions intended to level the playing field take hold, the disparities will largely disappear. The aim is not equality of outcome but equality of opportunity. Beyond these two foundational tax policies, additional interventions such as an increased federal minimum wage, augmented bargaining power for labor, and a streamlined (and loophole-free) corporate tax code are also needed. As President Obama argued, this is the "defining challenge of our time. . . . The basic bargain at the heart of our economy has frayed. . . . The combined trends of increased inequality and decreasing mobility pose a fundamental threat to the American dream, our way of life, and what we stand for around the globe. . . . It should compel us to action. We are a better country than this."[61]

As America engages in a coordinated sustainability revolution to mitigate the damage that has been caused during the Hydrocarbon Age, it must get its house in order. It must make democracy truly available to everyone. It must formulate smarter approaches in the national security sphere. It must seek stability and fairness in the economic arena. Only then will society be resilient enough to truly take on and overcome the challenges that it will face as the climate crisis worsens in the coming decades. The United States has the tools to make this happen. It is better situated than the rest of the world to address the problems we all face. It has much to gain by leading the global community as it moves forward. While the Fourth Political Age has experienced a rocky start, it is time to realize the promise of a new progressive cycle and make the changes that will restore America to greatness.

CHAPTER SIX

Climate Security and a Sustainability Revolution

On 2 April 2007 a policy window opened for aggressive action to solve the climate crisis. On that date, one of the most important in the history of American environmental policy, the US Supreme Court announced its opinion in *Massachusetts v. Environmental Protection Agency*. The case originated with the submission by twelve states, four local governments, and thirteen environmental groups of a petition for certiorari arguing that the EPA was failing to appropriately regulate emissions of four greenhouse gases, including carbon dioxide. The court was tasked with determining the meaning of section 202(a)(1) of the Clean Air Act, which states: "The [EPA] Administrator shall by regulation prescribe (and from time to time revise) in accordance with the provisions of this section, standards applicable to the emission of any air pollutant from any class or classes of new motor vehicles or new motor vehicle engines, which in his judgment cause, or contribute to, air pollution which may reasonably be anticipated to endanger public health or welfare." Carbon dioxide hadn't been included in the original act because at the time climate science was in its earliest stages; tracking atmospheric concentrations in Hawaii had only started a decade earlier. During the intervening thirty-five years, however, an overwhelming consensus had been reached within the scientific community that carbon dioxide was a key cause of global warming. Under President Bill Clinton, the EPA had determined that it had the authority under the Clean Air Act to regulate carbon dioxide emissions. Yet when President George W. Bush took office, the agency had reversed itself and refused to take action. The Bush administration contended that the EPA only had authority to control local emissions, which removed carbon dioxide from consideration because its *alleged* impacts were felt globally.[1]

Suit was initially filed at the US Court of Appeals for the District of Columbia, the second most important court in the nation. A three-judge panel found that the Bush EPA had been correct to sidestep rulemaking owing to the supposed uncertainty of the science. Based on this finding, the Supreme Court agreed to hear the case. Much of the case came down to standing, the legal capacity of a party to bring suit in court. A person cannot bring a suit challenging the constitutionality of a law without demonstrating imminent harm. In this case the question was whether Massachusetts had suffered some actual damage that would provide the federal courts with jurisdiction. The Bush EPA argued that since greenhouse gas emissions were purported to inflict widespread harm, standing presented an insurmountable jurisdictional hurdle. Justice John Paul Stevens, writing for a 5–4 majority, did not agree. Massachusetts met the three key requirements for standing: injury, causation, and remedy. With regard to injury, the state faced serious territorial damage to its coast from rising sea levels resulting from climate change. With regard to causation, the EPA's failure to take action contributed to the scope of this injury because of the cumulative impact of emissions over time. Finally, with regard to remedy, a "reduction in domestic emissions would slow the pace of global emissions increases." Thus, Massachusetts had standing, and the court had jurisdiction.[2]

With the jurisdiction hurdle cleared, Justice Stevens turned to the merits of the case. The majority found that Congress had unmistakably provided the EPA with the flexibility to recognize new pollutants as scientific understanding evolved. Furthermore, Stevens wrote, the agency could not avoid taking action if a pollutant "may reasonably be anticipated to endanger public health or welfare." Thus, the court found: "Because greenhouse gases fit well within the Clean Air Act's capacious definition of 'air pollutant,' we hold that EPA has the statutory authority to regulate the emission of such gases from new motor vehicles." Perhaps more importantly, the agency could only avoid taking action if it determined that carbon dioxide didn't contribute to climate change. While not requiring that the EPA make an endangerment finding, the finding left little wiggle room for the agency and the Bush administration to avoid doing so and at the same time meet its obligation to execute domestic laws. Thus, the majority reversed the lower-court decision and remanded the case for further proceedings consistent with the ruling. As the *New York Times* noted, the "decision was a strong rebuke to the Bush administration." Even so, the administration, under the leadership of EPA administrator Stephen Johnson, blocked an endangerment finding until Bush left office. It would be up to the newly elected president, Barack Obama, to carry out the Supreme Court mandate. In so doing, the new administration would have the great advantage of working with the sanction of the judicial branch, which would prove invaluable.[3]

Before tackling carbon emissions directly, President Obama was able to gain congressional approval for a substantial investment in clean energy technologies. Taking office at the height of the Great Recession, which had been prompted by the bursting of an $8 trillion housing bubble, the new president moved quickly to gain passage of a Keynesian stimulus package. Less than a month after taking office he signed into law the American Recovery and Reinvestment Act (ARRA), which would pump more than $800 billion into the economy. Approximately one-eighth of the money allocated was directed toward investments related to clean energy. This was unprecedented; funding had previously averaged only $2 billion to $3 billion annually. The stimulus helped create tens of thousands of jobs, but its long-term impacts were far larger. Solar energy generation rose thirtyfold in just seven years, with system costs decreasing more than 50 percent. Wind production rose threefold, but from a more significant base. Major investments were made in advanced vehicles, battery research and manufacturing, energy efficiency technologies, grid modernization, home weatherization, and public transit and high-speed rail. The Council of Economic Advisers estimated that these investments contributed to a 2–3 percent increase in national GDP within two years. Overall, giving priority to clean energy investments was unprecedented among previous administrations and gave hope to environmentalists.[4]

In the following year, as President Obama was expending considerable political capital to gain passage of both financial and health reform, congressional Democrats made a major push to gain approval of an American cap-and-trade system. This idea, which had emerged two decades earlier as a market-based solution from the conservative intelligentsia, was by this time anathema to Republican Party dogma. The Democratic House was barely able to pass the relatively weak Waxman-Markey bill, which was riddled with exceptions for powerful business interests, but the bill never gained much traction within the Democratic Senate, especially after conservatives somewhat successfully branded it as cap-and-tax, which met with a weak response from the environmental movement. President Obama decided that it wasn't worth pushing harder for the House bill or a potential Senate compromise, a decision that was largely vindicated when Senator Lindsey Graham (R-SC) abandoned his own legislation in the face of pressure from the Tea Party movement. After the Democrats' crushing defeat in the 2010 midterm elections, any climate legislation was a nonstarter. In a disconcerting development, President Obama was blamed by environmentalists for his failure to push the issue forward despite the political situation.[5]

Meanwhile, President Obama was quietly beginning to exercise his executive power to take aggressive action to mitigate climate change. Utilizing authority granted under the Clean Air Act and validated by the Supreme Court in *Massachusetts v. Environmental Protection Agency*, the president tasked the EPA with finding ways to reduce greenhouse gas emissions. This effort began on 7 December 2009 with an endangerment finding for carbon dioxide, methane, nitrous oxide, hydrofluorocarbons, perfluorocarbons, and sulfur hexafluoride. The agency found that these gases jeopardized public health. Under the leadership of Lisa Jackson and Gina McCarthy, the EPA developed and promulgated several new regulations that would place the nation on a path toward sizable emissions reductions. At the outset, the agency worked with the National Highway Traffic Safety Administration to adopt rules that would require light-duty carmakers to meet fleetwide averages of 54.5mpg, solely through improved fuel economy, by 2025. (At the end of the Obama administration, for the first time ever, the agency would also propose additional standards for heavy-duty vehicles.)[6]

While transportation regulations were important, the Obama EPA truly broke new ground with a Clean Power Plan (CPP), which tackled the largest source of carbon pollution in the American economy, power plants. Under the auspices of the Clean Air Act, section 111(d), the plan adopted performance standards for both coal- and gas-fired power plants to meet by 2030. One of the signature features of the rule was that each state could meet the objective in a way that best fit its unique circumstances. The states were required to develop implementation plans to be submitted no later than 2018. The EPA envisioned an initial effort to make coal-fired plants more efficient (partially by decommissioning older facilities), followed by a midterm transition to lower-emitting gas-fired plants, and concluding with the eventual replacement of coal-fired electricity production by renewable sources. These pathways were tailored to states in each of the three large electricity-grid interconnections, to provide customized performance targets within each region. Thus, each state had its own goal based upon existing circumstances and achievable energy mixes. The CPP encouraged the adoption of emissions trading as a cost-effective way to achieve state objectives, thus trying to reboot Waxman-Markey at the state level. The CPP was, without doubt, the most consequential climate policy in American history to that point in time; however, as I write it is under dire threat from the Trump administration.[7]

With the CPP in his back pocket, President Obama was able to play an outsized role in international negotiations aimed at adopting a treaty to replace the Kyoto Protocol. In December 2015 the international community met in Paris to finalize an agreement that had been under consideration for many years. Unlike earlier, largely unsuccessful efforts like Kyoto, the Paris Agreement doesn't legally require any action by signatory nations. Rather, it is a set of diplomatic tools aimed at leveraging international public opinion to encourage countries to fulfill their obligations. Among these tools are transparency measures to ensure that national emissions data are fully accessible, as well as funding for poor nations to transition to cleaner energy and adapt to changing conditions. America pledged to decrease its greenhouse gas emissions by 26 percent of 2005 levels by 2025; the CPP was central to achieving this goal. Europe and Brazil promised even larger cuts. China declared that its emissions would peak by 2030. India vowed to continue reducing its overall economic carbon intensity. While these goals are insufficient to keep global temperatures at anywhere close to safe levels, the treaty represented a major step forward in two specific ways. First, it included all of the major global economies. Second, in almost all cases the pledges exceeded previous pledges. Thus, while it is not a perfect agreement, there is hope that it will build needed momentum toward achieving more substantial emission reductions by midcentury. In August 2017, however, the Trump administration delivered notice to the United Nations that it planned to withdraw from the treaty. This represented a body blow to President Obama's legacy and more importantly to the environmental movement.[8]

From a climate security perspective, the Fourth Political Age had a better beginning than many recognize. Major investments were made in clean energy and key infrastructure. The transportation sector was given important emissions standards to meet. Vital regulations were promulgated to begin phasing out coal-fired power plants, and a major international agreement was reached, with every major economy signing on. Constant conservative sniping and active obstructionism limited what was politically possible, but that didn't prevent President Obama from pushing forward a pragmatic progressive agenda that was a sharp departure from the failure to act during the previous political age. Unfortunately, the Trump administration is in the process of gutting most of this regulatory progress. It is unlikely that this will change the general trajectory of the American economy toward the adoption of more sustainable forms of energy, but it will certainly slow progress meaningfully. Thus, should progressives retake control of the government in the coming years, there will be much to do to achieve the promise of this new political age. A straightforward sustainability revolution, laid out below, might serve as an organizing principle for the entire progressive movement. It would put America on a path to not only meet its own obligations with regard to mitigating climate change but also help guide the larger global community. As it has many times in its history, it is once again time for the United States to lead by example. But for this to happen, a new generation of leaders must take control and make climate security a central task of government. They should pursue a policy agenda with five primary columns to create the sustainable future Americans and the world need.

Sustainability Revolution, Column 1
Placing a Price on Carbon

For political reasons, it has long been considered unwise to speak truth to the American people about what will be required to fight the climate crisis. During the recent conservative cycle, Republicans and their business allies manufactured a campaign aimed at discrediting the environmental movement. The severe backlash against environmental regulations resulted in great timidity among progressives, not necessarily without good reason at the time. What was striking, however, was the length of time before key leaders determined they could start talking sense again without fear of election losses. In light of the rapidly changing demographics of the nation, there is in fact far more to be gained by leveling with the American people. Of course, there are some who will never be convinced. But their share within the electorate is diminishing quickly, as Generation X'ers and Millennials are voting more regularly. It has also helped that many environmentalists have moved away from the failed strategy of shaming the nation to take action. The United States is far more likely to take action if the climate crisis is framed as a collective societal problem, even one that might require modest personal sacrifices.[9] This is particularly true if the citizenry can be convinced that their standard of living will likely be little affected and their quality of life will likely improve. That said, solutions to the climate crisis do have costs. Progressives should be clear about what they are and who will pay.

At the outset, progressives should be perfectly honest about their goals. Without a doubt, the most important initial goal is to dispatch the coal industry. This is because of coal's large carbon concentration, making it the single largest contributor to national greenhouse gas emissions. While this argument may be unpopular in a few states where coal is economically important, there are few jobs left in this sector. In 2014 only about 75,000 coal-mining jobs remained in the United States (along with perhaps as many as 60,000 jobs at coal-fired plants). Far more people are employed directly in the solar and wind sectors. So there is really no rationale for not being frank about the need to stop burning coal. The second target should be the oil industry, but here the story is more complicated. There is a straightforward need to eliminate the use of petroleum in automobile transportation, which is where it is most used. And this should be relatively easy as electric-vehicle technologies become ever more viable. Likewise, Americans should stop using oil to heat both commercial and residential buildings. It may be more difficult to purge petroleum from other sectors. Medium- and heavy-duty trucking, marine transportation, and air travel are all still heavily reliant on dense liquid fuels. This is perhaps even truer for manufacturing, where high-heat processes require hydrocarbon combustion. Fossil fuels are also critical industrial feedstocks, particularly for petrochemicals and plastics. While solar fuels may provide an alternative within a couple of decades, in the near term it will be difficult to transition without harming the economy. Still, the nation can easily aim to eliminate a majority of the petroleum it currently uses in the near term before fully phasing it out by midcentury. The final objective will be to remove gas from the energy mix, but this will likely be a longer-term goal because its lower carbon concentration makes gas an attractive bridge fuel on the path toward a far cleaner energy future. So the intention is to largely dismantle both the coal and oil industries in the next decade or two. Let's be honest about the necessity to do this to preserve humanity's ability to survive on this planet. Progressives must be clear about the problem so the nation can engage in a real conversation about the best solutions.[10]

* * *

The single most important way to achieve the above goals is to place a price on carbon. Burning coal and oil must be more expensive, not only so we can be truthful about the actual costs of these fossil fuels but also to make renewable energy sources comparatively cheap. What economic sense would it make to continue using coal and oil if cleaner options were far less expensive? Although the prices of renewables are already coming down, wind energy is already broadly competitive with coal-fired plants, and solar energy costs are plummeting every year, making a quick transition away from hydrocarbons requires the long-term stability provided by an easily understood governmental intervention to make them more expensive. Cap-and-trade is only a good option for accomplishing this objective in a world where progressives are playing politics rather than directly tackling the problem. During a conservative cycle when environmentalists were engaged in this gamesmanship, it might have made sense to take emissions trading seriously, largely because as a market-based solution it could potentially provide conservative support for putting a price on carbon. The difficultly, however, is that emissions trading schemes are extremely complicated and hard to implement across an entire economy (although they have been somewhat more successful within very narrow sectors). One need only examine the experience in Europe to question the benefits of this approach. The good news is that a much better alternative exists: a carbon tax.

When addressing public policy problems, there aren't many truly new solutions. Humans have been governing themselves for several millennia, and modern governments have grown in size and intricacy for several centuries, so they have had to consider a wide variety of options for dealing with societal challenges. Thus, as John Kingdon and others have suggested, much of the policy process involves applying existing alternatives to new problems (or finding the right combination to achieve optimal results). Taxes are the oldest and probably best-understood policy intervention. The primary downside of using taxes to tackle the climate crisis is that during the most recent conservative cycle they became anathema to a large segment of the American populace even though paying taxes has long been the price of civilization, required to fund the programs that provided the United States with unprecedented levels of national and economic security. The abhorrence of taxes has begun to wane, as evidenced by the fact that Democratic presidential candidates actively campaigning to raise them on wealthier Americans have paid little political price. If a carbon tax is properly structured, it can take advantage of this fact without resulting in major political costs to the leaders pushing it forward.

What are the advantages of a carbon tax? Whereas cap-and-trade is a quantity approach, in which society can theoretically control the exact amount of annual emissions, a carbon tax is a price approach. Many economists favor a carbon tax for two primary reasons. First, it provides far more certainty for companies determining how best to make the transition to a low-carbon future. This is no small benefit, given that everyone should want to make this shift with as little economic trauma as possible. Solving the climate crisis should not be about punishing evil corporations; it should be about making the future better for all members of society. Second, and more importantly, a carbon tax would address the negative externalities that have always resulted from burning hydrocarbons. It would achieve this by taxing fossil fuels in proportion to their carbon content. Thus coal would face the highest tax, trailed by various oil by-products; natural gas, which produces roughly half the emissions of coal, would face the lowest tax. Relatively little new administrative machinery would be required, because numerous governmental agencies already have taxing authority. Since taxes lead to predictable changes in behavior, we can be assured that consumers will begin to shift away from increasingly expensive hydrocarbons. As renewables become the cheaper option, their adoption will accelerate even beyond the high rates experienced in recent years.

An interagency working group formed by President Obama found in 2016 that the current social cost of emitting one metric ton of carbon dioxide is roughly forty dollars. The goal should be to offset these damages relatively quickly with a tax rising to that level over a decade, with a specific emissions reduction target in mind. Some might argue for a more aggressive approach, but given that energy investment decisions are made based upon long time frames, the private sector would need this amount of time to make the transition to new power and feedstock sources while also avoiding major economic disruptions. Over the next few decades, a well-designed carbon tax could generate trillions of dollars in revenues. A portion of these proceeds could be directed toward far more robust technology development programs, local and regional transportation projects, and financial incentives to promote green urbanism. Another popular idea, at least among progressives, is to use much of the remainder to finance payroll tax relief for poor and working families, which might be popular among both white and nonwhite demographic groups. This would reduce the negative impacts of this regressive tax, while also ensuring that higher energy costs would be paid primarily by those members of society best able to handle them. In the end, a carbon tax is the most effective way to achieve climate security in the coming decades. It should be the primary policy objective of the progressive and environmental movements.[11]

Sustainability Revolution, Column 2
Urban Density and Livable Cities

While policy solutions such as energy efficiency and clean energy technology have long enjoyed widespread attention, until very recently urban density was ignored as a policy alternative. This snub resulted from the lingering belief among environmentalists that cities were the root of evil, the birthplace of pollution and climate change. Influential intellectuals from Jefferson to Thoreau to Muir subscribed to an ethic that placed dense urban areas at the center of all that was wrong with modernity. This was and is disconcerting because well-designed cities are the second most important climate security policy. To a considerable degree, this is because trying to replicate an idealized "life in the country" led to extraordinarily damaging forms of suburban, exurban, and rural living. The stupendous lack of energy efficiency observed in these arrangements has been very harmful. For example, those living in rural Vermont, which is held up as a sustainable exemplar, utilize far more hydrocarbons than those living in dense cities. In recent years, authors like Peter Calthorpe, David Owen, Andres Duany, Douglas Farr, and Edward Glaeser have advanced urbanism as a climate solution. We are now at the point, however, when the ideas they promote need to make their way onto the national agenda. This is vital to our collective future.

Peter Calthorpe offers a useful definition of urbanism: "Urbanism is always made from places that are mixed in uses, walkable, human scaled, and diverse in population; that balance cars with transit; that reinforce local history; that are adaptable; and that support a rich public life." David Owen writes that there are three overarching advantages to living in this type of dense city. First, urban dwellers live in much smaller spaces. Worldwide, buildings represent only about 15 percent of annual greenhouse gas emissions; in America, they represent more than 27 percent. The average size of a new home in America has grown to approximately 2,600 square feet (although household size has shrunk), but the homes of those living in the densest cities aren't nearly as large. As a result, the latter consume less energy, since there is less space to heat, cool, and illuminate. Even better, these folks benefit from the embodied efficiency of the larger buildings in which they often live. Most homes in suburban, exurban, and rural areas lose a great deal of energy simply because they have four walls and a roof exposed to the elements. Apartment occupants, on the other hand, share their interior walls with other tenants, so their heated and cooled air passes back and forth between their spaces rather than quickly moving into the ambient environment. Urbanites also tend to consume less, largely because they don't have nearly as much room to store things; even their appliances are smaller. Second, urban dwellers live closer to work, services, and leisure. The best neighborhoods are mixed use, so that all the ingredients of a high-quality life can be found within a short distance of home. Consequently, there is less need to build extensive energy, transport, and water infrastructures to support far-flung developments. Third, because city residents live closer to work, services, and leisure, there is far less need to drive. Worldwide, transportation represents only about 12 percent of annual greenhouse gas emissions; in America, it accounts for more than 27 percent. This is because those living suburban, exurban, and rural lives must drive to do most things in life. The design of these areas, with zoning often preventing mixed-use neighborhoods, forces residents to be car dependent. In cities, on the other hand, people commute to work by foot, bike, or public transit. Likewise, they don't need a car to conduct everyday errands; they can simply walk to the market, restaurant, or shop. Accordingly, many urbanites don't even own a car. In sum, Americans who live in dense urban areas use far less energy and therefore contribute fewer damaging greenhouse gas emissions. An important policy objective, therefore, should be to encourage Americans to live in more compact cities.[12]

A major problem is that America has too few truly dense cities. There are exceptions of course. Places like New York, Boston, and San Francisco. Perhaps more troubling than the raw density numbers, however, is poor zoning and insufficient public transit. Greater Los Angeles, for example, is surprisingly dense, even when compared with the above three urban areas. But it lacks the zoning and transport needed to genuinely benefit from the advantages found in other dense city centers. Progressives need to promote a collection of policy interventions, starting at the national level but quickly flowing to municipal regions, to redevelop America's cities to maximize density. A good start would be to acknowledge that both the United States and the world have an urban future. Even in the unlikely event that we Americans completely abandon industrial farming, most people are still going to be living in cities. Thus, determining how best to take advantage of this prospect should be at the top of any serious policy agenda. And there are many reasons to believe that this path offers tremendous opportunities for increasing national sustainability.

Americans must begin adopting more enlightened urban policy by embracing whole systems thinking, which in this context means planning on a regional scale. It is no longer acceptable to tolerate the inefficiencies that result from the fractured governance existing in every major American urban area, where the wealthiest attempt to isolate themselves from their fellow citizens. Not only does this lead to dramatic economic inequity and social dislocation but it makes it harder to design metropolitan areas that best shepherd limited natural resources, while eliminating the need to use hydrocarbons. According to Calthorpe, "53 percent [of GHG emissions] depend on the nature of our buildings and personal transportation system—the realm of urbanism." Traditional urbanism, focused on creating more densely populated and livable cities, can by itself reduce this number. It is a logical first step, because it would allow society to rightsize its energy use and thus determine what would be required moving forward. Green urbanism, in which advanced conservation techniques and clean-power systems are added to the mix, can then provide truly meaningful emissions reductions. A green townhome in a transit-oriented development consumes 58 percent less energy than the average suburban home, while a green urban condo consumes 73 percent less. And it provides noteworthy cost savings because of reduced transportation and housing costs, as well as reduced taxes because overall infrastructure spending would decrease as the need for far-flung systems disappears. One study conducted in California in 2010 found that each urban household could enjoy an average annual savings of more than ten thousand dollars. How can anyone argue with a policy solution that essentially pays for itself?[13]

The best way to foster urbanism is by adopting one of a variety of models of regional development that fit local conditions. Portland, Oregon, has adopted a top-down approach known as an *urban growth boundary*. It prohibits development outside a specific area, attempting to discourage sprawl while encouraging urban infill, whereby underused space inside the boundary is targeted for development. Seattle has a more intricate system based upon a *hierarchy of centers*, each with an individual growth boundary that takes into account local considerations. Maryland favors a system that allocates infrastructure funding to *priority funding areas*, each of which meets specific minimum-density standards. The Twin Cities have embraced an approach that combines aspects of the urban growth boundary and the priority funding area. The *urban service boundary* limits where government will invest in water, sewer, and road infrastructure, thus incentivizing developers to build in these areas so they don't have to internalize those costs. Finally, Salt Lake City benefitted from a bottom-up approach led by a civic group that sought to educate the public about the benefits of compact, transit-oriented development. Envision Utah showed that accommodating an additional million residents in 112 square miles instead of a sprawling 439 square miles would result in an approximately thirty-thousand-dollar reduction in infrastructure costs for each new home. Considerable public support for the density option, which was also more suitable given local trends, led to the passage of the Quality Growth Act and creation of a commission to select *smart growth areas* for redevelopment and new development. The federal government can actively promote these and other approaches to incentivizing compact, transit-oriented development by giving priority for national infrastructure spending to well-designed regional planning efforts. This will require that it stop subsidizing the suburbanization of the country, especially through wasteful highway spending and hydrocarbon subsidies. Instead, it must rapidly transition to providing financial incentives aimed at revolutionizing the urban built environment and also redirecting transportation funding to urban transit projects. Both are essential to promoting green urbanism.[14]

* * *

In order for green urbanism to become a reality, building design will need to change fundamentally. A comprehensive set of policy solutions must be adopted. The first step would be a highly visible educational outreach effort to publicize the benefits of sustainable buildings, which result in substantial long-. term cost savings and environmental advantages. It will be indispensable to provide stable financial incentives (tax credits, mortgage agreements that promote energy efficiency, low-interest loans, and rebates) for retrofitting existing buildings to meet robust energy efficiency standards. Research funding in this area has never been adequate, particularly when one considers the share of American emissions that come from commercial and residential buildings. The budgets for the Oak Ridge National Laboratory Buildings Technology Center and the Department of Energy Emerging Technologies Program should be dramatically enlarged. Finally, and almost certainly most importantly, better building codes are needed across the country. While there are many potential models, the International Living Future Institute's Living Building Challenge provides the most forward-thinking standards for sustainability in the built environment. The challenge requires buildings to meet twenty mandatory imperatives within seven performance categories—place, water, energy, health, materials, equity, and beauty.[15]

I was recently able to visit the first commercial building to receive a full certification from the International Living Future Institute, the Bullitt Center in Seattle. Inspired by Denis Hayes, a co-organizer for the original Earth Day and a former director of the National Renewable Energy Laboratory, it is a remarkable building. As the *New York Times* wrote, the building was intended to "demonstrate that a carbon-neutral office space can be commercially viable and aesthetically stunning without saddling its occupants with onerous demands." My conversations with tenants suggest that it has certainly achieved those objectives. The building met the challenge requirements in the following ways:

- Place: urban location supports a pedestrian-, bicycle-, and transit-friendly lifestyle.
- Water: rainwater is collected on the roof, stored in an underground cistern, and used throughout the building.
- Energy: a solar array generates 160 percent of electricity needs, geothermal energy heats and cools the building, and computers automatically adjust passive and active systems to keep the building comfortable and efficient.

- Health: the building includes inviting stairways and operable windows.
- Materials: the building does not contain Red List materials commonly found in building components.
- Equity: all work stations are within thirty feet of large operable windows, offering workers access to fresh air and natural daylight.
- Beauty: the striking design includes an innovative overhanging photovoltaic array, a green roof, large structural timbers, and a revitalized pocket park.

Perhaps what is most impressive about the Bullitt Center is that it cost only 23 percent more to construct than other commercial buildings in Seattle. And it was an experimental building. Broadly adopting the approaches listed above would almost certainly bring down capital outlays as architects and contractors become more familiar with the techniques employed. Furthermore, the owners of traditional office buildings still have to make monthly payments for electricity, heating and cooling, and water, costs that are essentially eliminated from living buildings (bringing their lifetime costs down radically). In addition, the building has a very limited environmental impact compared with similar structures. Although beauty might be in the eye of the beholder, the other six performance categories seem quite reasonable as the starting point for a model building code. It is likely that the national government will not want to mandate the adoption of such standards, but once again it could develop intelligent financial incentives to promote their acceptance by local governments. This will be an indispensable step in fostering green urbanism in American cities.[16]

* * *

In order to make widespread changes in the built environment, progressives must lead the charge with regard to a full-scale transformation of urban planning. Many aspects of city design must change; most importantly, zoning laws that prohibit the development of mixed-use neighborhoods must be eliminated. "Traditional zoning plays a significant role in the inefficiencies of low-density development by creating two distinct infrastructures in place of the traditional multipurpose town or city. With the home and the workplace separated, often by long car commutes, two well-serviced developments are created with duplicate retail, service, and parking institutions: the bedroom community and the office park." Given the huge size of the United States and the fact that most of its cities were planned after the introduction of the automobile, this division, particularly when it didn't appear to have major costs, is probably not too surprising. Today, however, we recognize that there are major problems associated with segregated auto suburbs. Not only have they driven the rise in greenhouse gas emissions but they have led to major social dislocation and rising partisan disharmonies. As Robert Putnam writes, "For the first two-thirds of the twentieth century a powerful tide bore Americans into ever deeper engagement in the life of their communities, but a few decades ago—silently, without warning—that tide reversed and we were overtaken by a treacherous rip current. Without at first noticing, we have been pulled apart from one another and from our communities over the last third of a century." If Americans desire to reduce overall vehicle miles traveled and begin to knit the citizenry back together, they must discard this bifurcation in favor of mixed-use and mixed-income districts.[17]

Americans must build human-scale cities in which cars are largely abandoned in favor of walking, biking, and public transit—places where community can be rebuilt, while the environment is protected. Calthorpe argues that putting these ideals into practice requires a reassessment of the fundamental components of regional design. These include the following:

- Neighborhoods—basic building blocks of community, walkable areas composed of a range of housing options with parks, schools, and local services;
- Centers—mixed-use destinations for a group of neighborhoods, where people find housing, jobs, services, and retail spaces;
- Districts—special use areas designated for a single use such as a university or cultural venue;
- Preserves—open space reserved for natural habitat, parkland, or agriculture; and,
- Corridors—transport connections between neighborhoods, centers, and districts.[18]

Although the evolution of neighborhoods, centers, and districts is a somewhat organic process, enlightened urban planning can help direct it in progressive ways. One way is to discourage sprawl, while encouraging vertical growth and urban infill (the redevelopment of derelict or underutilized urban spaces). Just as important is building transportation corridors that provide alternatives to driving. At the neighborhood level, this would involve a transition to "complete streets." Rather than being autocentric, these thoroughfares reduce not only the space available for cars but also the speeds at which they can travel.

Vehicles are replaced with generous sidewalks for pedestrians and clearly de-marcated lanes for bicycle traffic. Many European cities, most notably places like Amsterdam and Copenhagen, have been so successful in adopting these strategies that more than half of all daily trips are now made by bike. This can and is being done, and Americans have much to gain by bringing these meth-ods to their shores.

Public transit needs to be expanded in nearly every American city to con-nect centers and districts, creating networks that are far superior to the roads that currently dominate the urban landscape. The goal is yet another trans-formation in the ongoing INNATE revolution in transportation. Fortunately this process has already begun, with many metropolitan areas inaugurating limited light rail systems and rudimentary bus rapid transit projects. Still, the funding and coverage of these efforts has been wholly inadequate as political leaders have continued to focus on building more roads. It is time to fully embrace public transit for a sustainable future. It is the key solution for re-ducing car ownership and utilization. People will continue to drive unless two things happen. First, urban driving must become an expensive and unpleas-ant undertaking. Gasoline prices must be increased by adding a carbon tax. Cars should be charged tolls to access urban highways. Roadway expansions aimed at reducing traffic congestion should be eliminated, because they aim to make driving between city center and suburbs and exurbs easier—exactly the opposite of what is most desirable from a climate security perspective. Park-ing should be more expensive, and/or parking places should be more difficult to find (this will require changing zoning laws to reduce parking-space allot-ments for residential and commercial buildings). Purchasing an automobile should include a sizable tax, further pushing up the cost of vehicle ownership. Car share programs should be encouraged, so that the environmental costs of manufacturing an automobile are shared by many users, while still allowing residents to have infrequent access to a car. "A truly effective traffic program for [a] dense city," writes Owen, "would impose high fees for all automobile access and public parking (thereby discouraging all car use and raising money to support transit) while also gradually eliminating automobile road capacity (thereby reducing total car traffic volume without eliminating the environ-mentally beneficial burden of driver frustration and inefficiency) and increas-ing the capacity, frequency, and efficiency of public transit."[19]

Second, adequate funding must be provided to develop alternatives to auto-mobiles. The future of regional and local transportation lies in a mix of walk-ing, biking, and public transit such as bus rapid transit, light rail, streetcars, subways, and commuter trains. The federal government must provide funding mechanisms to not only create complete streets for walking and biking but also build public transit networks. Creation of a model similar to the current one for national highways—in which the federal government provides 90 percent of the funding for development and maintenance, paid for by a transportation tax—would be an excellent start. National financing is important because public transit has a chicken-and-egg problem. For transit systems to support themselves, the population density must be such that there is sufficient ridership. But currently the population density doesn't exist, because the transportation infrastructure is designed to spread people out. Thus, government capital is needed to bankroll the construction and provide the impetus for people to move to transit-oriented centers and surrounding neighborhoods, which could be connected to form an urban network. Calthorpe writes that "transit boulevards are at the heart of this new network. They are multifunctional streets designed to match the mixed-use urban development they support. They have dedicated lanes for transit—whether bus rapid transit, streetcars, or light rail." These are places where people eat at cafes and shop at small stores before retiring to nearby apartments or small-lot single-family homes on the avenues and streets in the surrounding neighborhoods. It will take decades to fully transform the American urban landscape to match this vision of dense, human-scaled cities designed around walking, biking, and transit. But now is the time to begin shifting federal funding and financial incentives to begin this exciting undertaking.[20]

Sustainability Revolution, Column 3
Renewed Energy Security

As humanity continues moving into the climate age, America has a distinct advantage compared with other nations: it has the natural resources, the technological capabilities, and the capital to ensure energy security. As we have seen, these are critical for maintaining modernity. A key question facing the United States, however, is how best to leverage its hydrocarbon largesse in the near term to preserve energy security while transitioning to the clean alternatives demanded by the times. As discussed above, the nation must quickly discard coal through aggressive carbon taxes. In the next ten to fifteen years, once battery-electric-vehicle technology has advanced sufficiently and gasoline prices have been driven up, it can also eliminate most oil use. The third hydrocarbon, gas, will see a much different near-term future. As the changeover to renewables gains momentum, gas will play an important role as a bridge fuel, for several reasons. First, over the past several decades gas has become the preferred hydrocarbon for electricity production, heating and cooling, and cooking. Second, unconventional sources are highly abundant and recoverable using hydraulic fracturing. This is particularly beneficial for the United States as a major producer itself, because there are limited geopolitical challenges associated with its employment. There are lingering environmental concerns about methane leakage from gas extraction, but many experts agree this could be significantly reduced through better regulation. Third, gas has roughly half the carbon content of the coal it is replacing. And it burns efficiently, especially when consumed in ultraefficient combined-cycle turbines that capture and use waste heat. In addition, a great deal of waste-heat generated in large manufacturing plants is not being used to produce electricity because the current regulatory framework doesn't allow commercial sales of this energy (clearly an example of a bureaucratic failure that must be corrected). Fourth, gas-fired generators cost approximately half as much as their coal-fired equivalent. It is therefore not surprising, as Paul Roberts writes, that "the power sector has emerged as the strongest impetus for the gas business, and the first piece in a gas 'bridge' to the next energy economy. Electrical power, even more than transportation fuel, is the critical resource for modern economies that are increasingly based on technology and services." Fifth, because gas-fired turbines can come online very quickly, they are an excellent backup to address any fluctuations in either renewable-based supply or consumer-driven demand. Finally, gas is far more scalable for electricity production than coal. Rather than requiring enormous boilers as industrial-scale coal-fired power plants do, gas-fired microturbines have much smaller boilers and can be used in a more decentralized fashion—a perfect match for green urbanism. If the political will existed for aggressive climate policies aimed at keeping carbon dioxide emissions in the atmosphere below 450 parts per million, gas would have a limited role, because the changeover to clean resources would need to be so rapid. But in a more politically realistic scenario, in which the goal is 550 parts per million, a gas bridge will play an essential role. Although there is no doubt that by midcentury we will seek to largely remove gas from the energy mix, in the next couple of decades it will remain an important resource to maintain modernity while transitioning to a clean energy future.[21]

* * *

During the 2008 election cycle, Barack Obama called for a crash clean energy research program in the spirit of the Manhattan and Apollo projects. Although the financial crisis provided him with the opportunity to champion a huge one-time infusion of research funding under the auspices of the American Recovery and Reinvestment Act, the Republican takeover of Congress after the 2010 midterm elections eliminated any possibility that a permanent research program would gain approval. As a result, Department of Energy resources have remained relatively limited. The annual budget for the Office of Energy Efficiency and Renewable Energy (EERE) is less than $2 billion. In 2015, the department received only $233 million for solar energy research, $107 million for wind energy research, $61 million for waterpower research, and $55 million for research on geothermal technologies. The entire Advanced Research Projects Agency-Energy, which funds high-potential, high-impact energy projects, only received $280 million that same year. The Renewable and Appropriate Energy Laboratory at the University of California at Berkeley has calculated that to truly have an impact on global warming, a crash research program would need an annual budget of $15 billion to $30 billion. This would require a five- to tenfold increase in federal spending, which is comparable to past big technology efforts. Adoption of a carbon tax could easily provide the funding for such a program; the key would be to dedicate a certain amount of the revenue to clean energy research. A stable budget would allow for the development of not just replacements for hydrocarbons but sources that will permit America to continue expanding energy supplies to support economic progress.[22]

The Department of Energy's EERE has had three broad research programs, which correspond almost perfectly with three INNATE revolutions, intended to develop a robust clean energy economy: these programs are in sustainable transportation, renewable power, and energy efficiency. It has also pursued crosscutting initiatives, most importantly facilitating the incorporation of clean electricity into a reliable national grid (which includes approaches for storing renewable energy). Sustainable transportation aims to replace gasoline in light-duty vehicles; diesel in trucks, rail, and marine applications; and jet fuel in aircraft. The EERE has prioritized several technologies to accomplish these goals, including fuel-cell electric vehicles, plug-in electric vehicles, advanced-combustion technologies, vehicle light-weighting, and large-scale use of biofuels. A recent National Research Council study found that this suite of approaches could reduce transportation-related petroleum use by 40 percent by 2030 and by 80 percent by 2050. Electrification of transportation is quite obviously a key priority, because it means the sector can directly make use of the energy provided by renewables. Electric vehicles are already highly price competitive, so the aim is to provide them with the same performance, convenience, and safety standards as cars with internal-combustion engines. Production of alternative fuels will likely be oriented toward passenger and cargo aircraft, which are far more difficult to electrify.[23]

As the nation works toward a green urbanist future, research is needed to advance energy efficiency technologies. Likewise, ways must be found to reduce waste in the manufacturing sector. As we have seen, energy used in homes, buildings, and industry accounts for 70 percent of American energy consumption. While traditional urbanism can eliminate much of the resulting emissions, sustainable technologies can offer further cuts. Market-ready products already exist that can reduce building and manufacturing energy requirements by as much as 30 percent, but research is required to bring those savings down even further. This is really important, because at "a time when the nation faces likely major energy system upgrades, energy efficiency can help reduce the need for additional energy system infrastructure improvements." This means potentially huge cost savings to taxpayers and ratepayers. With regard to buildings, the EERE has prioritized enhancements to the thermal capabilities of the structural envelope (including windows), as well as integrating capabilities into consumer products to self-diagnose, communicate, and schedule services to improve the electricity grid. On the industrial front, the "focus will include high efficiency motors and drive systems, reduced energy intensity thermal systems, combined heat and power (CHP) systems, advanced forming and fabrication technology, manufacturing design and sensors and process controls." As discussed above, the adoption of these approaches requires federal leadership to educate, regulate, and provide financial assistance. In most cases, however, adopting these technologies saves homeowners and companies money. Thus, the benefits of providing political support for rapid progress in these areas should be obvious.[24]

For many, the most interesting areas of technology development relate to clean energy. This is not surprising, because at the end of the day Americans must find ways to largely eliminate their hydrocarbon use. But they are unlikely to do so unless there are new resources available that will allow the nation to maintain modernity. Sustainable transportation and energy efficiency technologies will only take the country so far without replacing the dirty electrons currently in use with clean electrons. The United States has vast supplies of renewable resources. This is an area in which it is already making major strides, with wind energy already cost-competitive with hydrocarbons and solar energy prices falling dramatically. Placing a price on carbon will quickly change the economics of clean energy alternatives, making them the preferred choice; existing concerns about raising the capital necessary for both decentralized and industrial-scale projects will largely disappear. Still, progressives must support continued advancement in these technologies. While it is important to replace hydrocarbons as a solution to the growing climate crisis, in doing so progressives have the opportunity to dramatically grow the available supply of cheap energy, thus potentially helping to support economic growth in the coming century and beyond.

Developing technologies to utilize America's renewable energy resources offers a wide range of research opportunities. Because initial capital requirements represent the vast majority of the costs associated with renewables, the EERE has focused much attention on producing advanced materials and components. The objective is to reduce start-up costs. Another avenue for investigation is how to improve the capabilities of wind and solar technologies; although other renewable sources are also the subjects of important research, wind and solar power are receiving extra attention because they are more readily scalable. Traditional windmills used mechanical power to mill grain, grind spices and dyes, make paper, and saw wood. As we have seen, they were commonplace in Europe before the First Industrial Revolution. Modern horizontal-axis turbines are designed to take advantage of aerodynamic forces generated by airfoils to extract energy from the wind and turn it into electricity. The EERE has prioritized several research initiatives to increase the energy captured by designing taller towers, longer blades, higher tip speeds, and enhanced wind farm design. There are two primary ways to use solar energy: via thermal power or via photovoltaic cells. Solar thermal methods include daylighting (today this means designing buildings to limit the need for interior artificial lights), passive solar water heating in warm regions, active solar water heating in cooler regions, and concentrated solar collectors that boil water to drive steam engines and produce electricity. Photovoltaic cells utilize solar photons to produce electricity. They work like this: photons hit a cell and are absorbed by semiconducting materials (such as silicon); negatively charged electrons are knocked loose from their atoms; the electrons are allowed to move in a single direction to produce a current; and the combined current of many cells is fed into the electricity grid. The first cells were developed in the 1870s, but they had very low conversion rates (approximately 1%). In the 1950s, monocrystalline silicon cells achieved much greater efficiencies, reaching conversion rates of approximately 20 percent after decades of research. Recent breakthroughs in polycrystalline silicon and thin film photovoltaics have provided cells that are far cheaper to produce although not quite as efficient, so that bigger arrays are needed to produce the required electricity. Today, solar innovation is concentrated on advancing conversion for traditional silicon, polycrystalline thin-film, and multijunction photovoltaic cells. Researchers are also invested in improving concentrated solar power techniques, which could be critical to industrial-scale production.[25]

* * *

Among the challenges posed by wind and solar energy in particular is their intermittency, which has become ever more important as the prospect for rapidly ramping up their share of the energy mix is realized. The wind doesn't always blow. The sun doesn't always shine. Over a large region, however, the fluctuations in wind availably are somewhat less dramatic then one might assume because the wind is usually blowing somewhere. Evidence suggests relatively modest slew rates (the variation in energy production), well within the ranges currently handled by the grid. There are quite lengthy lulls at times, which is one reason why maintaining gas-fired and nuclear power plants as backups is a good idea, at least for now. Solar variability might be a bigger issue, particularly if the United States relies heavily on decentralized production. It becomes a more manageable problem if large amounts of electricity are produced in the American Southwest, where there are few cloudy days, although effectively transporting those clean electrons to population centers would be a challenge. Still, researchers will certainly have to find ways to use solar energy at night, when the sun is behind the planet. Leveraging geothermal and hydroelectric resources at those times is a partial solution, but they cannot fully fill the gap.[26]

Given the intermittency problems related to renewables, the most important area for technology development is energy storage. In one approach, called *pumped storage*, water is pushed uphill into a reservoir using cheap energy when the wind is blowing and the sun is shining; then it is discharged to flow back downhill to be used by standard hydroelectric turbines. A major benefit of this approach is its efficiency, which can reach 90 percent or even higher. Another benefit is that it can be turned on quickly to react to slews or lulls. The biggest challenge with this approach is to either locate geographic areas that provide suitable uphill and downhill reservoirs (which could be quite controversial among environmentalists) or build huge numbers of stand-alone facilities—imagine an artificial lake on the surface, with a penstock reaching to an underground chamber and thus creating the high head needed to generate power. Investigations into both possibilities are needed to estimate the potential for this solution to address slews and lulls. Another approach would be to store excess energy in the batteries of electric cars and draw electrons from them when required. David MacKay describes how this would work: "Electric cars could be plugged in to smart chargers, at home or at work. These smart chargers would be aware both of the value of electricity, and of the car user's requirements. . . . The charger would sensibly satisfy the user's requirements by guzzling electricity whenever the wind blows, and switching off when the wind drops, or when other forms of demand increase. . . . In times of national electricity shortage [they could] run their chargers in reverse and put power back into the grid." An even better solution would be exchangeable vehicle batteries; millions could be waiting at service stations and available to the grid. One potential issue with this alternative is that it assumes there will still be large car fleets, although these should be much smaller if green urbanism succeeds. In the near term, however, this is an idea worth pursuing. A final approach that is receiving a great deal of attention is demand management. This involves technologies that allow thermostats and water heaters and appliances to determine when and how to draw electricity from the grid. They could be programmed to temporarily reduce their use when there is a slew or lull in renewable production and to increase use when energy is cheap and abundant. In addition to these three approaches, there are scores of other potential storage technologies that need funding to investigate whether they are worthwhile—things like flywheels and supercapacitors and vanadium-flow batteries. Each of these approaches needs sizable research funding to determine which combination works best, both on a regional and a national scale. If a sustainability revolution is to be workable, the storage problem must be solved.[27]

Another challenge posed by wind and solar energy is the mismatch between the highest-quality resources and major population centers. The richest land-based wind resources are in mountain ranges or the Great Plains, far away from big coastal cities. Along the coasts it can be difficult to site turbines, because sprawl has left few places for major wind farms. More expensive offshore resources might be able to fill this gap, but the United States currently has a single such wind farm because of NIMBYism concerns; it has fallen far behind other advanced economies in this arena. The most dependable solar resources are in the American Southwest. While it is certainly possible to provide some energy through decentralized rooftop solar arrays elsewhere, it is highly unlikely that these can provide sufficient energy. Fortunately, the United States has a nationwide electricity grid—actually three large regional grids—that can transport electricity from resource-rich areas to areas with high population density. Unfortunately, this grid is horribly outdated. This is very sad. The American electricity grid was one of the great engineering feats of the twentieth century,

but it has suffered decades of neglect. It still relies mostly on technologies designed before the age of microprocessors. One result of this is that it is hugely
inefficient to send electricity over even relatively short distances on existing
power lines (as many as half of the electrons are lost as waste heat). So a top
priority in the coming years will be to provide funding to develop cutting-edge
local, regional, and national grids capable of better utilizing the nation's enormous wind and solar resources. What is needed is a supersmart grid.

A lot of attention has already been directed toward Department of Energy
and private sector efforts to design local and regional smart grids. These projects focus on developing the technologies needed to make the grid operate
more proactively, including integrated communications that provide real-time
information; monitoring for real-time pricing and demand-side management;
advanced components research in superconductivity, storage, electronics, and
diagnostics; advanced control methods to enable rapid diagnosis and reactions
to slews and lulls; and decision-support interfaces for grid operators and consumers. There has been less concentration on the need to research, develop,
and build extensive long-distance grids capable of moving clean electrons
over long distances (perhaps thousands of miles) without major losses on the
journey. There are several potential approaches for such a super grid. Some
research has been conducted on nitrogen-cooled superconducting cables that
could transmit energy over great distances with very few losses. While this is a
method that deserves serious attention and funding, in the near term the priority should be developing a few technologies that would allow high-voltage
direct current (HVDC) cables to convey energy over long distances far more
inexpensively (although not quite as efficiently). The key HVDC technology
is the converters at either end of the line, which switch high-voltage alternating current to high-voltage direct current and back again. A half century or
more ago, these converters could not make these changes without considerable
losses. But over the decades a great deal of progress has been made. Current
voltage-source converters have had losses of just 1 percent, although it is hoped
that additional research will lead to increased efficiency. Another important
technology need is fast-acting, large-capacity circuit breakers that can react
quickly to any system problems. Although building a super grid could ultimately cost hundreds of billions of dollars, it is likely that it will be required
if the United States hopes to replace hydrocarbons and continue to expand its
clean energy resources to meet growing demands.[28]

* * *

The final area for a concerted research effort is nuclear power. For reasons that entirely ignore the facts, this energy source remains controversial more than six decades after the first commercial plant became operational. It is contentious even though there has never been a full reactor meltdown in the United States and there have been only two elsewhere, one caused by a poor design and human error, the other the result of a natural disaster. The Three Mile Island accident in Pennsylvania involved a partial meltdown that was completely contained and resulted in no attributable deaths. During the 1980s, however, nuclear weapons and commercial nuclear power were lumped together by antinuclear and environmental activists. As a result, nuclear research budgets were gutted, and no new plants were constructed, although approximately one hundred facilities were safely producing roughly 20 percent of American electricity without releasing any greenhouse gas emissions. A common refrain among the small but vocal antinuke crowd was that even if nuclear plants were unlikely to suffer catastrophic failures, radioactive waste made them dangerous. Interestingly, similar arguments weren't made about coal-fired power plants, even though coal ash produces far more radioactive waste. In fact, nuclear power produces relatively small amounts of waste compared with the hydrocarbons it might replace, and 97 percent of the waste it generates is not highly radioactive. For a nation of 325 million people, a lifetime's worth of waste would only fill about 350 Olympic-sized swimming pools, or, put another way, one square kilometer at a depth of one meter for a nation nearly ten *million* square kilometers in size. Furthermore, this waste has been safely stored for many decades. There is no doubt that this is nasty stuff, but it is important to put the risks in context. Each year, approximately 30,000 people die on American roads, another 33,000 gun deaths occur, and there are more than 480,000 premature deaths from smoking-related causes. Compared with risks Americans accept every day and the enormous threats posed by the continued use of hydrocarbons, the risks posed by commercial nuclear power are infinitesimal. Yet the potential benefits are enormous.[29]

Nuclear power will never fully replace hydrocarbons, largely because of resource constraints. Still, there is enough land-based uranium (most uranium is in seawater) to provide a large share of global energy needs. That share would increase dramatically if Americans were to abandon their current inefficient fission reactors. Traditional reactors burn less than 1 percent of the available uranium; they are far too unproductive to provide a "sustainable" energy supply that would last for centuries. Thankfully, other nuclear power technologies exist already, and there are many other potential avenues of research. Fast breeder reactors burn the U-235 isotope (which makes up only 0.7 percent of mined uranium); more importantly, they transform the predominant U-238 isotope into "fissionable plutonium-239 [to] obtain 60 times as much energy from the uranium." Not surprisingly, these reactors produce far less highly radioactive waste. And even better, they can burn the uranium left in spent fuel rods from current power plants, further reducing the existing waste stream. Fast breeder reactors currently cost more to construct than traditional fission reactors, but this is only true if they are built as commonplace one-gigawatt behemoths. Fast breeders are particularly well suited, however, to being deployed as small modular reactors, which could be fully constructed in a manufacturing plant and then delivered to a specific site for installation. If this approach became popular, SMRs could be far cheaper because of economy-of-scale benefits. While fast breeder reactors seem like an obvious candidate for a rapid research and development effort, they are by no means the only technology the government should be investigating.[30]

In the next decade there should also be an aggressive program to examine the pebble-bed reactor, also known as the Next Generation Nuclear Plant. As Andrew Kadak writes, these "small, modular, inherently safe" reactors "use a demonstrated nuclear technology and can be competitive with fossil fuels. Pebble bed reactors are helium-cooled reactors that use small tennis ball size fuel balls consisting of only 9 grams of uranium per pebble to provide a low power density reactor." Low power density and a large graphite core deliver built-in safety by reducing the peak temperatures reached even in the case of a complete loss of coolant, essentially eliminating the possibility of a meltdown. Even better, the reactor could be coupled to a hydrogen-production facility that would transfer heat from the coolant system to produce the valuable gas effectively at no cost. Other reactor approaches are also worthy of serious consideration, particularly the supercritical water-cooled reactor, the molten-salt reactor, the gas-cooled fast reactor, the sodium-cooled fast reactor, and the lead-cooled fast reactor. In addition to considering alternative reactor designs, America should take the lead in exploring alternative nuclear fuels. Thorium is the main substitute that should be studied, primarily because it is a radioactive element that is far more readily available; there is at least three times as much land-based thorium as there is land-based uranium. TH-90 can be entirely burned up in simple reactors, but a modular breeder reactor using the isotope would be better, especially if it employed both a liquid core and an accelerator-driven system, in which particle beams are used to stimulate the fission process. The resulting liquid fluoride thorium reactor could conceivably be two hundred to three hundred times more fuel efficient than traditional reactors. According to Richard Martin, "They are safer, simpler, smaller, less expensive to build, and less expensive to run to produce electricity on a cost-per-kilowatt basis."[31]

The bottom line is this: there are many potential reactor designs that could safely and cheaply provide nuclear power to the grid with zero emissions and reduced waste. Electric utility operators have concerns about transitioning to a system that uses primarily renewable sources that experience slews and lulls. While gas can certainly provide standby power in the near term, nuclear power provides an alternative that will have far fewer environmental costs in the long term. And we haven't even discussed nuclear fusion. Stars are powered mostly by fusing hydrogen atoms, but replicating this process safely for commercial energy production is currently only in the exploratory phase. This is because fusion creates so much heat, more than can be contained by any Earth substance. Yet on the plus side, fusion can create seven times as much energy as fission, and the source materials are abundant (although the hydrogen isotope deuterium would have to be harvested from seawater, which could be costly). For decades, various nations have attempted to design fusion reactors that can contain the astonishingly high heat needed to fuse hydrogen nuclei. The focus has been on constructing machines that use magnetic fields to keep superheated plasma in check. The focus of research since the 1970s has been on the tokamak arrangement, in which particles move in a symmetrical doughnut-shaped ring surrounded by electric coils that produce a magnetic field. But the evolution of the technology has been slow, largely because investigators have been trying to determine how to control immense magnetic forces without damaging the reactor. The other problem with tokamaks is that they can only channel a current through the plasma in short bursts because it is difficult to keep particles from drifting off course, potentially limiting its usefulness for electricity production. Recently an old competitor, the stellarator, has reemerged. Although it could sidestep some of the issues facing tokamaks, the stellarator faces its own difficulties. These nonsymmetrical instruments are uncommonly hard to build, but because they overcome particle drift by twisting the plasma, they can maintain steady-state operations better suited for utility-scale energy generation. Regardless of approach, we are still many decades away from commercializing any fusion reactor. Yet providing funding for research in this area seems more than worthwhile, because if it were ultimately successful it would provide enormous amounts of energy.[32]

* * *

If a sustainability revolution is to succeed, the United States has a lot of work to do to eliminate the use of hydrocarbons and replace them with clean energy resources. A typical citizen consumes roughly 250 kilowatt-hours every day, an enormous amount compared with the consumption patterns in the rest of the world. Europeans use about half that amount of energy but still maintain an advanced economy that serves their citizens well. The main reason for this disparity is that Americans use energy so inefficiently, particularly in transportation and buildings, both related to large-scale urban sprawl and suburbanization. The nation should be able to cut these requirements in half by adopting policies that foster higher-density lifestyles, incentivize efficiency upgrades, and alter the transportation infrastructure. It would still have to provide 125 kilowatt-hours per day per person. David MacKay has modeled how much energy various renewable resources might offer, summarized (and slightly adjusted by the author) as follows:

- On-shore Wind: assuming wind farms covering an area the size of California, the nation could produce roughly 42 *kilowatt-hours per day* per person.
- Off-shore Wind: assuming wind farms covering an area the size of Delaware and Connecticut combined, the nation could produce roughly 5 *kilowatt-hours per day* per person.
- Geothermal: assuming enhanced systems are available and operating in ideal locations, the nation could produce roughly 8 *kilowatt-hours per day* per person.
- Hydroelectric: assuming current capacity plus some micro-hydro on suitable rivers, the nation could produce roughly 5 *kilowatt-hours per day* per person.
- Solar: assuming concentrated solar power facilities covering an area the size of Arizona in the American Southwest, the nation could produce roughly 400 *kilowatt-hours per day* per person.
- Total: *460 kilowatt-hours per day* per person.

Thus, relying on nonsolar renewables doesn't get the United States even close to replacing hydrocarbons. A major advance in solar power production is essential. And it seems very unlikely that this goal could be met by depending solely on distributed solar photovoltaic installations alone, although they could play an important role. Only by leveraging an enormous expansion of industrial-scale concentrated solar—which could take many decades to accomplish—would the nation be able to achieve its central objectives with regard to clean energy production. Given the time required to build up this almost entirely new energy economy and the problems associated with stabilizing the grid while relying so heavily on solar energy, America will almost certainly need an initial gas bridge and might need next-generation nuclear power in the long term (without radical innovations in storage capacity). Finally, all of these pathways require enormous investment in a twenty-first-century electricity grid. The government is faced with an enormous task. There is no doubt that the country has the financial resources to fund both the needed clean energy research and the infrastructure development. What is required is the collective will to make the sacrifices necessary to make it happen during the Fourth Political Age, as waiting would likely make it far more difficult for large numbers of humans to live on the planet.[33]

Sustainability Revolution, Column 4
Intercity Travel

Medium- and heavy-duty trucks are among the most polluting features of American life, accounting for approximately 6 percent of annual greenhouse gas emissions. They are still the primary means of shipping cargo from ports and factories to retailers and other customers across the nation. Tractor-trailers pollute three times as much as freight rail transport. Although there are plans to make them more fuel efficient, one must ask why the American government subsidizes the trucking industry to the tune of tens of billions of dollars annually? Beyond their obvious environmental effects, fully loaded eighteen-wheelers do most of the damage to US roads; it takes ninety-six hundred cars to do the damage that one eighteen-wheeler does. They cost taxpayers hundreds of millions of dollars in highway maintenance, yet trucking companies only pay a small share of these outlays. And it isn't as if there isn't an alternative. Freight trains can do the work of long-distance heavy-duty trucks at a much lower cost to society. So why does the government continue relying on the marketplace to determine the future of the rail industry and rail infrastructure?[34]

Commercial airliners are another component of America's high-emissions transportation sector, accounting for approximately 2 percent of annual greenhouse gas emissions. In comparing the climate impact of aircraft to that of trains, one must consider a variety of factors, including the distance traveled and average occupancy. In an apples-to-apples comparison, however, the trains are roughly twice as energy efficient as the airliners. This assumes, of course, that the planes and trains are both using hydrocarbons. While commercial airliners cannot easily be electrified, however, many trains already are, which means that when they are using renewable or nuclear power they produce essentially zero emissions. Current comparisons also don't take into account the potential adoption of more energy-efficient technologies for trains, such as regenerative braking, which would have a much bigger impact than anything on the horizon for aircraft. Yet, like trucking, commercial aviation has been subsidized to the tune of hundreds of billions of dollars in the past century. This might have made sense during the twentieth century, when air travel was central to economic growth and before we truly understood its subsidiary environmental impacts. But times have changed. It is likely that air transport will remain economically important for transcontinental and transoceanic trips (planes have lower emissions than do ocean liners). When it comes to regional passenger travel, however, high-speed rail can provide the same service at a much lower cost to society. So, once again, why is the government relying on the marketplace to determine the future of the rail industry and rail infrastructure?[35]

The primary answer has long had to do with jobs. But although trucking employs approximately 1.6 million people, the industry is having a harder time finding people that want to take on this demanding and inflexible job; in 2015 there was an estimated shortage of about 50,000 drivers nationwide. The Bureau of Labor Statistics reports that approximately 3.1 million people are already working in green industries, and one study estimates that an increased commitment to transit and rail could produce as many as 3.7 million more jobs, including 600,000 in manufacturing. These are jobs that cannot be offshored, particularly if it were mandated that American companies must provide the steel, trains, and other infrastructure components. So, the lack of government investment in rail transportation cannot really be about throwing people out of work. America could choose a transportation strategy that would employ far more people, while at the same time reducing overall emissions appreciably.[36]

*   *   *

Given that more than 90 percent of daily trips by car and light truck are local, putting a price on carbon and green urbanism is an appropriate way to reduce the overall number of vehicle miles traveled. Pricing carbon is also important for promoting intercity transport because fully accounting for the emissions coming from heavy-duty trucks and commercial airliners will make sustainable alternatives more competitive. But this isn't enough. The nation must also completely reorient current energy and transportation subsidies. Rather than supporting major emitters, the government must commit to a phased infrastructure development plan that favors cargo and passenger rail for long-distance travel. This would represent an indispensible new chapter for the INNATE revolution in transportation. As we saw in chapter 4, in recent decades there has already been something of a renaissance in the use of freight trains. Yet far more needs to be done to transition long-distance transport back onto the rails. (Local transport will likely always rely on trucks, although these can be engineered to be far more energy efficient, because they are traveling shorter distances.) By 2020, freight volume delivered by rail is expected to have doubled in just a dozen years, a trend that will continue into the future. This requires a national freight transportation strategy that returns rail to the forefront of planning, with trucking operating as a local and regional partner. The centerpiece of any new policy direction, not surprisingly, will be a sizable surge in infrastructure spending. After missing an entire generation of needed upgrades, it is time to fix existing problems and also expand capacity. The Association of American Railroads estimates that over the next three decades this will require at least $135 billion in investment. This estimate is based on the assumption that growth will continue at the current pace, although it would be desirable to increase the pace to foster the changeover from environmentally harmful long-distance trucking. While the private sector provides much of this needed capital today, the government will need to step forward to provide incentives and funding to ensure that the nation gets the system it needs. This is an area in which a national infrastructure bank would be an asset. Also useful would be an infrastructure tax credit similar to the one proposed in the failed Freight Rail Infrastructure Capacity Expansion Act, which would have provided a 25 percent credit on specific types of rail investments. Gil Carmichael, the head of the Federal Railroad Administration under President George H. W. Bush, argues that such a tax credit "would help [railroads] restore double track, add new triple track, and new GPS systems. . . . We need to build this new high-speed freight and passenger system, and we need to do it in the next 25 years." There is also potential for public-private partnerships to help develop key rail corridors, which would be more difficult to accomplish without federal backing. Finally, because rail must be electrified, the corridors being developed could also serve as the pathways for the super-grid discussed previously—an idea called Solutionary Rail. Much of the funding for these efforts can be obtained from private investors and by redirecting current subsidies that benefit the trucking industry. There would be a big environmental upside, and those still utilizing the highway network would enjoy better and safer roads.[37]

* * *

In view of the hefty amount of greenhouse gas emissions produced by commercial airliners, it is time for America to invest in a more environmentally friendly alternative, high-speed passenger trains. A century ago the United States had a rail network that was the envy of the world, but the slow recovery of freight networks has not been matched in the transport of intercity travelers. As we saw in chapter 2, during the mid-twentieth century rail corporations, which had once been at the forefront of American economic progress, began steadily losing passengers to cars and airliners. Train travel ultimately slid into near irrelevance. The government did little to help. In fact, the Interstate Commerce Commission helped push the industry over the cliff by prohibiting it from eliminating unprofitable routes. In 1970, after a number of major lines were forced into bankruptcy, President Nixon and a Democratic Congress attempted to revive the industry by creating a government corporation to operate passenger trains. For nearly half a century, Amtrak has operated a mediocre national system, with most of its routes losing money. It simply cannot compete with other highly subsidized forms of intercity transportation. While highways receive $69 billion annually in subsidies, Amtrak gets only $1.4 billion (in contrast to the rail system in China, which received $128 billion in 2014 alone). As the US rail infrastructure has slowly deteriorated and trains have fallen far behind the state of the art, it has been difficult to attract customers. Even the Acela, which operates along the Northeast Corridor and is America's fastest class of trains, faces challenges because it only averages speeds of sixty-eight miles per hour. Still, the service is quite well liked, and rail is now more popular than air travel between New York City and Washington. But elsewhere, where service is even worse, most citizens still prefer car and air travel for intercity trips. How do progressives fix this? It is time to forgo any sense of American exceptionalism and look to the example of other advanced economies, which have long operated superior networks.[38]

Japan was the early innovator in establishing a comprehensive national rail network with high-speed trains rushing along major trunk lines. In 1964, more than five decades ago, Shinkansen bullet trains began operating within the island nation at speeds ranging from 150 to 200 miles per hour; even trains on spur lines travel faster than the Acela. For trips of up to five hundred miles, the Shinkansen is cost competitive with air travel and gets passengers to their destinations more quickly, because stations are centrally located and service is frequent and punctual (trains are rarely late by even seconds). The trains themselves are extraordinarily comfortable and roomy, and the ride is smooth because they employ continuous welded rails with concrete beds. My own experience is that riding these trains is like floating, a pleasant departure from a bumpy flight (or the jarring experience of a typical American train trip). And this is all accomplished with an exemplary safety record. Over a half century, with more than ten billion passengers served, not a single fatality has been caused by a derailment or a collision. Because the traffic volume in Japan is so high, today the rail network is essentially self-financing, with six for-profit railways operating the system. These companies refused, however, to operate in rural areas where a lack of population density made it impossible to operate cost-effectively. To provide service in these regions, the government decided to create a government corporation that would build the needed infrastructure and then lease it to the companies. This has allowed for the development of a truly national passenger network, with all the inherent advantages provided by such an arrangement. Given its long success, it isn't surprising that Japan has become a model for the construction of high-speed rail around the world. It has been modeled throughout Europe, where TGV, ICE, AVE, and Eurostar trains rush passengers around the continent. China has gone even further, building by far the largest high-speed network, comprising roughly 60 percent of all such railways worldwide.[39]

The United States has fallen far behind the rest of the world in this key transportation sector, but there is a path forward to restore its position as a global leader. Progressives need to lead the way by changing the national mind-set with regard to passenger rail. Like other advanced nations, America should see trains as essential public goods. The goal is not profitability but economic connectivity. Neither highways nor airports make a profit; they connect the nation. Why should railways be treated any differently? The objective of a reimagined train network would be to economically connect densely populated cities to promote growth, while at the same time protecting the environment. Trains can do this better than cars and aircraft. That is why they are popular with political leaders outside the Washington bubble, where local interests still hold sway. Mayors are interested in economic development, and high-speed train connections have long been recognized as a key smart-growth policy. The problem is that cities don't have the money to build regional or national networks, although it is clear that if such systems were available passengers would use them. The Northeast Corridor is a prime domestic example, and there are scores of other models around the world. What is needed is a new way of thinking.[40]

The federal government should refrain from viewing high-speed rail (HSR) as a business, at least initially. Instead, it should be approached in the same way as any other infrastructure program used by individuals and companies to conduct the nation's business. A share of the proceeds from a carbon tax should be committed to the development of a twenty-first-century HSR network. The country needs a dedicated system of high-speed corridors with a line running in each direction and pass-through tracks in stations to allow express trains to overtake local trains that make more stops. Once the physical infrastructure is under construction, the Federal Railroad Administration could then be tasked with awarding contracts to for-profit companies to operate the system (although early subsidies might be required until operations and demand mature). It is fairly obvious where the United States should start. Critics have long argued that HSR cannot work in American because of the large size of the country and the relatively low population density. This argument ignores the fact that there are several regions where a large proportion of Americans are concentrated—the Northeast, California, the Pacific Northwest, the Great Lakes, the Piedmont, the Texas Triangle, and Florida. Any rational national plan would begin with networks in each of these regions, where the government should already be promoting greater population density. Once these networks are complete, lines can be added to allow travel between regions, and medium-speed lines can be added to connect smaller cities. Such a phased plan, focusing initially on the highest-density regions, would allow the steady development of HSR, while spreading the considerable costs (and job benefits) over several decades. By no later than midcentury, America could have a national rail network worthy of a great country.[41]

HSR has been discussed in the United States for decades. In the 1980s, six states created boards tasked with developing HSR projects. In the 1990s, Congress passed legislation charging the Federal Railroad Administration with identifying corridors where HSR might operate. In 2008, things really began to move when Californians approved a ballot measure issuing nearly $10 billion in bonds for the construction of a network stretching from San Diego to San Francisco and Sacramento. This was augmented the following year when President Obama championed the inclusion of $8 billion for the development of HSR in the American Recovery and Reinvestment Act. After several Republican governors rejected the money for their states, much of the funding was directed toward California and the Northeast Corridor. This was a positive outcome, because HSR will be easier to promote if it begins operations in these two corridors. Once political leaders see the benefits that accrue to these regions (e.g., reduced hydrocarbon use and greenhouse gas emissions, increased economic development, more reliable intercity service, and better transportation safety), they will be more likely to want HSR in their own regions.[42]

California will be the true test case for North America, mainly because construction has already started there and also because of the steadfast support it has received from the state's progressive-dominated political establishment. The first phase of construction will link Anaheim and San Francisco via California's Central Valley, and the second phase will extend service to San Diego and Sacramento. The initial phase will be complete by approximately 2030. It will propel passengers from the Los Angeles Basin to the Bay Area at speeds of up to 220 miles per hour in just two hours and forty minutes. While the infrastructure is being constructed by the state, a for-profit company will operate the actual train service. One of the reasons why the California HSR plan is so promising is that it only requires reaching political consensus within a single state, and there has been relatively little federal-government involvement beyond the ARRA funding. This would not be the situation in the Northeast Corridor, where a proposed HSR would almost certainly require federal financing and agreement from at least seven state governments (not to mention local governments, which will have hopes and concerns regarding the routing). Proponents believe that the political obstacles can be overcome since the population density in the region can support such a system. In 2010, Amtrak published its vision for an HSR service within this corridor, citing demand forecasts as the primary reason for development. Under this plan, express trains would rush between Washington, DC, and Boston in roughly three hours, making stops in Philadelphia and New York, with local trains stopping at stations in smaller cities. Amtrak found that HSR development would support 44,000 jobs annually during construction and approximately 120,000 permanent jobs. Most importantly, it would allow for at least a fourfold increase in ridership in the Northeast Corridor, providing eighty million passengers with an alternative to cars and planes. The national objective should be to use both the California and Northeast Corridor HSR systems to show that such services can thrive in North America and then begin creating additional regional networks with interregional connectors by midcentury. This will be an essential component of a sustainability revolution's intercity transportation strategy, and if successfully adopted, it will both reduce hydrocarbon use and promote urban density.[43]

Sustainability Revolution, Column 5
Genetically Modified Organisms and Reduced Meat Consumption

Climate change has three major types of impacts: extreme weather events, rising sea levels, and food insecurity. All three are extraordinarily distressing. Extreme weather events like hurricanes and floods or heat waves and droughts can have immediate impacts on sizable populations. Rising sea levels could ultimately result in tens of millions of climate refugees worldwide, while also requiring the investment of billions upon billions of dollars to protect imperiled cities. But food insecurity is by far the greatest long-term threat to humanity, because global warming jeopardizes our ability to feed more than seven billion people. The most vulnerable nations are mostly in Africa and Asia, although countries from Paraguay to Mexico in the Americas also face serious exposure. Mass hunger greatly increases the risk of conflict. We have already seen this as hunger riots have spread across the globe in recent years. The situation will only get worse unless we determine how to feed everyone, even as we face ever-harsher growing conditions. For America, this is both a security issue and a moral imperative. There is good reason to believe the country can continue to feed itself, but civil unrest around the world could create threats at home. It is almost certain that it would do immense damage to key economic partners. In the end, however, the main reason why the United States must take the threat seriously is that it has a moral obligation to assist the peoples of other nations. While it must act in its own self-interest, the country's political culture demands that its people assist those facing calamity. With regard to food security, Americans have an outsized ability to help. Not only can they provide surplus foodstuffs to endangered populations but they can share technology and expertise to foster self-sufficiency.

There are several reasons why the international community is faced with growing food insecurity. On one hand, there is rising demand for foodstuffs. Every year the global population increases by approximately eighty million people. Additionally, globalization has resulted in the emergence of a worldwide middle class of approximately three billion. This growing affluence has resulted in far greater consumption of meat, milk, and eggs. Approximately 30 percent of the planet's landmass is now dedicated to raising grains to feed livestock. This is troubling, because the consumption of red meat, in particular, accounts for ten to forty times as many greenhouse gas emissions as vegetables and grains. Finally, corn has been used increasingly to produce ethanol, an environmentally dubious undertaking that has increased the acreage under cultivation for corn production while reducing the amount of land being used to actually feed people. (This has led to major increases in corn prices in places like Mexico, where the government now has to subsidize tortillas.) On the other hand, there is the potential for dramatic reductions in the supply of foodstuffs. Across the globe, soil erosion is a major threat, and drought-related desertification is overwhelming previously arable land. Warmer weather is reducing agricultural yields, since many plants thrive within very specific temperature ranges. Aquifer depletion is an increasing worry, particularly in nations like the United States, where aggressive irrigation programs have been undertaken for many decades. Melting glaciers and smaller snowpacks are reducing agricultural resiliency, because farmers can no longer rely on a steady flow of water throughout the growing season. Rising sea levels are increasing the salinity of farmland close to oceans, decreasing the yields in these formerly productive regions. Finally, arable land is being lost to suburban and exurban sprawl. The combination of the demand and supply pressures will significantly reduce our ability to feed the global population, unless strategies are embraced to adapt to the new reality.[44]

Unfortunately, neither the local-foods nor the organic-agriculture movement is adequate to prevent rising food insecurity. There are certainly benefits to eating locally. It encourages economic resiliency, builds community, and sometimes delivers tastier food. The environmental benefits of locavorism, however, are questionable, primarily because transportation accounts for only 11 percent of industrial agriculture's energy budget, while farmers driving vehicles with internal-combustion engines to local markets can result in considerable emissions. Shipping food long distances can actually be quite positive if the food is grown in a climate that is particularly well suited to its cultivation, resulting in sizable reductions in production-related emissions, which accounts for 45 percent of total agricultural emissions. Certain regions are simply better suited to producing particular types of crops or even raising certain kinds of livestock. Food miles are a poor metric for determining the climate friendliness of the farming system. Organic agriculture likewise involves major challenges. Perhaps the biggest is that while it can theoretically provide yields similar to those of industrial agriculture, it requires far more farm labor to accomplish this outcome. At least in the developed world, though, there is little evidence that the citizenry wants to return to the farm. There is a reason why fewer than 1 percent of

Americans still work on farms: they don't want to perform such backbreaking labor. Without an abandonment of cities and a return to the farm, therefore, it seems unlikely that organic farming could feed seven to nine billion people. So although returning to some idealized vision of small family farms growing foods organically sounds nice, this is unlikely to happen unless society has no other options. Fortunately, we do have other options.[45]

While some form of industrial agriculture is probably necessary to feed the world, it doesn't have to be the system existing today. The government can work to make it better. It can adopt regulations and policies to improve energy efficiency, encourage more intelligent application of petrochemicals, and foster polyculture. It can fund soil-restoration programs that promote tree shelter-belts, strip cropping, and contour farming. It can raise the price of water so that farmers use this resource more efficiently, while funding the adoption of drip irrigation. It can incentivize local farming and animal husbandry where it makes the most sense. It can mandate that all animals being raised for consumption be treated humanely throughout their lives, including sustaining them with forage that ensures they are healthy. All of this will set the stage for the two biggest changes needed in agriculture to make a sustainability revolution successful.[46]

First, the environmental movement needs to abandon its foolhardy campaign against genetically modified (GM) crops. Green politics is guided by science. It should not be held hostage by emotion. Yet for too long environmentalists have allowed their distrust of large agribusinesses, which fund most of the research related to GM crops, blind them to scientific facts. And those facts are pretty straightforward. According to *Scientific American*, "Overwhelming scientific evidence . . . proves that GMOs are safe to eat, and that they bring environmental benefits by making agriculture more sustainable." Not surprisingly, these genetic modifications are consistent with the selective breeding that humans have been engaged in since the beginning of the Neolithic Revolution. The benefits of GM crops are overwhelmingly positive. They can be customized to need fewer pesticides. They can be tailored to reduce tillage. For example, Monsanto is developing Roundup-ready crops that are herbicide resistant. With reasonable application of herbicides, crops can be planted and weeds can be eliminated without the need for damaging plowing. The result is a noteworthy reduction in soil erosion, as well as reduced hydrocarbon use to propel tillers through the fields. GM crops can be adapted to tolerate more heat and to require less water, which will be vital as the impacts of climate change worsen. They can protect biodiversity, because higher yields mean that less land needs to be cultivated, which will be particularly important in places like Brazil and Africa. Rather than fighting against GM crops, environmentalists should be battling for federal research dollars to more quickly innovate in this important arena. Not only would such funding lead to breakthroughs that promise to improve society's chances of adapting to the climate changes that are already baked into the system but government would control the intellectual property. The United States could accomplish much good around the globe by sharing knowledge about GM crops widely, which would almost certainly be a cost-effective way to bolster friendships with other nations.[47]

Second, Americans must begin eating lower on the food chain, not only because it is environmentally sounder but also because it is better for the collective health of the population. Progressives should build a national policy around Michael Pollan's quite elegant food rule: "Eat food, mostly plants, not too much." This mantra suggests that everyone must focus their diet on real food, not the processed food that has come to overwhelm supermarkets. It also argues that most calories should come from fruits, vegetables, and whole grains. This does not mean that people should not eat any meat or dairy products; it simply means that they should limit their consumption of these products, particularly grain-fed livestock. Not only would this have important health benefits but the environmental advantages would be enormous. As we have already seen, a lot of energy is expended to raise corn-finished beef, about ten times the energy required to raise grass-fed animals. Ruminants are responsible for major methane releases in the form of burps and farts. Even conventionally raised cattle require considerable land. Finally, they are huge consumers of fresh water, putting further pressure on already stressed reserves of this precious resource. If people would eat less meat, particularly beef and lamb, society would enjoy reductions in the use of hydrocarbons, land, and water, as well as fewer direct methane emissions.[48]

How might government accomplish this objective, when most people like to eat meat? Progressives should adopt a three-step approach to reduce emissions in this area. First, government can invest in public-education campaigns about the link between meat consumption and environmental problems. While such campaigns would likely only partially succeed in reducing how much people consume, they would help explain why more aggressive policies are needed. Second, government must provide alternatives to the most resource-intense meat. It is worth saying again: people really enjoy eating animal protein. Americans eat 125 pounds of beef, pork, lamb, chicken, and turkey annually. They also consume 250 eggs, roughly 15 pounds of fish and shellfish, and 630 pounds of milk, yogurt, cheese, and ice cream. The public is likely to resist any approach that seeks to completely eliminate these staples from their diet. But not all meat is created equal. Christopher Weber and Scott Matthews have found that a "dietary shift can be a more effective means of lowering an average household's food-related climate footprint than 'buying local.' Shifting less than one day per week's worth of calories from red meat and dairy products to chicken, fish, eggs, or a vegetable-based diet achieves more GHG reduction than buying all locally sourced food." The bottom line is that people should simply be eating far less red meat, even from animals fed only grass. Pork is somewhat more environmentally friendly, although it too should be consumed only rarely. Dairy products will probably continue to be part of the daily diet, but people should seek to reduce their overall consumption of these foodstuffs. The key alternative to red meat should be humanely raised poultry, eggs, and fish and shellfish.[49]

Third, government must provide financial disincentives to eat the most resource-intense animal protein, while also attempting to reduce overall meat consumption. The first step would be to require higher standards for the humane treatment of animals and the protection of local and regional ecosystems. Because complying with these new regulations would make meat production more expensive, the cost to the consumer would increase. Although this would get the ball rolling, it probably wouldn't be sufficient. Therefore, carbon taxes might have to be levied on livestock operations. If meat costs more, people will eat less. The highest taxes would be imposed on beef and lamb, with more moderate taxes on pork and dairy and the lowest on chicken and eggs. Fish and shellfish, particularly highly sustainable bivalves (e.g., oysters, clams, and mussels), should not be taxed as long as they meet appropriate ecosystem-protection standards. In fact, government might want to promote the development of a more robust aquaculture industry. Raising fish and shellfish is far more sustainable than raising land-based animals. An aquaponics operation, in which plants and fish are in a contained system, can produce fifty thousand pounds of fish and one hundred thousand pounds of vegetables annually on just one acre (one grass-fed cow, producing 75 pounds of meat, requires 8 acres). And it does not take up valuable arable land. There are already successful aquaculture examples, particularly the rice-tilapia-carp approach in Asia and catfish farming in America. But the scale needs to be increased, and government can provide financial incentives to make this happen. Some might argue that decreasing meat consumption and promoting a diet focused lower on the food chain is wishful thinking. Yet researchers at Chatham House, a London-based think tank, have found that when citizens understood the rationale for such interventions they were generally receptive. Given the substantial global emissions resulting from meat production, this sector can no longer be ignored. America must seek to persuade people that by making modest adjustments to their diet they can dramatically reduce agriculture's climate impacts. Then it must put in place specific policies to promote this reorientation toward more plants and less red meat. Combined with more reliance on GM crops, this will allow a sustainability revolution to provide food security for the world population.[50]

A Sustainability Revolution in the Fourth Political Age
An Organizing Framework for the Progressive Movement

The arc of American history bends toward progress, but it doesn't bend on its own. Hydrocarbons played a crucial role in this national advancement, delivering the energy resources needed to provide broad-based security. The INNATE revolutions allowed the growing country to leverage its vast reserves of fossil fuels to become an economic and geopolitical behemoth. The United States was birthed as Great Britain's First Industrial Revolution was beginning, but the new nation's economy was dominated by farming and relatively small-scale sources of renewable energy for many decades. From America's earliest days, providing national and economic security was at the heart of its experiment. The colonists sought independence because they believed their future security was threatened by being part of the British Empire. The First Political Age focused on ensuring long-term national security, which was achieved largely via westward movement and territorial expansion. Under Jefferson and Polk a truly continental nation was created. This was vital to the American Rise, because it not only made the country difficult to invade but also provided access to both the Atlantic and the Pacific trade. While national security was a common focus, however, Federalists and northern states were putting in place a controversial program to advance macroeconomic security. A centerpiece of the resulting push toward economic modernization was the development of water mills, which provided the American System of Manufactures with its primary motive force for nearly a century. The beginning of the First Transportation Revolution provided another piece of the puzzle, as steamboats and canals began reshaping the economic landscape. This transformation was accelerated by the arrival of the railroads, although their initial impact was largely limited to the Northeast. In less than a century the United States had become one of the most vibrant nations on the planet. It was secure at home, and it was quickly developing a model for macroeconomic security that could make it an important commercial power. All this almost entirely without using its immense hydrocarbon reserves. Bringing those resources to bear quickly vaulted the country into the international pantheon.

The Second Political Age began with a civil war that tore the United States apart, although the conflict ended with the union preserved and slavery abolished. Behind the scenes of this wrenching struggle, an underappreciated subplot was the firm establishment of a governmental role in providing macroeconomic security. The First Agricultural Revolution, during the previous political age, had resulted in a more productive farming sector, which meant that surplus labor was available for the development of a radically different economy—a transition that would take more than a half century. Coal powered this new system. It transformed the nation economically, geopolitically, and politically. It established the Hydrocarbon Nation. In the postwar years it would fire smelters in the burgeoning steel industry. Starting with the railroads and continuing with the development of modern urban skylines, this foundational consumer product fundamentally changed the national landscape. It created a country with the infrastructure, cities, and factories capable of dominating the Second Industrial Revolution. With coal use doubling every decade during this period, it was the resource that made everything else possible. Railroads crisscrossed the country from coast to coast, providing it with the most impressive transportation network in the world. Coal-fired steam turbines launched an Electrification Revolution that soon provided the power to illuminate cities, power assembly lines, and run innovative household appliances, all of which dramatically increased the productivity of American workers. All at once, America had an energy source that could power simultaneous revolutions in manufacturing, transportation, and electrification, offering the nation the energy security needed to successfully add the provision of macroeconomic security to the government portfolio.

Without such a rapidly growing economy, it would have been more difficult to continue to pursue the Lincolnian Platform, which had established a system of national banks and a national currency, inaugurated a progressive income tax, passed protective tariff legislation, encouraged western settlement, transformed higher education, and enabled construction of transcontinental railroads. It would also have made it more difficult to begin tackling the worst excesses of the Gilded Age during the Progressive Era, which was done by further advancing macroeconomic security and, perhaps more importantly, by taking the initial steps in providing microeconomic security to all Americans, particularly those left behind by the transition from rural agriculture to urban manufacturing. This two-pronged advance included Theodore Roosevelt's efforts to control corporations, protect consumers, and conserve the natural environment, as well as Woodrow Wilson's reinstatement of a federal income tax and lowering of the basic tariff rate, creation of the Federal Reserve and the Federal Trade Commission, and efforts to increase the voice of labor unions in economic decision making. Although a short-lived but highly effective conservative cycle undid some of these advances temporarily, the foundation had been created for a far more progressive era to come. What the economic and political system needed to make this happen, however, was another hydrocarbon that could expand energy security even further.

The Third Political Age would be dominated by oil. Coal maintained its foundational role in both manufacturing and electrification, particularly as grid power was expanded into rural America, leading to new migration from the Northeast and Midwest to the Sunbelt. But it was petroleum that provided expanded energy security and opened the door for more national economic and geopolitical assertiveness. Oil fueled the Second Transportation Revolution and set in motion the Second Agricultural Revolution. Within the transportation sector, the rise of gasoline-powered automobiles led to a radical makeover of the national environment. Public transit and railroads steadily lost ground as suburbs sprawled across the countryside and highways became the central arteries of the economy. In the postwar years, factories were dedicated to producing the cars and building supplies and appliances needed for this reimagination of the American Dream, while the rapidly expanding electricity grid supplied the suburbs with the power they needed. But all of these developments revolved around the cheap and abundant petroleum reserves that made car-centric suburbanization possible. Meanwhile, in farming, a series of innovations fundamentally altered humans' ability to grow food. This began with the mechanization of tractors and combines, continued with the introduction of petrochemicals, progressed with the development of miracle seeds, and was reinforced by large irrigation projects. The extraordinarily increased yields delivered by these advances allowed America not only to easily feed a rapidly growing population at home but also to become the world's top food exporter. Hydrocarbons were the essential component of this change, as the energy intensity of the farming sector increased eightyfold during the Second Agricultural Revolution.

The maturation of the INNATE revolutions in the Third Political Age established the preconditions for America's emergence as the indispensible world power. The nation had the resources to do something unique in human history. It could provide not only national security but also global security for the process of globalization that it had initiated in the aftermath of the Second World War. And it was able to do this with a startlingly small investment as a share of the overall economy. During the Cold War the nation faced a significant nuclear threat from the Soviet Union, but it was still essentially uninvadable. On the home front, progressive leaders were able to pursue simultaneously both macroeconomic and microeconomic security. Franklin Roosevelt's New Deal launched major new financial regulations (most importantly creating a firewall between investment and commercial banking), gave the Federal Reserve a more active role, cemented the decisive role of organized labor, and made vast investments in infrastructure. At the same time, major microeconomic security policies were being adopted. Higher marginal tax rates provided the needed funding for Social Security, unemployment insurance, and welfare programs like Aid to Families with Dependent Children. In the coming decades, Roosevelt's successors would build upon this foundation, most notably when Lyndon Johnson gained approval for Medicare, Medicaid, and various antipoverty programs. The result of these progressive programs was the Great Compression, the longest period of economic stability and equality in US history. Hydrocarbons, and oil in particular, had made this achievable. These crucial resources provided the energy security that made it possible for the nation to expand security in so many different ways at the same time.

The peaking of American conventional petroleum reserves disrupted this national advance, making it more expensive to maintain energy security. The redirection of US geopolitics to ensure the flow of cheap petroleum from the Middle East cost taxpayers many trillions of dollars. Direct economic losses were just as troubling, with reliance on foreign markets costing the national economy approximately $10 trillion. Meanwhile, direct subsidies, tax breaks, and research funding for oil companies have added billions in additional costs. And this doesn't even include the immense indirect environmental and health consequences of the nation's continued reliance on oil. These combined societal outlays placed major fiscal pressures on the national government, leading to an erosion of the gains made during the progressive cycles of the Third Political Age. As a result, the conservative cycle that began with Reagan's election was characterized by a number of developments that steadily weakened America both at home and abroad. Energy insecurity and fundamental constitutional weaknesses combined to create highly damaging political dysfunction, which was exemplified by the inability of the federal government to find solutions to big problems facing the nation. Energy insecurity and an extreme conservative backlash against progressivism led to an unprecedented financialization of the economy, which caused a deep financial crisis. Energy insecurity, extreme conservatism, and financialization reversed the Great Compression, giving rise to destabilizing levels of income and wealth inequality. Energy insecurity, extreme conservatism, and income and wealth inequality brought about severe inequity in an educational system rooted in federalism, thus destroying the key ladder to realizing the American Dream. Energy insecurity and extreme conservatism gave birth to fiscal constraints that led to the abandonment of the national commitment to maintaining and growing a cutting-edge infrastructure. Energy insecurity and political dysfunction occasioned the collapse of national prestige abroad, with extensive geopolitical and economic consequences. Finally, extreme conservatism and political dysfunction lead to an inability to address the growing climate crisis by dramatically reducing the use of hydrocarbons and finding alternatives to restore the energy security that was critical to America's economic success. These postpeak developments were highly damaging to the country and its citizens.

We have now entered the Fourth Political Age, brought about by dramatic generational and demographic shifts that changed the nature of the American electorate. It was also caused by a return of energy security, delivered by increased production of unconventional oil and gas. Once elected, President Barack Obama gained approval of the most ambitious collection of progressive legislation in a half century. This included a Keynesian stimulus package, regulation of the financial sector, educational reforms, and a historic health care law. And Obama took unprecedented action to fight the climate crisis. This included championing extraordinary new investments in sustainability research, as well as executive actions to dramatically reduce the nation's use of coal and petroleum. Despite the momentous nature of these accomplishments, compared with the inauguration of past progressive cycles they were drastically constrained. This was a consequence of the deep political dysfunction of the previous conservative cycle, which proved so difficult to quickly reverse that the 2016 presidential election resulted in the elevation of a leader who seems committed to erasing most of his predecessor's accomplishments—although it is far from clear whether these efforts will be successful. The outcome of the most recent election has made it clear that American democratic and republican principles must be restored, which will likely require changes to the US Constitution. America needs a Democracy Amendment, which would limit citizenship rights to natural persons, provide government with clear authority to regulate campaign financing, mandate nonpartisan congressional redistricting, eliminate the Electoral College, and guarantee a fundamental right to vote. America also needs a Republican Amendment, which would make it easier to amend the Constitution, eliminate the US Senate, reduce the size of the constituency for each US representative (while lengthening representatives' terms to four years), mandate a more rigorous annual schedule for Congress, require quadrennial budget resolutions, and compel Congress to confirm or reject all presidential nominations within ninety days. While gaining passage of these constitutional amendments would be extraordinarily difficult, they would be a useful component of any organizing framework for the progressive movement.

* * *

The central component of a progressive organizing framework in the Fourth Political Age is a sustainability revolution. As we have seen throughout this book, nothing is more important to continued American prosperity than ensuring long-term energy security. Thanks to new extraction technologies, for the first time in nearly a half century the nation is in a position to provide this security almost entirely through the use of hydrocarbons. But at what cost? The fact that the United States possesses these resources doesn't necessarily mean it has to use all of them, particularly when they are known to be the key contributors to the growing climate crisis. While it might be true that America is in a particularly good position to adapt to global warming, this does not mean that it is in the nation's best interest to go down a path where it seeks to flourish as the rest of the international community struggles to survive. The world would be a better place if the country not only dramatically reduced its own use of hydrocarbons but also helped proliferate the technologies and approaches needed to reduce global greenhouse gas emissions. Taking on this challenge would enhance America's national, macroeconomic, microeconomic, and climate security. It would make Americans safer at home by reducing their need to engage geopolitically in parts of the world where they have few interests beyond promoting the stability of hydrocarbon energy markets. It would make the national economy more stable by restoring and expanding the national infrastructure. It would reduce the economic fears faced by tens of millions of Americans by providing the resources needed to make fairness the cornerstone of political life. Finally, it would mitigate the negative environmental impacts of the country's 150 years of nearly unfettered exploitation of hydrocarbons. Why wouldn't progressives lead the way to a more sustainable future? Isn't the whole point of the American government to make life better for We the People? Whether based in the Democratic or the Republican Party, or some combination of the two, a progressive movement centered on promoting sustainability would be an unstoppable force if its leaders could educate the citizenry about the overwhelming benefits to the nation that would result. This should be the great task of the United States in the coming decades.

As we have seen, there are five key steps to executing such a sustainability revolution. First, and most important, progressives should push for the adoption of an aggressive carbon tax to make coal and oil (and eventually gas) more expensive, while opening market space for solar, wind, and nuclear power. The resulting revenues would help fund other measures within the larger revolution, while also providing enhanced microeconomic security in the form of payroll tax relief for the poorest Americans. Second, progressives should support policies that foster greater urban density. Using regulations, financial incentives, and growth models, the nation can transform how residential and commercial buildings are constructed and also encourage regional and urban planning that favors mixed-use development and public transit. Third, progressives should expand national energy security with a crash clean energy research program, a dedicated energy efficiency effort, and a new mix of energy resources. As government eliminates the use of coal and oil, it should build a temporary gas bridge, renew a commitment to next-generation nuclear power, construct a national supersmart grid, and invest heavily in solar and wind energy. Fourth, progressive should support the construction of a twenty-first-century transportation infrastructure that favors a reinvigorated freight rail network and an intercity high-speed passenger rail. Finally, progressives must work to incentivize a transition to eating lower on the food chain by making resource-intense animal protein sources more expensive while at the same time providing ethically raised alternatives. Along with more reliance on GM crops, this will allow the nation to continue to play an outsized role in providing food security for the global population while at the same time reducing the negative impacts of the current industrial agriculture system. Not only would this straightforward and eminently feasible plan make the United States far more sustainable but it would enable it to confidently tackle other problems that face humanity in the modern age. If pursued successfully, this plan would help guarantee the nation prosperity within a global order that seeks to emulate its model for economic and political success. This would represent the kind of American Century of which all its citizens could be proud.

Napatree Point

Sand between my toes, waves breaking at my feet, sun on my face, and my dog Izzy trotting by my side. This is how most of my summer days begin. After tumbling sleepily out of bed just as the sun creeps over the eastern horizon, I drive a few miles for a morning run at my favorite place on this small planet we all call home—Napatree Point. This sandy ribbon of land juts out from the Rhode Island coast, stretching from the village of Watch Hill toward Fishers Island and providing shelter for Little Narragansett Bay. It is a place of great beauty, with bent grasses dancing in the coastal breezes, endangered piping plovers darting into the ebbing waves, and dramatic sea views across Block Island Sound. While there are many places around the world that are more endangered with regard to the potential impacts of global warming, this natural gem has become my own frame of reference for the rapidly developing crisis. This is mostly because Napatree is so much a part of my lived experience but also because this small spit of land has such an interesting history. Ancient geological processes shaped it, and more recently it has been reshaped by both human activities and extreme weather events. For me it is not only a setting for physical and spiritual restoration but also a place where I can consider the fragility of our circumstances here on Earth.

During the academic year, I live with my partner, Kate, and my son, Sam, in a charming little college town in southwestern Ohio. In the early nineteenth century, the state decided to locate its second public university in an almost uninhabited wilderness area on the border with Indiana. Miami University was named after an indigenous tribe that had been forcibly removed from the area, while the village that was founded as a home for the university was given the classic name Oxford. More than two hundred years later, the university and town are still thriving. Like our fellow Oxonians, we enjoy the town's lively historic district, pleasant neighborhoods, and wide range of cultural, community, and athletic events. One of our favorite features is the extensive university-owned nature preserve, with seventeen miles of hiking trails located on more than one thousand acres of leafy forests. This is a place where we are teaching Sam to love and value the natural world. It is here where we tramp along Harker's Run Creek in search of ancient marine mammals deposited in the region's limestone and dolomite bedrock millions of years ago, when a shallow sea covered this part of the globe. It is here where we explore the glacial deposits along Collins Creek, where large ice sheets advanced and retreated twenty-five thousand years ago. And it is here where we explore the seasonal changes of a deciduous forest filled with beech, maple, and oaks. It is a wonderful learning laboratory, particularly for one who studies climate policy. It enhances life in an already lovely town. Yet our coastal upbringing continues to pull Kate and me back to the ocean.

As this yearning to return to the shore begins tugging at my heart after a few months in our small corner of the American Midwest, I often think of President Kennedy's explanation for this phenomenon. In a speech delivered in Newport, just forty miles from Napatree, JFK said:

> I really don't know why it is that all of us are so committed to the sea, except, I think, it is because in addition to the fact that the sea changes, and the light changes, and ships change, it is because we all came from the sea. And it is an interesting biological fact that all of us have, in our veins, the exact same percentage of salt in our blood that exists in the ocean, and, therefore, we have salt in our blood, in our sweat, in our tears. We are tied to the ocean. And when we go back to the sea, whether it is to sail, or to watch it, we are going back from whence we came.[1]

Every summer and winter break, my family answers this call by returning to the place where we feel most at home. But even as we make this journey every year, we are aware that this gorgeous coastline is in great danger from storm surges and hurricanes and sea level rise. And we are reminded that there are more critical locations around the world where large portions of humanity face far greater dangers.

* * *

Approximately 115,000 years ago, the Wisconsin Glaciation began as polar ice sheets advanced toward the equator. Over the next hundred thousand years the Laurentide Ice Sheet, which covered five million square miles, including almost all of modern Canada, fundamentally altered the continental landscape. Roughly twenty-five thousand years ago, it briefly stretched as far south as New England and the Upper Midwest. In some areas it was a couple of miles thick, although Rhode Island was under a thinner fringe of the ice sheet. Napatree is part of what Dorothy Sterling calls the "Outer Lands," the remains of the shoreline that existed at the time of the last glacial maximum, when the global sea level was nearly four hundred feet lower. "When the glacier reached the soft margins of the coastal plain," writes Sterling, "it scooped out basins of bays and sounds and pushed on toward the sea. Halted by currents of warm air, it dumped the load of rocks and pebbles it had carried from the mainland. These glacial dump heaps, torn from once-solid bedrock, formed ridges hundreds of feet high. Known as moraines, they are the framework of the Outer Lands." The southernmost terminus of the Laurentide can be seen today in the hills that run from Queens to Montauk on Long Island before reemerging on Block Island, Martha's Vineyard, and Nantucket. A recessional moraine runs parallel to this boundary and traverses northern Long Island, crosses Block Island Sound to Fishers Island and Napatree, and encompasses the shore and saltwater ponds of Rhode Island's Washington County before jumping across Buzzard's Bay to pass through Upper Cape Cod.[2]

Napatree was created by glacial deposition from the great ice sheet as it paused in its northern retreat approximately twenty thousand years ago. Over many millennia the sea battered the shores of the Outer Lands and turned boulders into rocks, rocks into pebbles, and pebbles into sand, ultimately forming the beaches that came to border these realms. When Europeans arrived to colonize North America in the seventeenth century, Napatree was a three-mile-long, sickle-shaped peninsula that was thickly wooded (thus a Nape of Trees). In 1745 a lighthouse was built just a few hundred yards away, at the southernmost point of the Rhode Island colony, and the village of Watch Hill slowly developed just to the east of the neck. Eighty years after the lighthouse was established, the infamous Great September Gale of 1815 struck. Estimated to have been a category four hurricane, the storm wreaked havoc from Long Island to New Hampshire. The storm killed nearly forty people all told, but the primary victims on Napatree were the trees. The entire spit was totally deforested in a matter of hours.[3]

In the late nineteenth century, Watch Hill and Napatree began to be slowly developed. Watch Hill evolved into an exclusive summer resort for families who were either priced out of or uninterested in fast-paced Newport. At the same time, the American military became interested in Napatree. A joint army-navy task force building a network of artillery batteries from Galveston to Maine selected the neck as a good location for guns that would protect the entry to Long Island Sound. In 1898, Fort Mansfield and its three gun emplacements were constructed at the elbow of the peninsula, and a road was built from Watch Hill. When it was determined after an extensive military exercise that the guns were too vulnerable, however, the entire fort was decommissioned, after just eight years. Two decades later the land was sold to a developer, and thirty-nine summer homes were built along Fort Road. R. A. Scotti describes these elegant dwellings as "two and three stories of weathered shingles, their broad front porches a few strides from the Atlantic. Cement walls, three, maybe four feet high, protected them from the sea's darker moods." On 21 September 1938 these barricades would be put to the test.[4]

On that final day of summer, a colossal hurricane galloped up the Atlantic coast toward New England. It was the worst natural disaster in the region's storied history, with the destructive power of several nuclear bombs. And unlike those living in Watch Hill today, the local residents had no way to know that it was coming. "Rampaging through seven states in seven hours, it would rip up the famous boardwalk in Atlantic City, flood the Connecticut River Valley, and turn downtown Providence into a seventeen-foot lake." Just before four o'clock in the afternoon the hurricane slammed into Napatree and Watch Hill, when a thirty-foot wave rushed ashore. In the coming minutes wave after powerful wave bypassed the seawall and pounded into the cottages on Fort Road, where there were forty-two souls inside. In short order, several of the houses were swept off their foundations and deposited in the middle of Little Narragansett Bay. The wind was gusting at up to 150 miles per hour, lashing the coast with sand and water and debris. As residents attempted to evacuate, the violent sea carried them away. As the minutes ticked by, more and more of the solidly built houses followed them, bursting apart in the face of the unimaginable powers of the storm. Finally only one home remained, with Jeff and Catherine Moore, their four children, a close friend, and three servants taking cover inside. In a panic, the family ran upstairs to the second floor and then the third floor and then the attic as the structure was enveloped by the angry Atlantic Ocean. Then most of the roof was torn off by a giant gust of wind, and the attic floor was cast into the stormy waters with the family clinging on for dear life. Wave after wave rushed over their heads, but the ersatz raft kept them above water before it providentially delivered them to safety on Barn Island, on the far side of the bay. They had been saved, but 15 people had been killed on Napatree that fateful day; 433 were killed in Rhode Island alone, 682 overall.[5]

By the time the storm blew itself out, Napatree had quite literally been turned into several small islands. Two breachways had been ripped through the peninsula by the hurricane, one just northeast of Watch Hill and the second separating the sand spit that had reached up toward Stonington, Connecticut. The former would eventually be healed, but the latter was never recovered, and today there is a thirty-five-acre island called Sandy Point in its place. The peninsula has remained uninhabited by humans ever since the great storm, and visitors are often incredulous when told that gracious cottages once lined the beach. Watch Hill ultimately bounced back from the disaster, with most of the hotels that were destroyed being replaced by large summerhouses built by affluent families from New York, Philadelphia, and Cincinnati. As Stephen Birmingham wrote, Watch Hill became "an Andorra of Victoriana on the New England shore."[6] This sense has been recently augmented by the restoration of both the Ocean House Hotel and the Watch Hill Inn, which have attracted new vacationers to the resort town. Since 1938, storms and weathering have pushed Napatree Point more than two hundred feet toward the mainland, a testament to nature's immense power. Hurricane Sandy essentially subsumed the spit just a few years ago, but it largely survived intact. Nearby Misquamicut, a more modest village where my mothers-in-law own a house, was not so lucky. The storm surge destroyed the Hotel Andrea and numerous seaside businesses, while tearing a number of beach houses off their foundations. While not nearly as damaging as previous extreme weather events, the storm served as a cautionary note for those concerned about the vulnerability of this entire coastline to climate change.

I hope to retire to the Rhode Island coast with Kate in a couple of decades, so I am very cognizant of the potential impacts storm surges and sea level rises might have in the area. While the best-case scenarios are troubling, the worst-case scenarios could see entire villages subsumed by the Atlantic Ocean. Not surprisingly, this gets me thinking about the potential effects of extreme weather events and sea level rise and food insecurity in far more densely populated regions. Many Americans, and many citizens of the world, share concerns for places they love and depend upon. They share concerns for the people they love. They share concerns about the ability of our species to continue to thrive on this planet. The strength of the scientific consensus has convinced me that we must all support political parties that will place this issue at the top of the national agenda. I am particularly keen about leaders who can articulate a vision in which climate security acts as an organizing principle for repairing democratic and republican government while maintaining national security and restoring economic security. This is the type of program that would attract Millennials and nonwhites, not just because it sounds good but because it represents the type of progressive change they yearn for in this new political age. There is no time to waste. While the pause button has been pushed on the current cycle, the return of energy security and the changing demographics of the nation mean that the opportunity for major political action remains. But we need leaders who can overcome the dysfunction of the previous political age and inaugurate a sustainability revolution that can ensure that America will continue to serve as a model for the international community.

* * *

Napatree Point is a protected conservation area today, although the headland remains strikingly vulnerable. The strand facing Block Island Sound is quite popular, with beachgoers trekking over from Watch Hill on sunny days to enjoy the sand and surf. Trails transport birdwatchers to the small saltmarsh, where they can spy a variety of species (including piping plovers, least terns, oystercatchers, and ospreys) that stop over on their migrations both north and south. During summer, the protected bayside hosts dozens of watercraft moored in the shallow water, while on a foggy morning one can see clammers hunting bivalves. At dawn or dusk one might also spy a magnificent local fox that lives on the point. For nearly two decades I have been taking morning runs on this gorgeous peninsula. When Sam was a bit younger, he and I would often return in the afternoon because the protected beach was easier for him to navigate. When the surf is too big in Misquamicut, the whole extended family will often pile into a car for the quick drive to Watch Hill in search of more manageable waves to bodysurf, an activity about which we are all passionate. Napatree has become deeply entangled in our lives and occupies a special place in our hearts.

One clear piece of evidence for Napatree's storied past can still sometimes be seen on this stunning beach. Traversing the shoreline after a storm, you will find small chunks of coal liberally sprinkling the sand. A couple hundred yards out from the shore, buried in the seabed, the foundations of the elegant cottages that once graced this sandy spit can still be found. As the breakers work the ocean bottom, they dislodge this hearty hydrocarbon that had been used by the former residents to heat their homes. As I run past these small pieces of history on a sunny morning, I often find myself thinking about the role prehistoric organisms have played in the rise of the American nation. At the same time, I wonder whether humanity can muster the will to keep similar fossil fuels buried deep inside the Earth to prevent the worst impacts of climate change. If not, the entire planet could face the same level of destruction wrought on this small strip of land eighty years ago. Yet I am still filled with hope that a new generation of leaders can turn the tide. I am confident that my descendants will be able to return to this lovely spot to enjoy the sand between their toes, the waves breaking at their feet, and the warm sun on their faces.

Prologue

1. Riki Ott, *Not One Drop: Betrayal and Courage in the Wake of the Exxon Valdez Oil Spill* (White River Junction, VT: Chelsea Green, 2007), 21 [quote], 24–25.

2. Ibid., 35–36.

3. *Oil Spill Case Histories, 1967–1991: Summaries of Significant U.S. and International Spills* (Seattle: NOAA Hazardous Materials Response and Assessment Division, 1992); "Oil Spill Facts," *Exxon Valdez* Oil Spill Trustee Council, accessed 13 July 2011, http://www.evostc.state.ak.usindex .cfm?FA=facts.home.

4. Ibid.

5. "Oil Spill Facts."

6. Ott, 46.

7. "Oil Spill Facts."

8. Ott, 52.

9. Ibid., 77–78.

10. Ibid., 78.

11. "Exxon Valdez Case Timeline," *Anchorage Daily News*, 25 June 2008.

12. Editorial Board, "Exxon Verdict: Supreme Court Makes Life Easier for Corporate Wrongdoers," ibid.

13. Ibid.

14. Charles H. Peterson, Stanley D. Rice, Jeffrey W. Short, Daniel Esler, James L. Bodkin, Brenda E. Ballachey, and David B. Irons, "Long-Term Ecosystem Response to the Exxon Valdez Oil Spill," *Science*, 19 December 2003, 2082–86.

15. Richard E. Thorne and Gary L. Thomas, "Herring and the 'Exxon Valdez' Oil Spill: An Investigation into Historical Data Conflicts," *ICES Journal of Marine Science*, 1 January 2008, 44.

Introduction

1. Timothy Mitchell, *Carbon Democracy: Political Power in the Age of Oil* (London: Verso, 2011).

2. V. O. Key Jr., "A Theory of Critical Elections," *Journal of Politics* 17, no. 1 (February 1955): 3.

3. E. E. Schattschneider, *The Semi-Sovereign People: A Realist's View of Democracy in America* (Chicago: Holt, Rinehart & Winston, 1960).

4. David R. Mayhew, *Electoral Realignments: A Critique of an American Genre* (New Haven, CT: Yale University Press, 2004); A. James Reichley, "The Future of the American Two-Party System in the Twenty-First Century," in *The State of the Parties: The Changing Role of Contemporary American Politics*, ed. John Clifford Green and Daniel J. Coffey (Lanham, MD: Rowman & Littlefield, 2007).

5. Frank R. Baumgartner and Bryan D. Jones, *Agendas and Instability in American Politics* (Chicago: University of Chicago Press, 1993), 3–24.

6. Karl Manheim, "The Problem of Generations," in *Essays on the Sociology of Knowledge*, by Karl Manheim, ed. Paul Kecskemeti (New York: Routledge, 1952), 286.

7. Ibid., 292.

8. Alan Spitzer, "The Historical Problem of Generations," *American Historical Review* 78, no. 5 (December 1973).

9. Ibid.

10. Arthur M. Schlesinger Jr., *The Cycles of American History* (New York: Houghton Mifflin, 1999), vii.

11. Ibid., 3–50.

12. William Strauss and Neil Howe, *Generations: The History of America's Future, 1584 to 2069* (New York: Quill, 1991).

13. Ibid., 74.

14. Ibid., 35.

15. John Kenneth Galbraith, *The Affluent Society* (New York: Houghton Mifflin, 1998), chap. 8.

16. Ibid.

17. Ian Morris, *Why the West Rules—For Now: The Patterns of History, and What They Reveal About the Future* (New York: Picador, 2010), 161, 281, 583.

18. Ibid., 621–45; J. R. McNeill, *Something New Under the Sun: An Environmental History of the Twentieth Century* (New York: Norton, 2001), chap. 1.

Chapter 1 · Steam, National Security, and the First Political Age

1. Richard Manning, *Against the Grain: How Agriculture Has Hijacked Civilization* (New York: North Point, 2004), 3–21.

2. Richard Wrangham, *Catching Fire: How Cooking Made Us Human* (London: Profile Books, 2009), 1–14.

3. Manning, 3–21 [quote on 20].

4. Tom Standage, *An Edible History of Humanity* (New York: Walker, 2009), 16–17 [quote on 16].

5. Ibid., 17–23.

6. Manning, 37–45 [quote on 37]; Standage, 25–59 [quote on 32]; Jared Diamond, *Guns, Germs, and Steel: The Fates of Human Societies* (New York: Norton, 1997), 169–74.

7. Manning, 54–56 [quote on 54]; Carolyn Steel, *Hungry City: How Food Shapes Our Lives* (London: Random House, 2009).

8. Standage, 63–104 [quotes on 67, 82]; Robert Curley, ed., *Fossil Fuels* (New York: Britannica Educational, 2012), 103–12.

9. Robert C. Allen, *Global Economic History: A Very Short Introduction* (Oxford: Oxford University Press, 2011), 1–8; Daron Acemoglu, Simon Johnson, and James Robinson, "The Rise of Europe: Atlantic Trade, Institutional Change, and Economic Growth," *American Economic Review* 95, no. 3 (June 2005): 546–51; Niall Ferguson, *Empire: The Rise and Demise of the British World Order and the Lessons for Global Power* (New York: Basic Books, 2002), 15–16; Paul Kennedy, *The Rise and Fall of the Great Powers: Economic Change and Military Conflict from 1500 to 2000* (New York: Vintage Books, 1987), 76–85; Kevin Phillips, *American Theocracy: The Peril and Politics of Radical Religion, Oil, and Borrowed Money in the 21st Century* (New York: Viking, 2006).

10. Ferguson, 16–19; Alfred Thayer Mahan, *The Influence of Sea Power upon History, 1660–1783* (Cambridge, MA: John Wilson and Son, University Press, 1890), 29.

11. E. J. Hobsbawm, *Industry and Empire: From 1750 to the Present Day* (New York: New Press, 1968), 3, 46 [quote]; David S. Landes, *The Unbound Prometheus: Technological Change 1750 to the Present* (Cambridge: Cambridge University Press, 1969), 1 [quote], 41–51; Allen, 1, 27 [quote].

12. Paul Roberts, *The End of Oil: On the Edge of a Perilous New World* (New York: Houghton Mifflin, 2004), 24–26 [quote on 25]; Daniel Yergin, *The Prize: The Epic Quest for Oil, Money and Power* (New York: Free Press, 1991), 8; Barbara Freese, *Coal: A Human History* (New York: Penguin, 2004), 1, 15–16; Norbert Berkowitz, *Fossil Hydrocarbons: Chemistry and Technology* (San Diego, CA: Academic, 1997).

13. Thom Hartmann, *The Last Hours of Ancient Sunlight: The Fate of the World and What We Can Do Before It's Too Late* (New York: Three Rivers, 2004), 2; Curley, xiii–xiv; Jeff Goodell, *Big Coal: The Dirty Secret behind America's Energy Future* (New York: Mariner Books, 2006), 8–11; Freese, 17–21.

14. Roberts, 26–27; Freese, 53–56.

15. Roberts, 22–24 [quote on 22–23]; Vaclav Smil, *Energy in World History* (Boulder, CO:

Westview, 1994), 165 [quote]; Allen, 32–37; Alfred W. Crosby, *Children of the Sun: A History of Humanity's Unappeasable Appetite for Energy* (New York: Norton, 2006), 70–76.

16. Ferguson, 43.

17. John R. Alden, *A History of the American Revolution* (New York: Knopf, 1969), 3–16; Ferguson, 49–69.

18. Edmund S. Morgan, *The Birth of the Republic: 1763–1789* (Chicago: University of Chicago Press, 1977), 14–60; Ferguson, 73–74 [quote on 73]; Samuel Adams, "The Rights of the Colonists" (1772), in *American Political Thought: A Norton Anthology*, ed. Isaac Kramnick and Theodore Lowi (New York: Norton, 2008), 108 [quote]; Alden, 13.

19. Morgan, 68–86; Ron Chernow, *Washington: A Life* (New York: Penguin Books, 2011).

20. Morgan, 128–43.

21. "From Thomas Jefferson to James Madison, 20 December 1787," National Archives, accessed 10 February 2017, tinyurl.com/z8fetq9.

22. Chernow, 770–71.

23. Alexander Hamilton, *The Report on a National Bank*, 14 December 1790, tinyurl.com/yd z64fup; Hamilton, *Opinion on the Constitutionality of the Bank*, 1791, tinyurl.com/5ug8n4j; Chernow, 650.

24. George Washington, Farewell Address, 19 September 1796.

25. John H. Aldrich, *Why Parties? A Second Look* (Chicago: University of Chicago Press, 2011), 72–83; Ron Chernow, *Alexander Hamilton* (New York: Penguin Books, 2004), 270–91; Louis Johnston and Samuel H. Williamson, "What Was the U.S. GDP Then?," MeasuringWorth, 2011, tinyurl .com/yd73ha50; David McCullough, *John Adams* (New York: Touchstone, 2001), 467–567.

26. Thomas Jefferson, first inaugural address, 4 March 1801.

27. Stephen E. Ambrose, *Undaunted Courage: Meriwether Lewis, Thomas Jefferson, and the Opening of the American West* (New York: Touchstone, 1996), 52–57 [quotes on 52, 56].

28. Albert Gallatin to Thomas Jefferson, 18 December 1807, tinyurl.com/ycovfomv; Gordon S. Wood, *Empire of Liberty: History of the Early Republic, 1789–1815* (Oxford: Oxford University Press, 2011), 174–208.

29. Henry Adams, *History of the United States: Volume IV* (New York: Scribner & Sons, 1890), 288; Donald R. Hickey, *The War of 1812: A Forgotten Conflict* (Urbana-Champaign: University of Illinois Press, 1989), 21; Francis D. Cogliano, *Thomas Jefferson: Reputation and Legacy* (Charlottesville: University of Virginia Press, 2008), 250; Robert W. Tucker and David C. Hendrickson, *Empire of Liberty: The Statecraft of Thomas Jefferson* (Oxford: Oxford University Press, 1990), 204–10, 222; Douglas A. Irwin, "The Welfare Cost of Autarky: Evidence from the Jeffersonian Trade Embargo, 1807–09," *Review of International Economics* 13 (2005): 631–45.

30. Ibid.

31. Hickey, 304–5; Daniel Walker Howe, *What Hath God Wrought: The Transformation of America, 1815–1848* (Oxford: Oxford University Press, 2007), 79–90.

32. Michael Lind, *Land of Promise: An Economic History of the United States* (New York: Harper, 2012), 51 [quote], 85–87 [quote on 87]; Chaim M. Rosenberg, *The Life and Times of Francis Cabot Lowell: 1775–1817* (Lanham, MD: Lexington Books, 2011), 297–308 [quote on 300]; Barbara M. Tucker, "The Merchant, the Manufacturer, and the Factory Manager: The Case of Samuel Slater," *Business History Review* 55, no. 3 (Autumn 1981): 297–313.

33. David A. Hounshell, *From the American System to Mass Production, 1800–1932* (Baltimore: Johns Hopkins University Press, 1984), 17 [Joseph Wickham Roe quote], 25; Thomas Jefferson to John Jay, 30 August 1785, tinyurl.com/ybfg42kr.

34. Hounshell, 33–46 [quote on 33]; Merritt Roe Smith, "Military Entrepreneurship," in *Yankee Enterprise: The Rise of the American System of Manufactures*, ed. Otto Mayr and Robert C. Post (Washington, DC: Smithsonian Institution Press, 1981), 65–80.

35. Hounshell, 46–123.

36. John Lauritz Larson, *Internal Improvement: National Public Works and the Promise of Popular Government in the Early United States* (Chapel Hill: University of North Carolina Press, 2001), 2–10; Christian Wolmar, *The Great Railroad Revolution: The History of Trains in America* (New York: Public Affairs, 2012), 5–6.

37. George Rogers Taylor, *Economic History of the United States*, vol. 4, *The Transportation Revolution: 1815–1860* (New York: Rinehart, 1951), 56–73; Louis C. Hunter, *Steamboats on the Western Rivers: An Economic and Technological History* (Cambridge, MA: Harvard University Press, 1949), 4; Wolmar, 6–7 [quote on 7]; Frederick Moore Binder, *Coal Age Empire: Pennsylvania Coal and Its Utilization to 1860* (Harrisburg: Pennsylvania Historical and Museum Commission, 1974), 85–110.

38. *Gibbons v. Ogden* (1824).

39. Peter L. Bernstein, *Wedding of the Waters: The Erie Canal and the Making of a Great Nation* (New York: Norton, 2005), 22 [quote], 208; John F. Stover, *American Railroads*, 2nd ed. (Chicago: University of Chicago Press, 1997), 5–8; Howe, 117–20; Wolmar, 8–9; Taylor, 15–31.

40. Wolmar, 9–13 [quote on 12]; Taylor, 15–31.

41. Wolmar, 2–4 [quote on 2]; Taylor, 74.

42. James Dilts, *The Great Road: The Building of the Baltimore & Ohio, The Nation's First Railroad, 1828–1853* (Stanford, CA: Stanford University Press, 1996), 1–7 [quote on 1]; Wolmar, 18–23; Taylor, 79; Stover, 20.

43. Stover, 28.

44. Samuel H. Williamson, "Seven Ways to Compute the Relative Value of a U.S. Dollar Amount, 1774 to Present," MeasuringWorth, 2017, tinyurl.com/yd73ha5o; Wolmar, 28–50; Taylor, 80–102; Stover, 24–34; Binder, 111–32.

45. Andrew Jackson, bank veto message, 1832.

46. Howe, 505.

47. Robert W. Merry, *A Country of Vast Designs: James K. Polk, The Mexican War, and the Conquest of the American Continent* (New York: Simon & Schuster, 2009), 19, 28–29.

48. Ibid., 184–94 [quote on 194].

49. Frederick Douglass, "What to the Slave Is the Fourth of July," 1852.

50. John C. Calhoun, speech on the reception of abolition petitions, 1837, tinyurl.com/y9t sn84q; Alexis de Tocqueville, *Democracy in American*, ed. Bruce Frohnen (Washington, DC: Regnery, 2002), 285; Daron Acemoglu and James A. Robinson, *Why Nations Fail: The Origins of Power, Prosperity, and Poverty* (New York: Crown, 2012).

Chapter 2 · *Coal, Macroeconomic Security, and the Second Political Age*

1. James M. McPherson, *Battle Cry of Freedom: The Civil War Era* (Oxford: Oxford University Press, 1988), 6–7; Ira Katznelson, Mark Kesselman, and Alan Draper, *The Politics of Power: A Critical Introduction to American Government* (New York: Norton, 2011), 67–68.

2. Ibid.

3. William Strauss and Neil Howe, *Generations: The History of America's Future, 1584 to 2069* (New York: Quill, 1991); Sidney M. Milkis and Michael Nelson, *The American Presidency: Origins and Development, 1776–2011* (Washington, DC: CQ Press, 2012), 177.

4. Abraham Lincoln, first inaugural address, 1861.

5. Abraham Lincoln, Gettysburg Address, 1863.

6. Michael Lind, *Land of Promise: An Economic History of the United States* (New York: Harper, 2012), 131.

7. Abraham Lincoln, quoted in Lind, 140; Milkis and Nelson, 168–69.

8. *Santa Clara County v. Southern Pacific Railroad*, 118 U.S. 394 (1886); Robert Engler, *The Politics of Oil: A Study of Private Power and Democratic Directions* (New York: Macmillan, 1961), 34–64.

9. Henry George, *The Annotated Works of Henry George*, ed. Francis Peddle and William Peirce, vol. 2, *Progress and Poverty* (Madison, NJ: Fairleigh Dickinson University Press, 2017), 55.

10. Alfred D. Chandler Jr., "Anthracite Coal and the Beginnings of the Industrial Revolution in the United States," *Business History Review* 46, no. 2 (Summer 1972): 141–45; Christopher F. Jones, *Routes of Power: Energy and Modern America* (Cambridge, MA: Harvard University Press, 2014), 59–87; Sean Patrick Adams, "The U.S. Coal Industry in the Nineteenth Century," EH.net, accessed 15 October 2016, http://tinyurl.com/jeo99q8.

11. Chandler, 150–54; Adams, sec. 3; Bruce Podobnik, *Global Energy Shifts: Fostering Sustainability in a Turbulent Age* (Philadelphia: Temple University Press, 2006), 33–34.

12. Chandler, 154–75; Podobnik, 34; Frederick Moore Binder, *Coal Age Empire: Pennsylvania Coal and Its Utilization to 1860* (Harrisburg: Pennsylvania Historical and Museum Commission, 1974).

13. Chandler, 175–77; Barbara Freese, *Coal: A Human History* (New York: Penguin, 2004), 124–28 [quote on 126]; Jones, 85–87; Binder, 41–60.

14. Joel Mokyr, "The Second Industrial Revolution," in *Storia dell'economia mondiale*, ed. Valerio Castronovo (Rome: Laterza, 1999), tinyurl.com/y8c98sty; Alfred D. Chandler Jr., "The American System and Modern Management," in *Yankee Enterprise: The Rise of the American System of Manufactures*, ed. Otto Mayr and Robert C. Post (Washington, DC: Smithsonian Institution Press, 1981), 153; David A. Hounshell, *From the American System to Mass Production, 1800–1932* (Baltimore: Johns Hopkins University Press, 1984), 46–50; E. J. Hobsbawm, *Industry and Empire: From 1750 to the Present Day* (New York: New Press, 1968), 87.

15. Thomas J. Misa, *A Nation of Steel: The Making of Modern America, 1865–1925* (Baltimore: Johns Hopkins University Press, 1995), xix–xxiii, 5–14 [quotes on xix, xx].

16. Freese, 137–42 [quotes on 137, 141]; Bob Johnson, *Carbon Nation: Fossil Fuels in the Making of American Culture* (Lawrence: University Press of Kansas, 2014), 63–104.

17. Engler, 5; Podobnik, 35.

18. Christian Wolmar, *The Great Railroad Revolution: The History of Trains in America* (New York: Public Affairs, 2012), 89–119 [quotes on 90, 94].

19. Vaclav Smil, *Energy in World History* (Boulder, CO: Westview, 1994), 195.

20. Stephen Ambrose, *Nothing Like It in the World: The Men Who Built the Transcontinental Railroad, 1863–1869* (New York: Simon & Schuster, 2000), 17–22, 369–71 [quotes on 369].

21. Wolmar, 159, 215–57 [quote on 216]; John F. Stover, *American Railroads*, 2nd ed. (Chicago: University of Chicago Press, 1997), 96–133.

22. Wolmar, 259–71 [quote on 271].

23. Wolmar, 272–324 [Albro Martin quote on 273–74]; Stover, 192–225.

24. Maury Klein, *The Power Makers: Steam, Electricity, and the Men Who Invented Modern America* (New York: Bloomsbury, 2008), 71–80 [Nikola Tesla quote on 71]; Walter Isaacson, *Benjamin Franklin: An American Life* (New York: Simon & Schuster, 2003), 133–45.

25. Klein, 84–97 [quote on 88–89]; Alan Hirshfeld, *The Electric Life of Michael Faraday* (New York: Walker, 2006), 111–24.

26. Lewis Coe, *The Telegraph: A History of Morse's Invention and Its Predecessors in the United States* (Jefferson, NC: McFarland, 1993), 1–13.

27. Michael Schiffer, *Power Struggles: Scientific Authority and the Creation of Practical Electricity before Edison* (Cambridge, MA: MIT Press, 2008), 41–47, 91–103, 137–54.

28. Thomas Edison, US Patent 0,223,898, issued 27 January 1880; Schiffer, 299–316.

29. Klein, 169–76 [quotes on 152, 176].

30. Richard Munson, *From Edison to Enron: The Business of Power and What It Means for the Future of Electricity* (Westport, CT: Praeger, 2005), 35–40.

31. Henry Prout, *A Life of George Westinghouse* (New York: Scribner's, 1921), 89; Jill Jonnes, *Empires of Light: Edison, Tesla, Westinghouse, and the Race to Electrify the World* (New York: Random House, 2003), 117–41.

32. Jonnes, 141–64.

33. Klein, 236–55 [quote on 251].

34. Ibid., 179–299 [quote on 295].

35. William Irwin, *The New Niagara: Tourism, Technology, and the Landscape of Niagara Falls, 1776–1917* (University Park: Pennsylvania State University Press, 1996), 99–101; Ginger Strand, *Inventing Niagara: Beauty, Power, and Lies* (New York: Simon & Schuster, 2008), 149–50; Klein, 333 [quote].

36. Klein, 349–52, 370–71 [quote on 350].

37. Josiah Strong, *Our Country: Its Possible Future and Its Present Crisis* (New York: Baker & Taylor, 1885); Alfred Thayer Mahan, "The United States Looking Outward," *Atlantic Monthly*, December 1890, 816–24 [quote on 821].

38. Paul Kennedy, *The Rise and Fall of the Great Powers: Economic Change and Military Conflict from 1500 to 2000* (New York: Vintage Books, 1987), 195–97, 241–48.

39. Arthur M. Schlesinger Jr., *The Cycles of American History* (New York: Houghton Mifflin, 1999), 82–87 [quotes on 82]; Milkis and Nelson, 211–13.

40. Lester Ward, "Plutocracy and Paternalism," *Forum*, November 1895, 300–310; Thorstein Veblen, *The Theory of the Leisure Class* (New York: Macmillan, 1899).

41. Theodore Roosevelt, speech on the nation's trusts, 3 December 1901, tinyurl.com/yd7ywo8a.

42. Doris Kearns Goodwin, *The Bully Pulpit: Theodore Roosevelt, William Howard Taft, and the Golden Age of Journalism* (New York: Simon & Schuster, 2013), 12; Katznelson, Kesselman, and Draper, 69–70; Milkis and Nelson, 218–26; Edmund Morris, *Theodore Rex* (New York: Random House, 2002), 151–69, 205–9, 430–38, 447–48, 485–87, 506–9.

43. Herbert Croly, *The Promise of American Life*, in *American Political Thought: A Norton Anthology*, ed. Isaac Kramnick and Theodore Lowi (New York: Norton, 2008), 1073 [quote]; Theodore Roosevelt, speech on the New Nationalism, 31 August 1910, tinyurl.com/y7sslqzv.

44. Woodrow Wilson, speech on the New Freedom, 5 October 1913, tinyurl.com/y87resbw; John Milton Cooper, *Woodrow Wilson: A Biography* (New York: Vintage Books, 2009); Louis D. Brandeis, "Industrial Absolutism and Democracy," in Kramnick and Lowi.

45. Morris, 301–2, 326, 402–13.

46. Milkis and Nelson, 253 [quote]; Cooper.

47. Freese, 147.

48. Richard L. Hills, *Power from Steam: A History of the Stationary Steam Engine* (Cambridge: Cambridge University Press, 1989), 283–88.

49. Ibid., 288–91; Klein, 374, 415 [quote].

50. Thomas Jefferson to Benjamin Rush, 23 September 1800, tinyurl.com/zf5p3xl; Joel Klotkin, *The City: A Global History* (New York: Modern Library, 2006), 92–95.

51. Klein, 375 [quote], 432–33.

52. John F. Wasik, *The Merchant of Power: Sam Insull, Thomas Edison, and the Creation of the Modern Metropolis* (New York: Palgrave Macmillan, 2006), 4 [quote], 5–16.

53. Phillip F. Schewe, *The Grid: A Journey through the Heart of Our Electrified World* (Washington, DC: Joseph Henry, 2007), 61 [quote], 67; Wasik, 33–54.

54. Schewe, 63–67; Wasik, 55–79.

55. Schewe, 67–70; Peter Fox-Penner, *Smart Power: Climate Change, the Smart Grid, and the Future of Electric Utilities* (Washington, DC: Island, 2010), 15.

56. Schewe, 70–71; Fox-Penner, 15–16 [quote on 15].

57. Schewe, 79–84; Forrest McDonald, *Insull: The Rise and Fall of a Billionaire Utility Tycoon* (Washington, DC: Beard Books, 1962), 333; Klein, 433; Fox-Penner, 16.

58. Hounshell, 217–21; Brian Black, *Crude Reality: Petroleum in World History* (Lanham, MD: Rowman & Littlefield, 2012), 113–16.

59. Hounshell, 221–34 [quote on 230]; Black, 119–21.

60. Hounshell, 237–61 [quotes on 237], 263.

61. Ibid., 268; Smil, 194 [quote].

62. Hounshell, 8 [quote], 214.

63. Earl Swift, *The Big Roads: The Untold Story of the Engineers, Visionaries, and Trailblazers Who Created the American Superhighways* (Boston: Mariner Books, 2011), 13–16 [Stone quote on 15]; Bruce Seely, *Building the American Highway System: Engineers as Policy Makers* (Philadelphia: Temple University Press, 1987), chap. 1; Owen Gutfreund, *Twentieth-Century Sprawl: Highways and the Reshaping of the American Landscape* (Oxford: Oxford University Press, 2004), 7–60.

64. Lind, 199–200; Smil, 168; Daniel Yergin, *The Prize: The Epic Quest for Oil, Money and Power* (New York: Free Press, 1991), 80; David E. Nye, *Consumer Power: A Social History of American Energies* (Cambridge, MA: MIT Press, 1999), 175–77; Alfred W. Crosby, *Children of the Sun: A History of Humanity's Unappeasable Appetite for Energy* (New York: Norton, 2006), 93–98; Black, 110–21.

65. Seely, chap. 2 [quote on 25]; Gutfreund, 16–19, 20 [quote].

66. Swift, 46; Seely, chap. 2; Gutfreund, 16–19.

67. Seely, 83; Swift, 71.

68. Roger Bilstein, *Flight in America: From the Wrights to the Astronauts* (Baltimore: Johns Hopkins University Press, 2001), 9–10.

69. Fred Howard, *Wilbur and Orville: A Biography of the Wright Brothers* (Mineola, NY: Dover, 1998), 15–16 [quote on 16].

70. Ibid., 32–140.

71. Ibid., 32–306; Tom D. Crouch, *The Bishop's Boys: A Life of Wilbur and Orville Wright* (New York: Norton, 1989), 133–394; Bilstein, 12.

72. Howard, 307–409; Crouch, 395–530.

73. Alex Roland, *Model Research* (Washington, DC: Government Printing Office, 1985), 22; Roger Bilstein, *Orders of Magnitude: A History of the NACA and NASA, 1915–1990* (Washington, DC: Government Printing Office, 1989), 1–14.

74. Nick A. Komons, *Bonfires to Beacons: Federal Civil Aviation Policy under the Air Commerce Act, 1926–1938* (Washington, DC: Smithsonian Institution Press, 1989).

75. T. A. Heppenheimer, *Turbulent Skies: The History of Commercial Aviation* (New York: John Wiley & Sons, 1995), 108; Bilstein, *Flight in America*, 85–92.

76. Brooks Adams, "The American Democratic Ideal," in Kramnick and Lowi, 884–90.

77. Milkis and Nelson, 265–71.

78. Ibid., 278–83 [quote on 278]; Katznelson, Kesselman, and Draper, 70–71.

Chapter 3 · *Oil, Microeconomic Security, and the Third Political Age*

1. Robert Curley, ed., *Fossil Fuels* (New York: Britannica Educational, 2012), xi–xiii; Brian Black, *Crude Reality: Petroleum in World History* (Lanham, MD: Rowman & Littlefield, 2012), 5–8.

2. Daniel Yergin, *The Prize: The Epic Quest for Oil, Money and Power* (New York: Free Press, 1991), 20–55 [quotes on 31, 37, 43]; Black, 45–48.

3. Yergin, 56–77; Timothy Mitchell, *Carbon Democracy: Political Power in the Age of Oil* (London: Verso, 2011), 31–36; Black, 48–50.

4. Yergin, 78–98 [quotes on 82, 87]; Black, 37–40.

5. Yergin, 106–13 [quote on 107]; Black, 73–77.

6. Peter Shulman, *Coal and Empire: The Birth of Energy Security in Industrial America* (Baltimore: Johns Hopkins University Press, 2015), 1–14; Black, 58–64; Yergin, 150–64 [Winston Churchill quote on 160].

7. Lord Curzon, quoted in *Crude: The Story of Oil*, by Sonia Shah (New York: Seven Stories,

2004), 10; Yergin, 170–83, 185 [quote]; David S. Painter, "Oil and the American Century," *Journal of American History* 99, no. 1 (2012): 25.

8. Yergin, 184–206.

9. Ibid., 207–24.

10. Ibid., 229–36; Painter, 26.

11. Yergin, 246–59 [quote on 247]; Robert Engler, *The Politics of Oil: A Study of Private Power and Democratic Directions* (New York: Macmillan, 1961), 86–91; David F. Prindle, *Petroleum Politics and the Texas Railroad Commission* (Austin: University of Texas Press, 1984), 41–70; Tyler Priest, "The Dilemmas of Oil Empire," *Journal of American History* 99, no. 1 (2012): 239.

12. Yergin, 280–302 [quote on 292]; Painter, 28; Priest, 237.

13. Herbert Hoover, *American Individualism* (New York: Doubleday, 1922).

14. Charles Beard, "Myth of Rugged American Individualism," *Harper's Monthly*, December 1931, 13.

15. Franklin D. Roosevelt, speech at Ogelthorpe University, 1932, tinyurl.com/y8owu5vt.

16. Franklin D. Roosevelt, Commonwealth Club speech, 1932, tinyurl.com/y88brhm2.

17. William Strauss and Neil Howe, *Generations: The History of America's Future, 1584 to 2069* (New York: Quill, 1991), 249.

18. Franklin D. Roosevelt, first inaugural address, 1933, tinyurl.com/y8l2mk2k.

19. Jonathan Alter, *The Defining Moment: FDR's Hundred Days and the Triumph of Hope* (New York: Simon & Schuster, 2006), 207–338; Patrick J. Maney, *The Roosevelt Presence: The Life and Legacy of FDR* (Berkeley: University of California Press, 1992), 47–87.

20. Franklin D. Roosevelt, White House press conference, 17 December 1940, tinyurl.com/y9lek2m7.

21. Franklin D. Roosevelt, annual message to Congress, 6 January 1941, tinyurl.com/ycqywlg3.

22. Yergin, 305–27; Black, 137–38; Painter, 27.

23. Yergin, 328–50; Black, 138–39.

24. Yergin, 368–88 [quote on 379]; Black, 139–41.

25. Claudia Goldin and Robert A. Margo, "The Great Compression: The Wage Structure in the United States at Mid-Century," *Quarterly Journal of Economics* 107 (1992): 1–34; Paul Krugman, *The Conscience of a Liberal* (New York: Norton, 2009), 37–56 [quote on 54].

26. John Kenneth Galbraith, *The Affluent Society* (New York: Houghton Mifflin, 1998), chap. 17.

27. Milton Friedman, *Capitalism and Freedom* (Chicago: University of Chicago Press, 1962).

28. John F. Kennedy, Democratic Convention acceptance speech, 15 July 1960, tinyurl.com/yafb8dx6; Theodore H. White, *The Making of the President, 1964* (New York: Harper Perennial Political Classics, 2010).

29. Lyndon B. Johnson, commencement address at the University of Michigan, 22 May 1964, tinyurl.com/y77vvz2v.

30. Martin Luther King Jr., "I Have a Dream" speech, August 1963.

31. Johnson.

32. Painter, 28–33 [quote on 28–29]; Harry Truman, address before a joint session of Congress, 12 March 1947, tinyurl.com/678rja; telegram, George Kennan to George Marshall, 22 February 1946, Harry S. Truman Administration File, Elsey Papers, Harry S. Truman Presidential Library, Independence, MO.

33. Priest, 238; Paul Kennedy, *The Rise and Fall of the Great Powers: Economic Change and Military Conflict from 1500 to 2000* (New York: Vintage Books, 1987), chap. 7; Dwight D. Eisenhower, farewell address, January 1961, tinyurl.com/k2l3xnj.

34. Paul Starobin, *After America: Narratives for the Next Global Age* (New York: Viking, 2009), 54.

35. Ibid.

36. David E. Nye, *Electrifying America: Social Meanings of a New Technology* (Cambridge, MA: MIT Press, 1990), 287–91 [quote on 291].

37. Ibid., 292–99 [quote on 293]; Laurence J. Malone, "Commonalities: The R.E.A. and High-Speed Rural Internet Access," *Hartwick College*, 2001, 3, accessed 24 June 2013, arxiv.org/pdf/cs/0109064.pdf.

38. Nye, 299–300 [quote on 300].

39. Ibid., 304–7; Malone, 3–4.

40. Nye, 307–14 [quote on 307].

41. Ibid., 314–21; Ronald R. Kline, "Resisting Development, Reinventing Modernity: Rural Electrification in the United States before WWII," *Environmental Values* 11, no. 3 (August 2002): 327; Malone 4–6 [quote on 6].

42. Nye, 321–35.

43. Ibid., 327.

44. James Howard Kunstler, *The Long Emergency: Surviving the End of Oil, Climate Change, and Other Converging Catastrophes of the Twenty-First Century* (New York: Grove, 2005), 280–89; Daron Acemoglu and James Robinson, *Why Nations Fail: The Origins of Power, Prosperity, and Poverty* (New York: Crown, 2012), 335–67.

45. Vaclav Smil, *Energy in World History* (Boulder, CO: Westview, 1994), 169.

46. US Energy Information Administration, *Annual Energy Review 2011*, tinyurl.comycb7s b20; US Environmental Protection Agency, *Inventory of U.S. Greenhouse Gas Emissions and Sinks: 1990–2011*, tinyurl.com/ybnqps4a.

47. Bruce Seely, *Building the America Highway System: Engineers as Policy Makers* (Philadelphia: Temple University Press, 1987), chap. 7 [quote on 149]; David Jones, *Mass Motorization and Mass Transit: An American History and Policy Analysis* (Bloomington: Indiana University Press, 2008), chaps. 3 and 4.

48. Earl Swift, *The Big Roads: The Untold Story of the Engineers, Visionaries, and Trailblazers Who Created the American Superhighways* (Boston: Mariner Books, 2011), 126.

49. Federal-Aid Highway Act of 1944, *United States Statutes at Large* 58 (1944): 838; Mark Rose, *Interstate: Express Highway Politics, 1939–1989* (Knoxville: University of Tennessee Press, 1990), 31 [quote]; Swift, 157–61; "Address of Vice President Richard Nixon to the Governors Conference, Lake George, New York, July 12, 1954," Federal Highway Administration, accessed 5 December 2012, www.fhwa.dot.gov/infrastructure/rw96m.cfm.

50. Rose, chaps. 5 and 6; Christopher W. Wells, "Fueling the Boom: Gasoline Taxes, Invisibility, and the Growth of the American Highway Infrastructure, 1919–1956," *Journal of American History* 99, no. 1 (2012): 72–81.

51. Swift, 183–87, 315; Wells, 78–81 [quote on 80–81]; Congressional Budget Office, *Spending and Funding for Highways* (Washington, DC: Government Printing Office, 2011).

52. Smil, 197–99 [quotes on 197, 198]; Kunstler, 248 [quote]; S. George Philander, ed., *Encyclopedia of Global Warming and Climate Change, Volume 1* (Thousand Oaks, CA: Sage, 2008), 103 [quote].

53. Nick A. Komons, *Cutting Air Crash: A Case Study in Early Federal Aviation Policy* (Washington, DC: Government Printing Office, 1984).

54. John R. M. Wilson, *Turbulence Aloft: The Civil Aeronautics Administration amid Wars and Rumors of Wars, 1938–1953* (Washington, DC: Government Printing Office, 1979).

55. Roger Bilstein, *Flight in America: From the Wrights to the Astronauts* (Baltimore: Johns Hopkins University Press, 2001), 125–65; T. A. Heppenheimer, *Turbulent Skies: The History of Commercial Aviation* (New York: John Wiley & Sons, 1995), 137–69.

56. Roger Bilstein, *Orders of Magnitude: A History of the NACA and NASA, 1915–1990* (Washington, DC: Government Printing Office, 1989), chap. 3.

57. Bilstein, *Flight in America*, 27–32.

58. Smil, 199.

59. Joseph A. Devine, "Transportation Policy," in *The American Economy: A Historical Encyclopedia*, ed. Cynthia Clark (Santa Barbara, CA: ABC-CLIO, 2011), 833.

60. R. Douglas Hurt, *American Agriculture: A Brief History* (West Lafayette, IN: Purdue University Press, 2002), 3–33; Charles C. Mann, *1491: New Revelations of the Americas before Columbus* (New York: Knopf, 2005), 243–338.

61. Robert C. Allen, *Global Economic History: A Very Short Introduction* (Oxford: Oxford University Press, 2011), 18–20; Niall Ferguson, *Empire: The Rise and Demise of the British World Order and the Lessons for Global Power* (New York: Basic Books, 2002), 2–49.

62. Tom Standage, *An Edible History of Humanity* (New York: Walker, 2009), 111 [quote].

63. John Winthrop, quoted in *American Political Speeches*, ed. Richard Beeman (New York: Penguin Civic Classics, 2012), 1–3; Hurt, 34–37.

64. Ferguson, 69.

65. Ibid., 37–50; Paul K. Conkin, *A Revolution Down on the Farm: The Transformation of American Agriculture since 1929* (Lexington: University Press of Kentucky, 2008), 3.

66. Hurt, 50–56.

67. Richard Manning, *Against the Grain: How Agriculture Has Hijacked Civilization* (New York: North Point, 2004), 85–88 [quote on 86]; Sidney M. Milkis and Michael Nelson, *The American Presidency: Origins and Development, 1776–2011* (Washington, DC: CQ Press, 2012), 56–61; Peter D. McClelland, *Sowing Modernity: America's First Agricultural Revolution* (Ithaca, NY: Cornell University Press, 1997), 10; Conkin, 19 and 3 [quotes].

68. Conkin, 17 [quote], 100–101; Manning, 88–89.

69. Conkin, 51–76 [quote on 75].

70. Ibid., 100–103 [quotes on 100, 103]; Jayson Beckman, Allison Borchers, and Carol A. Jones, "Agriculture's Supply and Demand for Energy and Energy Products," USDA Economic Research Service, May 2013, tinyurl.com/y97zjxbv.

71. Standage, 200–201; Black, 164–68.

72. Standage, 206–13; Conkin, 110.

73. Conkin, 111–12 [quote on 111].

74. Ibid., 98–99, 119–21; Manning, 90–91.

75. Standage, 214–19 [Borlaug quote on 219]; Manning, 93 [quote].

76. Conkin, 112–14.

77. Ibid.

78. Ibid., 115.

79. Stephen Grace, *Dam Nation: How Water Shaped the West and Will Determine Its Future* (Guilford, CT: Globe Pequot, 2012), introduction [Edwin James quote on 3].

80. Ibid., 10 [quote]; Marc Reisner, *Cadillac Desert: The American West and Its Disappearing Water* (New York: Penguin Books, 1986), 1–4.

81. Timothy Egan, *The Worst Hard Time: The Untold Story of Those Who Survived the Great American Dust Bowl* (Boston: Mariner Books, 2006), 9–90; John Opie, *Ogallala: Water for a Dry Land* (Lincoln: University of Nebraska Press, 1995), 1–14 [quotes on 4, 5].

82. Reisner, 120 [quote]; Gary Paul Nabhan, "The Beginning and the End of the Colorado River: Protecting the Sources, Ensuring Its Courses," Northern Arizona University Center for Sustainability Environments, accessed 1 November 2013, tinyurl.com/y8k9r9ac.

83. Reisner, 120–44.

84. Ibid. [quote on 132].

85. Conkin, 98–99 [quote on 98]; Standage, 200–201 [quote on 200].

86. Smil, 188–91; Patrick Canning, Ainsley Charles, Sonya Huang, Karen R. Polenske, and Arnold Waters, "Energy Use in the U.S. Food System," USDA Economic Research Service, March 2010, tinyurl.com/ycj6zwfp.

87. Tom Burke, quoted in *Hot, Flat, and Crowded: Why We Need a Green Revolution—And How We Can Renew America*, by Thomas Friedman (New York: Picador, 2009), 88.

88. L. T. Evans, *Feeding the Ten Billion: Plants and Population Growth* (Cambridge: Cambridge University Press, 1998), 7–43 [quote on 8].

89. Ibid., 44–71.

90. Ibid., 72–89.

91. Standage, 112–25; Evans, 74.

92. Evans, 90–113.

93. Ibid., 114–33.

94. Ibid., 134–64.

95. Ibid.

96. Michel G. J. den Elzen, Jos G. J. Olivier, Niklas Höhne, and Greet Janssens-Maenhout, "Countries' Contributions to Climate Change: Effect of Accounting for all Greenhouse Gases, Recent Trends, Basic Needs and Technological Progress," *Climatic Change*, 31 October 2013; Carlos Lopes, statement at Third Annual Conference on Climate Change and Development in Africa, Addis Ababa, Ethiopia, 21 October 2013, tinyurl.com/y8r29h9n.

97. Walter Isaacson, *The Innovators: How a Group of Hackers, Geniuses, and Geeks Created the Digital Revolution* (New York: Simon & Schuster, 2014), Kindle edition, chap. 1.

98. Ibid.

99. Ibid.

100. Ibid., chap. 2.

101. Ibid., chaps. 3 and 4.

102. Isaacson, chap. 5 [Kilby quote on 173].

103. Ibid.

104. Ibid.

105. David A. Hounshell, "The Evolution of Industrial Research in the United States," in *Engines of Innovation: U.S. Industrial Research at the End of an Era*, ed. Richard S. Rosenbloom and William J. Spencer (Boston: Harvard Business School Press, 1996), 30 [quote].

106. Vannevar Bush, *Science: The Endless Frontier* (Washington, DC: Government Printing Office, 1945); G. Pascal Zachary, *Endless Frontier: Vannevar Bush, Engineer of the American Century* (New York: Free Press, 1997), 218–40 [Bush quote on 233]; William A. Blanpied, "Inventing U.S. Science Policy," *Physics Today*, February 1998, 1–40; Nathan Rosenberg and Richard R. Nelson, "American Universities and Technological Advance in Industry," *Research Policy*, May 1994, 323–48; Walter W. Powell and Jason Owen-Smith, "Universities and the Market for Intellectual Property in the Life Sciences," *Journal of Policy Analysis and Management* 17, no. 2 (Spring 1998): 253–77; Richard Florida, "The Role of the University: Leveraging Talent, Not Technology," *Issues in Science and Technology*, Summer 1999, 67–73; John H. Gibbons, "Has the Time of Witches Passed Over?," in *AAAS Science and Technology Policy Yearbook 2000*, ed. Stephen D. Nelson, Celia McEnaney, and Stephen J. Lita (Washington, DC: American Association for the Advancement of Science, 2000).

107. Blanpied; Gibbons; John A. Alic, *Beyond Spinoff—Military and Commercial Technologies in a Changing World* (Boston: Harvard Business School Press, 1992).

108. Isaacson, chap. 7.

109. Ibid.

110. Ibid.

111. Ibid.; John Markoff, *What the Doormouse Said: How the Sixties Counterculture Shaped the Personal Computer Industry* (New York: Penguin Books, 2005), 165 [quote].

112. Isaacson, chap. 8.

113. Ibid.

114. Ibid., chap. 9.

115. Ibid.

116. J. Brooks Flippen, *Nixon and the Environment* (Albuquerque: University of New Mexico Press, 2000).

117. Richard Nixon, address to the nation about policies to deal with the energy shortages, 7 November 1973, tinyurl.com/yd9n6vdw; Yergin, 617.

118. Erwin C. Hargrove, *Jimmy Carter as President: Leadership and the Politics of the Public Good* (Baton Rouge: Louisiana State University Press, 1988), 48–50; Dominic Sandbrook, *Mad as Hell: The Crisis of the 1970s and the Rise of the Populist Right* (New York: Anchor Books, 2011), 139–58.

119. Jimmy Carter, national address on energy policy, 18 April 1977, tinyurl.com/ydx577he.

120. Jimmy Carter, environment message to Congress, 23 May 1977, tinyurl.com/kasgsnt.

Chapter 4 · Energy Insecurity and the American Decline

1. Daniel Yergin, *The Quest: Energy, Security, and the Remaking of the Modern World* (New York: Penguin Books, 2011), 233 [quote].

2. David J. Murphy, "The Implications of the Declining Energy Return on Investment of Oil Production," *Philosophical Transactions of the Royal Society*, 2013, 126.

3. Yergin, 239 [quote].

4. Murphy, 126 [quote]; Paul Roberts, *The End of Oil: On the Edge of a Perilous New World* (New York: Houghton Mifflin, 2004), 47 [quote]; International Energy Agency, *World Energy Outlook, 2013* (Paris, 2013).

5. Yergin, 227–41; James Howard Kunstler, *The Long Emergency: Surviving the End of Oil, Climate Change, and Other Converging Catastrophes of the Twenty-First Century* (New York: Grove, 2005), 41–44; David S. Painter, "Oil and the American Century," *Journal of American History* 99, no. 1 (2012): 33; Kevin Phillips, *American Theocracy: The Peril and Politics of Radical Religion, Oil, and Borrowed Money in the 21st Century* (New York: Viking, 2006), 41 [quote]; Painter, 24 [quote].

6. Yergin, 227–41 [quote on 235]; Kunstler, 41–48.

7. Painter, 34 [quote]; Tyler Priest, "The Dilemmas of Oil Empire," *Journal of American History* 99, no. 1 (2012): 240–41; John Duffield, *Over a Barrel: The Costs of U.S. Foreign Oil Dependence* (Stanford, CA: Stanford University Press, 2008), 1, 6 [quotes].

8. Duffield, 96–182 [quote on 97]; Bill Richardson, foreword to *Energy and Security: Toward a New Foreign Policy Strategy*, ed. Jan Kalicki and David Goldwyn (Washington, DC: Woodrow Wilson Center Press, 2005), xviii [quote]; Linda J. Bilmes, "The Financial Legacy of Iraq and Afghanistan: How Wartime Spending Decisions Will Constrain Future National Security Budgets," Harvard Kennedy School Faculty Research Working Paper Series RWP13-006, March 2013; Roger J. Stern, "United States Cost of Military Force Projection in the Persian Gulf, 1976–2007," *Energy Policy*, 2010, 9; Priest, 246; Keith Crane, Andreas Goldthau, Michael Toman, Thomas Light, Stuart E. Johnson, Alireza Nader, Angel Rabasa, and Harun Dogo, *Imported Oil and U.S. National Security* (Santa Monica, CA: RAND Corporation, 2009), 59–76.

9. David L. Greene and Sanjana Ahmad, "Costs of U.S. Oil Dependence: 2005 Update," *Oak Ridge National Laboratory*, February 2005; Duffield, 30–61.

10. US Energy Information Administration, "Annual Energy Review 2011," tinyurl.com/ycb7sb20.

11. Lester Brown, *Plan B 4.0: Mobilizing to Save Civilization* (New York: Norton, 2009), 245 [quote]; Robert Engler, *The Politics of Oil: A Study of Private Power and Democratic Directions* (New York: Macmillan, 1961), 158 [quote]; Andy Kroll, "Triumph of the Drill," *Mother Jones*, 1 November 2013; Shakuntala Makhijani, *Cashing in on All of the Above: U.S. Fossil Fuel Production Subsidies under Obama* (Washington, DC: Oil Change International, 2014); Engler, 151–61.

12. Committee on Health, Environmental, and Other External Costs and Benefits of Energy Production and Consumption, *Hidden Costs of Energy: Unpriced Consequences of Energy Production and Use* (Washington, DC: National Research Council, 2009).

13. Bob Johnson, *Carbon Nation: Fossil Fuels in the Making of American Culture* (Lawrence: University Press of Kansas, 2014), 167.

14. Kevin Mattson, *"What the Heck Are You Up To, Mr. President?" Jimmy Carter, America's "Malaise," and the Speech That Should Have Changed the Country* (New York: Bloomsbury, 2009), 13–14 [quotes on 13, 14]; Dominic Sandbrook, *Mad as Hell: The Crisis of the 1970s and the Rise of the Populist Right* (New York: Anchor Books, 2011), 291–95; Painter, 34; Priest, 242; Duffield, 104–12.

15. Mattson, 13–128.

16. Ibid.

17. Jimmy Carter, "A Crisis of Confidence," speech delivered 15 July 1979, tinyurl.com/yccv96va.

18. Mattson, 7.

19. Ibid., 172.

20. John Rawls, *A Theory of Justice* (Cambridge, MA: Harvard University Press, 1971).

21. Ryan Sager, *The Elephant in the Room: Evangelicals, Libertarians and the Battle to Control the Republican Party* (Hoboken, NJ: Wiley Books, 2006), 1–24.

22. Ibid.; Phillips, vii [quote].

23. Ronald Reagan, first inaugural address, 20 January 1981, tinyurl.com/z2qknjn.

24. Sidney M. Milkis and Michael Nelson, *The American Presidency: Origins and Development, 1776–2011* (Washington, DC: CQ Press, 2012), 377–90 [quote on 381].

25. Yergin, 715–44 [quote on 719]; Painter, 35–36; Priest, 245.

26. Jimmy Carter, State of the Union address, 23 January 1980, tinyurl.com/mbq554r.

27. Painter, 35; Duffield, 112–16.

28. Seth Allcorn and Howard F. Stein, "What, Me Worry? Deregulation and Its Discontents: Accurate Reality Testing Reveals Flaws to Deregulation," in *Towards a Socioanalysis of Money, Finance and Capitalism: Beneath the Surface of the Financial Industry*, ed. Susan Long and Burkard Sievers (New York: Routledge, 2012), 123 [quote]; Bruce R. Scott, *Capitalism: Its Origins and Evolution as a System of Governance* (New York: Springer, 2011), 553 [quote].

29. Milkis and Nelson, 397–405; Painter, 37; Priest, 245; Michael Klare, *Blood and Oil: The Dangers and Consequences of America's Growing Petroleum Dependency* (New York: Holt, 2005), 7; Phillips, 78 [quote].

30. Milkis and Nelson, 410–27; Jaime Fuller, "The 3rd Most Memorable State of the Union Address: Bye Bye Big Government," *Washington Post*, 26 January 2014, tinyurl.com/ya9evpmx.

31. Paul Wapner, "Clinton's Environmental Legacy," *Tikkun* 16, no. 2 (2001): 11 [quote]; Engler, 332–33, 485.

32. Zachary A. Goldfarb, "The Legacy of the Bush Tax Cuts, in Four Charts," *Washington Post*, 2 January 2013, tinyurl.com/b9l5vca.

33. Eric Shinseki, testimony before the US Senate Armed Services Committee, 25 February 2003.

34. George W. Bush, "Mission Accomplished" speech, 1 May 2003, tinyurl.com/hwf7w.

35. Dick Cheney, on *Meet the Press*, 16 March 2003.

36. *No End in Sight: The American Occupation of Iraq; the Inside Story From the Ultimate Insiders*, directed by Charles Ferguson, (Magnolia Home Entertainment, Los Angeles, 2007).

37. Thomas E. Mann and Norman J. Ornstein, *It's Even Worse Than It Looks: How the American Constitutional System Collided with the New Politics of Extremism* (New York: Basic Books, 2012), xxiv [quote].

38. Ibid., 31–43 [quote on 33].

39. Ibid., 44.

40. Ibid., 44–58.

41. Ibid., 58–67.

42. Ibid., 133–43.

43. Martin Gilens and Benjamin I. Page, "Testing Theories of American Politics: Elites, Interest Groups, and Average Citizens," *Perspectives on Politics*, September 2014, 564–81.

44. T. Alexander Aleinikoff and Samuel Issacharoff, "Race and Redistricting: Drawing Constitutional Lines after *Shaw v. Reno*," *Michigan Law Review* 92 (1993): 588 [quote]; Nate Silver, "As Swing Districts Dwindle, Can a Divided House Stand?," *New York Times*, 27 December 2012.

45. Sanford Levinson, *Our Undemocratic Constitution: Where the Constitution Goes Wrong—And How We the People Can Correct It* (Oxford: Oxford University Press, 2008), 81–97; Burdett Loomis, "Resolved, The president should be elected directly by the people / Pro," in *Debating the Presidency: Conflicting Perspectives on the American Executive,* ed. Richard J. Ellis and Michael Nelson (Washington, DC: CQ Press, 2010), 33–39.

46. Levinson, 29–38.

47. Ibid., 29–38, 58 [quote].

48. Ezra Klein, "Is the Filibuster Unconstitutional?," *Washington Post*, 15 May 2012, tinyurl .com/cdk64lr; Kevin Drum, "3 Charts Explain Why Democrats Went Nuclear on the Filibuster," *Mother Jones*, 22 November 2013, http://tinyurl.com/lmzxxof; Mann and Ornstein, 84–91.

49. Mark Warren, "Help, We're in a Living Hell and Don't Know How to Get Out," *Esquire*, 15 October 2014.

50. Mann and Ornstein, 101 [quote].

51. Carmen M. Reinhart and Kenneth Rogoff, *This Time Is Different: Eight Centuries of Financial Folly* (Princeton, NJ: Princeton University Press, 2009), xxxiv [quote].

52. Priest, 244 [quote]; David Leonhardt, *Here's the Deal: How Washington Can Solve the Deficit and Spur Growth* (New York: New York Times–Byliner, 2013), Kindle edition, chap. 1.

53. Historical Census of Housing Tables, US Census Bureau, accessed 17 August 2017, tinyurl .com/yan72zzy.

54. Jeff Madrick, *Age of Greed: The Triumph of Finance and the Decline of America, 1970 to the Present* (New York: Vintage Books, 2011), 43.

55. Ibid., 25 [quote].

56. Ibid., 86–95, 125–43 [quote on 81].

57. Ibid., 185–200 [quote on 185].

58. Ibid., 222–47, 318–350 [quote on 320].

59. Phillips, 281 [quote].

60. Priest, 248 [quote]; Madrick, 371 [quote]; Sheila Bair, "The Road to Safer Banks Runs through Basel," *Financial Times*, 23 August 2010.

61. Phillips, 266 [quote]; Jordan Weissman, "How Wall Street Devoured Corporate America," *Atlantic*, 5 March 2013; Jonathan House, "Five Takeaways From New GDP-by-Industry Report," *Wall Street Journal*, 25 April 2014.

62. Thomas Piketty, *Capital in the 21st Century* (Cambridge, MA: Harvard University Press, 2014); Paul Krugman, "Why We're in a New Gilded Age," *New York Times*, 8 May 2014.

63. Paul Krugman, *The Conscience of a Liberal* (New York: Norton, 2009), 38, 54 [quotes].

64. Drew Desilver, "For Most Workers, Real Wages Have Barely Budged for Decades," Pew Research Center, 9 October 2014, tinyurl.com/qxuhn2o.

65. Michael A. Fletcher, "Income Inequality Hurts Economic Growth, Researchers Say," *Washington Post*, 14 January 2014, tinyurl.com/kgss7kl.

66. Jared Bernstein, "Income Inequality: It's a Problem—Here's Why," *Christian Science Monitor*, 15 January 2012, tinyurl.com/pqjkb5c; Piketty.

67. William Forbath, "Workingman's Constitution," *New York Times*, 5 July 2012; Rajashri Chakrabarti and Matt Mazewski, "The Capitol since the Nineteenth Century: Political Polarization and Income Inequality in the United States," *Liberty Street Economics*, 23 June 2014, tinyurl .com/qczh99r.

68. Piketty; Nick Hanauer, "The Pitchforks Are Coming . . . For Us Plutocrats," *Politico Magazine*, July/August 2014, tinyurl.com/pgbca57.

69. Robert C. Allen, *Global Economic History: A Very Short Introduction* (Oxford: Oxford University Press, 2011), 15 [quote].

70. Patrick J. McGuinn, *No Child Left Behind and the Transformation of Federal Education Policy, 1965–2005* (Lawrence: University Press of Kansas, 2006), 1 [quote], 25–50.

71. *Brown v. Board of Education of Topeka*, 347 U.S. 483 (1954).

72. Lyndon Johnson, "Toward Full Educational Opportunity," special message to Congress, 12 January 1965, tinyurl.com/yahb6yek; McGuinn, 25–50.

73. McGuinn, 25–74.

74. Ibid., 75–165.

75. Thomas Friedman and Michael Mandelbaum, *That Used to Be Us: How America Fell Behind in the World It Invented and How We Can Come Back* (New York: Picador, 2012).

76. Sean F. Reardon, "No Rich Child Left Behind," *New York Times*, 27 April 2013; Rebecca Strauss, "Schooling Ourselves in an Unequal America," ibid., 16 June 2013.

77. Barry Goldwater, *The Conscience of a Conservative* (Glen Oaks, NY: LG Classics, 2014), chap. 8.

78. David Alan Aschauer, "Why is Infrastructure Important?," in *Is There a Shortfall in Public Capital Investment? Proceedings of a Conference* (Boston: Federal Reserve Bank of Boston, 1990), 21–63 [quote on 31].

79. Felix Rohatyn, *Bold Endeavors: How Our Government Built America, and Why It Must Rebuild Now* (New York: Simon & Schuster, 2009), 3 [quote].

80. 2013 Report Card for America's Infrastructure, American Society of Civil Engineers, 45–47 [quote on 46], accessed 16 March 2015, tinyurl.com/y8rwe3yq.

81. Ibid., 51–54 [quote on 51].

82. Ibid., 14–16, 23–25, 29–31.

83. Ibid., 60–63; Rohatyn, 3–5.

84. Klare, xiii [quote]; Phillips, 70–74.

85. Roberts, 40 [quote]; Yergin, 391–430; Priest, 237–38 [quote].

86. Daniel Yergin, "Energy Security and Markets," in Kalicki and Goldwyn, 53.

87. Edward M. Kennedy, *America Back on Track* (New York: Penguin Books, 2006), chap. 2.

88. "Global Attitudes Survey: U.S. Favorability," Pew Research Center, Spring 2014, tinyurl.com/yay8sopf.

89. Richardson, xviii [quote]; Leon Fuerth, "Energy, Homeland, and National Security," in Kalicki and Goldwyn, 411 [quote].

90. Brian Fagan, *The Long Summer: How Climate Changed Civilization* (New York: Basic Books, 2004), 99–252 [quote on 125].

91. Svante Arrhenius, *Worlds in the Making: The Evolution of the Universe* (New York: Harper & Brothers, 1906), 63 [quote]; Mark Maslin, *Global Warming: A Very Short Introduction* (Oxford: Oxford University Press, 2009), 23–24.

92. Maslin, 24–26.

93. Climate Central, *Global Weirdness: Severe Storms, Deadly Heat Waves, Relentless Drought, Rising Seas, and the Weather of the Future* (New York: Vintage Books, 2012), 3–28.

94. Ibid., 37 [quote].

95. Ibid., 81–164.

Chapter 5 · Gas and National Renewal in the Fourth Political Age

1. Paul Starobin, *After America: Narratives for the Next Global Age* (New York: Viking, 2009), 105; Cullen Murphy, *Are We Rome? The Fall of an Empire and the Fate of America* (New York: Mariner Books, 2007); Niall Ferguson, *Colossus: The Rise and Fall of the American Empire* (New York: Penguin Books, 2004).

2. Peter Heather, *The Fall of the Roman Empire: A New History of Rome and the Barbarians* (Oxford: Oxford University Press, 2006), 3–250 [quotes on 115, 190].

3. Ibid., 251–460 [quote on 288].

4. "Agricultural Productivity in the U.S.," USDA Economic Research Service, accessed 13 April 2015, tinyurl.com/6pjy4fr.

5. Kevin Phillips, *American Theocracy: The Peril and Politics of Radical Religion, Oil, and Borrowed Money in the 21st Century* (New York: Viking, 2006), 3–98.

6. Ferguson, 248 [quote].

7. Starobin, 53–72.

8. Paul Kennedy, *The Rise and Fall of the Great Powers: Economic Change and Military Conflict from 1500 to 2000* (New York: Vintage Books, 1987), 433 [quote].

9. Fareed Zakaria, *The Post-American World* (New York: Norton any, 2008), 1–6, 8 [quote], 167–260.

10. Starobin, 181–82; Zakaria, 258 [quote].

11. Paul Taylor, "The Next America," Pew Research Center, 10 April 2014, www.pewresearch .org/next-america/.

12. Seth Motel, "JFK Torchbearers Now Vote More Republican," Pew Research Center, 21 November 2013, tinyurl.com/mxsp42t.

13. Jeff Gordiner, *X Saves the World: How Generation X Got the Shaft But Can Still Keep Everything From Sucking* (New York: Penguin Books, 2008), xxi–66 [quotes on xxi, 23 (Richard Linklater), 52, 66].

14. Neil Howe and Bill Strauss, *13th Generation: Abort, Retry, Ignore, Fail?* (New York: Vintage Books, 1993), 224 [quote].

15. Paul Taylor and Scott Keeter, eds., "Millennials: A Portrait of Generation Next," Pew Research Center, February 2010, tinyurl.com/7s8pmgx.

16. Ibid.; Chris Mooney, Scott Clement, and Steven Mufson, "There's a Global Warming Generation Gap in the GOP, Like on the Issue of Gay Marriage," *Washington Post*, 19 November 2014.

17. George Packer, "The Choice: The Clinton-Obama Battle Reveals Two Very Different Ideas of the Presidency," *New Yorker*, 28 January 2008, tinyurl.com/yae66quj; Chuck Todd, *The Stranger: Barack Obama in the White House* (New York: Little, Brown, 2014), 19 [quote].

18. Barack Obama, Iowa Caucus victory speech, 3 January 2008, tinyurl.com/yclgn4w3.

19. Barack Obama, "A More Perfect Union" speech, 18 March 2008, tinyurl.com/ybuhbmkw.

20. Josh Bivens, Elise Gould, Lawrence Mishel, and Heidi Shierholz, "Raising America's Pay: Why It's Our Central Economic Policy Challenge," Economic Policy Institute, 4 June 2014, tiny url.com/ydbk5p8t.

21. Barack Obama, first inaugural address, 20 January 2009; Jonathan Chait, "Obama Promised to Do 4 Big Things As President: Now He's Done Them All," *New York Magazine*, 8 June 2014.

22. Jeffrey Sachs, *The Price of Civilization: Reawakening American Virtue and Prosperity* (New York: Random House, 2011), 27–46, 161–84, 251–64.

23. Paul Krugman, "In Defense of Obama," *Rolling Stone*, 8 October 2014.

24. Paul Krugman, "Nobody Said That," *New York Times*, 27 April 2015; Steven Rattner, "For Tens of Millions, Obamacare Is Working," ibid., 21 February 2015.

25. D. Robert Worley, "No Strategy, No Doctrine, No Organizing Principle?," *Huffington Post*, 27 August 2014, tinyurl.com/ma35an5.

26. Bill Scher, "Can Obama Save the Left from Isolationism?," *The Week*, 13 September 2013, tinyurl.com/knbbw3a; Robert M. Gates, *Duty: Memoirs of a Secretary at War* (New York: Knopf, 2014).

27. Gary Sernovitz, *The Green and the Black: The Complete Story of the Shale Revolution, the Fight over Fracking, and the Future of Energy* (New York: St. Martin's, 2016), 13–16.

28. Robert Kolb, *The Natural Gas Revolution: At the Pivot of the World's Energy Future* (Upper Saddle River, NJ: Pearson, 2014), 46–65.

29. Russell Gold, *The Boom: How Fracking Ignited the American Energy Revolution and Changed the World* (New York: Simon & Schuster, 2014), 63–85 [quote on 77].

30. Ibid., 86–132 [quotes on 7, 117]; Sernovitz, 20–23; Kolb, 54.

31. Gold, 133–52; Kolb, 56–58.

32. Ibid., 158–269 [quote on 190].

33. US Energy Information Agency, *Annual Energy Outlook 2015*, 14 April 2015; Tyler Priest, "The Dilemmas of Oil Empire," *Journal of American History* 99, no. 1 (2012): 236–37 [quote]; Kolb, 64–66.

34. Sanford Levinson, *Our Undemocratic Constitution: Where the Constitution Goes Wrong—And How We the People Can Correct It* (Oxford: Oxford University Press, 2008), 9 [quote], 11–24.

35. Sanford Levinson, "Our Imbecilic Constitution," *New York Times*, 28 May 2012.

36. H.J.Res.48, Proposing an Amendment to the Constitution of the United States Providing that the Rights Extended by the Constitution are the Rights of Natural Persons Only, 114th Cong. (2015).

37. John Paul Stevens, *Six Amendments: How and Why We Should Change the Constitution* (New York: Little, Brown, 2014), chap. 2.

38. George C. Edwards III, *Why the Electoral College Is Bad for America* (New Haven, CT: Yale University Press, 2011).

39. Ibid., 39–40 [quote].

40. Patrick Henry, speech before Virginia Ratifying Convention, 5–6 June 1788, tinyurl.com/ydgu8acm.

41. Levinson, *Our Undemocratic Constitution*, 159–66; Sanford Levinson, "Article V should be revised to make it easier to amend the Constitution and to call a constitutional convention / Pro," in *Debating Reform: Conflicting Perspectives on How to Fix the American Political System*, ed. Richard J. Ellis and Michel Nelson (Washington, DC: CQ Press, 2014), 5–10; Timothy Lynch, "Amending Article V to Make the Constitutional Amendment Process Itself Less Onerous," *Tennessee Law Review* 78 (2011): 823.

42. Daniel Lazare, "Abolish the Senate," *Jacobin*, 2 December 2014.

43. Bruce Bartlett, "Enlarging the House of Representatives," *New York Times*, 7 January 2014; Larry Sabato, *A More Perfect Constitution: 23 Proposals to Revitalize Our Constitution and Make America a Fairer Country* (New York: Walker, 2007).

44. Harold Meyerson, "Did the Founding Fathers Screw Up?," *American Prospect*, 26 September 2011, tinyurl.com/ybk2yvvf.

45. Priest, 249 [quote]; Kennedy, 514–15 [quote].

46. Zakaria, 1 [quote].

47. Peter Zeihan, *The Accidental Superpower: The Next Generation of American Preeminence and the Coming Global Disorder* (New York: Twelve, 2014), 46–91 [quote on 77]; Zakaria, 238 [quote]; Timothy Mitchell, *Carbon Democracy: Political Power in the Age of Oil* (London: Verso, 2011), 6 [quote].

48. Joseph Nye, *The Future Power* (Philadelphia: PublicAffairs, 2011), 20–24; Portland Media Group, "The Soft Power 30," accessed 20 September 2015, tinyurl.com/y8ydbjut.

49. Ibid.

50. Nye, 221 [quote].

51. Ibid., 230 [quote].

52. Thomas Barnett, *The Pentagon's New Map: War and Peace in the Twenty-First Century* (New York: Putnam & Sons, 2005), 295–340.

53. Barack Obama, quoted in Elizabeth Kolbert, "Congress Moves to Sabotage the Paris Climate Summit," *New Yorker*, 4 December 2015, tinyurl.com/jgxw7wr.

54. Sheila Bair, "The Road to Safer Banks Runs through Basel," *Financial Times*, 23 August 2010.

55. Elise Gould, "U.S. Lags Behind Peer Countries in Mobility," *Economic Policy Institute*, 10 October 2012, tinyurl.com/mfxymae.

56. 2013 Report Card for America's Infrastructure, American Society of Civil Engineers, accessed 16 March 2015, tinyurl.com/y8rwe3yq; Kenn Fisher, "Building Better Outcomes: The Impact of School Infrastructure on Student Outcomes and Behaviour," *Schooling Issues Digest*, 2001, tinyurl.com/jut9q84.

57. Jenny Anderson, "From Finland, an Intriguing School-Reform Model," *New York Times*, 12 December 2012; OECD, *Lessons from PISA for the United States: Strong Performers and Successful Reformers in Education* (Paris, 2011), 159–75 [quote on 169].

58. OECD, International Migration Database, accessed 15 November 2012, tinyurl.com/ychbzgra.

59. Keith Miller, Kristina Costa, and Donna Cooper, "Creating a National Infrastructure Bank and Infrastructure Planning Council: How Better Planning and Financing Options Can Fix Our Infrastructure and Improve Economic Competitiveness," Center for American Progress, September 2012, tinyurl.com/ya9h9n56.

60. "The Third Industrial Revolution," *Economist*, 21 April 2012; "Historical Trends in Federal R&D Policy," *American Association for the Advancement of Science*, accessed 13 March 2016, tinyurl.com/yblwt5dw; "Which Countries Spend the Most on Research and Development?," World Economic Forum, 9 July 2015, tinyurl.com/ybvuy2u6.

61. John Kenneth Galbraith, *The Affluent Society* (New York: Houghton Mifflin, 1998), 223–34; Barack Obama, remarks on economic mobility, 4 December 2013, tinyurl.com/yaepvxn8.

Chapter 6 · *Climate Security and a Sustainability Revolution*

1. *Massachusetts v. Environmental Protection Agency*, 127 S.Ct. 1438 (2007).

2. Ibid.

3. Ibid.; Linda Greenhouse, "Justices Say E.P.A. Has Power to Act on Harmful Gases," *New York Times*, 3 April 2007.

4. Executive Office of the President, "A Retrospective Assessment of Clean Energy Investments in the Recovery Act," *White House*, February 2016, tinyurl.com/y8ty8nfr.

5. John M. Broder, "'Cap and Trade' Loses Its Standing as Energy Policy of Choice," *New York Times*, 25 March 2010.

6. US Office of Transportation and Air Quality, "EPA and NHTSA Set Standards to Reduce Greenhouse Gases and Improve Fuel Economy for Model Years 2017–2025 Cars and Light Trucks," US Environmental Protection Agency, August 2012, tinyurl.com/y92cujat; US Office of Transportation and Air Quality, "EPA and NHTSA Propose Standards to Reduce Greenhouse Gas Emissions and Improve Fuel Efficiency of Medium- and Heavy-Duty Vehicles for Model Year 2018 and Beyond," US Environmental Protection Agency, June 2015.

7. US Environmental Protection Agency, Final Rule, "Carbon Pollution Emission Guidelines for Existing Stationary Sources: Electric Utility Generating Units," *Federal Register* 80, no. 205 (23 October 2015): 64661–65120.

8. Brad Plumer, "The World Just Agreed to a Major Climate Deal in Paris: Now Comes the Hard Part," *Vox*, 12 December 2015, http://www.vox.com/2015/12/12/9981020/paris-climate-deal.

9. Nick Obradovich and Scott Guenther, "Collective Responsibility Amplifies Mitigation Behaviors," *Climatic Change*, May 2016, 1–13.

10. "Annual Coal Report 2015," *Energy Information Agency*, November 2016, tinyurl.com/y7ws5tu8w.

11. Interagency Working Group on Social Cost of Carbon, "Technical Update of the Social Cost of Carbon for Regulatory Impact Analysis," August 2016, tinyurl.com/z8v3kjk.

12. Peter Calthorpe, *Urbanism in the Age of Climate Change* (Washington, DC: Island Press, 2011), 3 [quote]; David Owen, *Green Metropolis: Why Living Smaller, Living Closer, and Driving Less Are the Keys to Sustainability* (New York: Riverhead Books, 2009), 1–48; Tim Herzog, "World Greenhouse Gas Emissions in 2005," World Resources Institute, July 2009, tinyurl.com/ybv6x6pw.

13. Calthorpe, 7–25 [quote on 10].

14. Ibid., 58–61.

15. "Living Building Challenge 3.0," International Living Future Institute, tinyurl.com/ycx58x8e.

16. "The Greenest Commercial Building in the World," Bullitt Center, www.bullittcenter.org; Bryn Nelson, "A Building Not Just Green, but Practically Self-Sustaining," *New York Times*, 2 April 2013.

17. Owen, 111 [quote]; Robert Putnam, *Bowling Alone: The Collapse and Revival of American Community* (New York: Simon & Schuster, 2000), 27 [quote].

18. Based on Calthorpe, 64 [quote].

19. Owen, 149 [quote].

20. Calthorpe, 81 [quote].

21. Paul Roberts, *The End of Oil: On the Edge of a Perilous New World* (New York: Houghton Mifflin, 2004), 165–87 [quote on 175]; Michael Levi, "Climate Consequences of Natural Gas as a Bridge Fuel," *Climatic Change*, June 2013, 609–23.

22. US Department of Energy, Office of Energy Efficiency and Renewable Energy, "EERE FY 2016 Budget Request," accessed 27 June 2016, tinyurl.com/ycro6esw; Daniel M. Kammen, "The Rise of Renewable Energy," *Scientific American*, September 2006, 84–93.

23. Office of Energy Efficiency and Renewable Energy, "2016–2020 Strategic Plan and Implementing Framework," US Department of Energy, 2015, 9–11; Committee on Transitions to Alternative Vehicles and Fuels, National Research Council "Transitions to Alternative Vehicles and Fuels" (Washington, DC: National Academies Press, 2013).

24. US Department of Energy, Office of Energy Efficiency and Renewable Energy, "2016–2020 Strategic Plan and Implementing Framework," 15–18 [quote on 16].

25. Ibid., 12–14.

26. David MacKay, *Sustainable Energy: Without the Hot Air* (Cambridge: UIT Cambridge, 2009), 186–202.

27. Ibid., 186–202 [quotes on 195, 198].

28. Clark W. Gellings, "A Globe-Spanning Supergrid," *IEEE Spectrum*, July 2015, 48–54.

29. Mara Hvistendahl, "Coal Ash Is More Radioactive Than Nuclear Waste," *Scientific American*, 13 December 2007; MacKay, 161–76.

30. MacKay, 161–76 [quote on 162].

31. Andrew C. Kadak, "A Future for Nuclear Energy: Pebble Bed Reactors," *International Journal of Critical Infrastructures*, 2005, 330–45; Richard Martin, *Super Fuel: Thorium, the Green Energy Source of the Future* (New York: Palgrave Macmillan, 2012), 70–80, 232–40 [quote on 79].

32. MacKay, 161–76; Daniel Clery, "The Bizarre Reactor That Might Save Nuclear Fusion," *Science*, 21 October 2015, http://www.sciencemag.org/news/2015/10/bizarre-reactor-might-save-nuclear-fusion.

33. MacKay, 234–36.

34. US Office of Transportation and Air Quality, "Fast Facts: U.S. Transportation Sector Greenhouse Gas Emissions 1990–2014," US Environmental Protection Agency, June 2016, 1; Controller General, *Excessive Truck Weight: An Expensive Burden We Can No Longer Support* (Washington, DC: GAO, 1979), 23.

35. US Office of Transportation and Air Quality, "Fast Facts," 1; Jim Motavalli, "Plane, Train or Automobile: Which Has the Biggest Footprint?," Mother Nature Network, 17 February 2014, tinyurl.com/z9bsdkg.

36. Patrick Gillespie, "Truck Drivers Wanted: Pay—$73,000," CNN Money, 9 October 2015, tinyurl.com/jk5njj4; Bureau of Labor Statistics, US Department of Labor, "Employment in Green Good and Services," news release, 19 March 2013, tinyurl.com/6ocr8jm; Apollo Alliance, *Make It in America: The Apollo Clean Transportation Manufacturing Action Plan* (San Francisco, 2010), 4.

37. Bill Moyer and Patrick Mazza, *Solutionary Rail: A People-Powered Campaign to Electrify America's Railroads and Open Corridors to a Clean Energy Future* (Vashon, WA: Backbone Campaign, 2016); Rob van Harren, "Assessment of Electric Cars' Range Requirements and Usage Patterns based on Driving Behavior recorded in the National Household Travel Survey of 2009," July 2012, conducted as part of Solar Journey USA project, tinyurl.com/ya4g7g6k; Par Foran, "The Rail Renaissance Era is Here to Stay, Presuming Stakeholders Figure Out How to Partner Up (and Pony Up) to Ensure There is Enough Rail Capacity," *Progressive Railroading*, June 2008, tinyurl .com/zvgm6h5.

38. Simon van Zuylen-Wood, "Why Can't American Have Great Trains? A Washington Mystery," *National Journal*, 18 April 2015, 25–31; Tony Dutzik, Gideon Weissman, and Phineas Baxandall, *Who Pays for Roads? How the "Users Pay" Myth Gets in the Way of Solving America's Transportation Problems* (Boston: US PIRG Education Fund, 2015).

39. Daniel Albalate and Germa Bel, *The Economics and Politics of High-Speed Rail* (New York: Lexington Books, 2014); David Randall Peterman, John Frittelli, and William J. Mallett, *High Speed Rail (HSR) in the United States* (Washington, DC: Congressional Budget Office, 2009); Erik R. Pages, Brian Lombardozzi, and Lindsey Woolsey, *The Emerging U.S. Rail Industry: Opportunities to Support American Manufacturing and Spur Regional Development* (Washington, DC: National Institute of Standards and Technology, 2013); "Safety," under "About the Shinkansen," Central Japan Railway Company, tinyurl.com/ybv5foru; Yoshitaka Fukui and Kyoji Oda, "Is Public Private Partnership a Panacea for Transportation Infrastructure? A Lesson from the Funding of Japanese High-Speed Train Network," 14 September 2007, tinyurl.com/ya6lzqzc.

40. Albalate and Bel; Peterman, Frittelli, and Mallett; Pages, Lombardozzi, and Woolsey.

41. Ibid.

42. Ibid.

43. Ibid.; AMTRAK, "A Vision for High-Speed Rail in the Northeast Corridor," September 2010, tinyurl.com/hsg3qqv.

44. Lester R. Brown, *World on the Edge: How to Prevent Environmental and Economic Collapse* (New York: Norton, 2011), 3–58; Roddy Scheer and Doug Moss, "How Does Meat in the Diet Take an Environmental Toll?," *Scientific American*, 28 September 2011, tinyurl.com/m9fa55q.

45. James E. McWilliams, *Just Food: Where Locavores Get It Wrong and How We Can Truly Eat Responsibly* (New York: Little, Brown, 2009), 1–80.

46. Ibid.; Brown, 165–82.

47. Stefaan Blancke, "Why People Oppose GMOs Even Though Science Says They Are Safe," *Scientific American*, 18 August 2015, tinyurl.com/nzmpubg; McWilliams, 81–116.

48. Michael Pollan, *Food Rules: An Eater's Manual* (New York: Penguin Books, 2009).

49. Christopher L. Weber and H. Scott Matthews, "Food-Miles and the Relative Climate Impacts of Food Choices in the United States," *Environmental Science and Technology*, 2008, 3508 [quote].

50. McWilliams, 155–84; Laura Wellesley, Catherine Happer, and Antony Froggatt, *Changing Climate, Changing Diets: Pathways to Lower Meat Consumption* (London: Royal Institute of International Affairs, Chatham House, 2015).

Epilogue

1. John F. Kennedy, "Remarks at the America's Cup Dinner Given by the Australian Ambassador," 14 September 1962, tinyurl.com/y8mxwckh.

2. Dorothy Sterling, *The Outer Lands: A Natural History Guide to Cape Cod, Martha's Vineyard, Nantucket, Block Island, and Long Island* (New York: Norton, 1992), 9–21 [quote on 10].

3. Ray Cavanaugh, "Providence Besieged by Great Gale in 1815," *Providence Journal*, 20 September 2015.

4. R. A. Scotti, *Sudden Sea: The Great Hurricane of 1938* (New York: Back Bay Books, 2003), 18 [quote].

5. Ibid., 17–26, 147–58, 176–98.

6. Roy Bongartza, "Watch Hill to Point Judith: 20 Miles of Summer Fun," *New York Times*, 5 July 1981, tinyurl.com/yajqne6d.

Abizaid, John, 207

Acemoglu, Daron, 55, 133

Adams, Brooks, 101–2

Adams, Edward Dean, 77

Adams, Henry, 40–41

Adams, John, 34, 38

Adams, Sam, 32

Advanced Research Projects Agency (ARPA), 168–69, 352

Affordable Care Act, 288–89

Afghanistan War, 186, 205–6

African Americans, 125–26, 275, 303

AFRICOM, 320

Agricultural Revolution: First, 142–46, 155, 375; Second, 149–60, 376

agriculture: in colonial America, 143–45; in early societies, 24–25, 142; and fertilizers, 147–49, 159; government policies for, 145, 146, 147; and hybridization, 149–50; and hydrocarbons, 18–19, 155–56, 261, 262, 265, 370, 376; and irrigation, 25–26, 151–55; mechanization of, 145–47, 376; organic, 370–71; pesticides and herbicides in, 150–51, 154; and population growth, 156, 157–60; productivity of, 146, 155–56, 159; and urbanization, 26, 44, 132–33, 260, 270

Ahmad, Sanjana, 187

Aid to Families with Dependent Children, 116, 377

Aiken, Howard, 161, 162

Air Commerce Act, 100

air conditioning, 133, 212

Aircraft Noise Abatement Act, 126

Alaskan National Interest Lands Conservation Act of 1980, 177

Alaska Pipeline, 1–2, 4–5, 174

Aleinikoff, Alexander, 216–17

Allcorn, Seth, 202

Allen, Paul, 170, 171

Allen, Robert, 239

Al Qaeda, 205, 206, 252

Ambrose, Stephen, 39, 69–70

American Association of State Highway Officials (AASHO), 96

American Challenge, The (Peters and Waterman), 229

American Environmental History (Merchant), 11

American Geopolitical Cross (AGC), 320, 322; four pillars of, 317–19

American Individualism (Hoover), 111–12

American Israel Public Affairs Committee (AIPAC), 314

American Recovery and Reinvestment Act (ARRA), 336, 352, 367–68

American Society of Civil Engineers (ASCE), 247–48

America Online (AOL), 172

Americum, term, 156

Ampere, Andre-Marie, 72

Andreessen, Marc, 172

animal domestication, 25, 26

Antarctic Conservation Act, 177

anthracite, 63–65

anti-Americanism, 253, 254

Apple, 171

aquifers, 153, 259, 370

Arafat, Yasser, 253

Arkwright, Richard, 43

arms industry, 44–46, 128

Arnold, Henry "Hap," 141

Arrhenius, Svante, 257

Articles of Confederation, 33

Aschauer, David Alan, 245

assembly line, 92

Atanasoff, John Vincent, 161

Attila the Hun, 269

Australia, New Zealand, United States Security Treaty (ANZUS), 128

automobiles, 133, 135, 138–39, 142, 265, 349–50; invention of, 95; and oil shortages, 109, 185–86

aviation and airlines, 107–8, 140, 142, 246, 362–63, 366; airplane invention and early production, 97–101; high-speed trains as alternative to, 365; and jet propulsion, 140–41; navigation systems for, 100, 246; safety regulations for, 139–40

Axelrod, David, 280–81

Babbage, Charles, 161

Baby Boomers, 15, 260–61, 277; and elections, 197–98, 280, 282, 298

Bair, Sheila, 231, 325

Baker v. Carr, 218

Baldwin, Matthias, 50

banks and banking: in Great Depression, 114; investment firms' ties to, 115, 204, 230, 287, 324; US national bank, 35–37, 39, 41, 51–52, 375

Baran, Paul, 169

Bardeen, John, 163–64

Baumgartner, Frank, 13

Beard, Charles, 112

Begich, Nick, 2

Bell, Alexander Graham, 99

Benz, Karl, 95

Berners-Lee, Timothy, 172

Bernstein, Peter, 48

Beschloss, Michael, 204

Bessemer process, 66

bicameralism, 219–20, 308–9

bicycle, 94

Bilas, Frances, 162

Bilstein, Roger, 98

Bina, Eric, 172

bin Laden, Osama, 205, 206, 252, 290–91

Bipartisan Campaign Reform Act, 214

Bipartisan Policy Center, 302, 311

Birmingham, Stephen, 385

Bissell, George, 104

bituminous coal, 29–30, 63–64, 67

Boeing 707, 141

Boesky, Ivan, 229, 230

Bolt, Beranek and Newman (BBN), 168

Bomford, George, 44

Borlaug, Norman, 150

Bosch, Carl, 148

Boxer Rebellion, 80

Brattain, Walter, 163–64

Brazil, 160, 315, 316, 338, 372

Bremer, L. Paul, 207

Bretton Woods Agreement, 128

Brin, Sergey, 172–73

Britain, 40, 41, 53, 107–8, 120, 162; empire of, 27–28, 37, 143, 144, 270–72; and First Industrial Revolution, 28–31, 330, 374; and Middle East, 110, 250–51; and railroads, 49–50; US relationship with, 317–18

Brown, Lester, 188

Brown v. Board of Education, 125, 240

budget process, federal, 311–12

building design, 346

Bullitt Center, 346, 347

Bureau of Air Commerce, 100

Bureau of Public Roads, 96, 135

Bureau of Reclamation, 152

Burnham, Walter Dean, 12

Burr, Aaron, 221

Burton, William, 106

Bush, George H. W., 188, 202–3, 241, 364

Bush, George W., 188, 205, 218, 225, 231, 242, 275, 286; environmental policy of, 334, 335; and Iraq War, 205–7, 253, 254

Bush, Vannevar, 161, 166, 167, 168

cable news networks, 212–14

Caddell, Pat, 192

CAFE (corporate average fuel efficiency) standards, 196

Calhoun, John, 54–55

California corridor, 368

Calthorpe, Peter, 343, 344, 348, 350

campaign finance, 214–15, 218, 223, 301–2

Canada, 317

canals and waterways, 48, 63, 141–42, 246

Cantwell, Maria, 324–25

cap-and-trade, 340–42

Capital in the 21st Century (Piketty), 233, 234, 235, 237, 238

capitalism, 185, 202, 232–33, 301

Carbon Democracy (Mitchell), 10–11

carbon dioxide, 29, 139, 255; and global warming, 191, 257–58, 261–62, 265–66, 334. *See also* greenhouse gas emissions

carbon tax, 350, 353–54, 367, 373; need for, 340–42, 380

Cárdenas, Lázaro, 110

Carmichael, Gil, 364

Carson, Rachel, 151

Carter, Jimmy, 196–97, 201–2, 241; energy security efforts of, 175–77, 191–96, 254, 284

Carter Doctrine, 201–2, 252

Case, Steve, 172

Castro, Joaquin, 223

Cataract Construction, 77

Cayley, Sir George, 97

Cenozoic Era, 255

CENTCOM (US Central Command), 201, 314, 320

Central Treaty Organization (CENTO), 128

Cerf, Vint, 169–70, 172

certificates of deposit (CDs), 228

Chakrabarti, Rajashree, 238

Chamberlin, Thomas Chrowder, 257

Chandler, Alfred, Jr., 64, 65–66

Chappe, Claude, 73

Cheney, Dick, 207, 209, 210

Chernow, Ron, 34, 36

Chicago, IL, 89–90, 91

China, 80, 158, 318, 338, 366; agriculture in, 155, 157; and US geopolitics, 316–17, 318–19

Churchill, Winston, 107, 118

cities, 260, 261, 342–50, 380; and highways, 135–36, 138; and public transit, 349, 350; and Second Industrial Revolution, 64–65, 88; and zoning, 347–48. *See also* urbanization

Citizens United v. Federal Election Commission, 214, 301

Civil Aeronautics Act, 139

Civil Aeronautics Authority, 139

Civil Aeronautics Board, 140

Civil Aeronautics Regulations, 139–40

Civilian Conservation Corps, 115

Civil Rights Act of 1964, 125–26, 133, 173

Civil Rights Act of 1991, 203

civil rights movement, 125

Civil War, 56, 58–59, 65, 68–69, 145–46

Civil Works Administration, 115

Clark, Wesley A., 168–69

Clay, Henry, 51, 53, 56, 59

Clayton Anti-Trust Act, 84

Clean Air Act, 124, 337; 1970 extension of, 174; 1977 amendments to, 177, 203; court decision around, 334, 335

Clean Power Plan (CPP), 337

Clean Water Act of 1972, 174

Climate Central, 265

climate change, 156, 248–49, 255–67, 339; America's contribution to, 262, 265; in earth's history, 24–25, 255–56; failure to address, 258–59, 260–61; impact of, 259–60, 266, 369; Obama on, 324, 336–38. *See also* global warming

climate science, 257–58

climate security, 183, 272, 339–80; Clinton policies on, 204–5; and Fourth Political Age, 19, 284–87, 338–39; and geopolitics, 316, 319, 321; Obama policies on, 207, 285–86, 324, 336–38, 342; Supreme Court decision on, 334–35. *See also* sustainability revolution

Clinton, Bill, 188, 218, 242, 296, 334; and climate security, 204–5; and financial deregulation, 204, 230; and health care reform, 203–4, 210, 287

Clinton, DeWitt, 48

Clinton, Hillary, 280–82, 297–98

coal: anthracite, 63–65; bituminous, 29–30, 63–64, 67; and carbon emissions, 339; and electricity, 86–87, 134, 375; and First Industrial Revolution, 29–30; as primary US energy source, 62–65, 67–68, 69

Coal: A Human History (Freese), 10

Cold War, 127–29, 185, 250, 377

Cole, Tom, 222

collaterilized debt obligations, 230–31

Colorado River, 153–55

Colt, Samuel, 45

Columbian Exchange, 143, 158

Columbus, Christopher, 143

combined heat and power (CHP) systems, 353

Comey, James, 297–98

Commodity Futures Modernization Act, 231

Common Cause, 222

Common Core State Standards, 289

Commonwealth Edison, 91

compulsory voting, 305–6

computers: invention of, 161–62; personal, 165–66, 170–71; programming of, 162–63; transistors and microchips for, 163–66

congressional sessions, 311

Conkin, Paul, 145–46, 147, 149, 151

Conscience of a Conservative (Goldwater), 244

Constitution, US: amendment process in, 300–301, 307–8; Bill of Rights in, 299–300; Commerce Clause of, 47–48, 219, 299–300, 418; Constitution of Conversation in, 299–300; Constitution of Settlement in, 300–301, 307; drafting of, 33–34; Equal Protection Clause of, 61; and federalism, 218–19; Necessary and Proper Clause of, 36, 219, 299–300; proposals for reforming, 299–313; and Senate filibusters, 220–22; weaknesses in, 216, 298

Consumer Financial Protection Bureau, 324

Contract Air Mail Act, 100

Coolidge, Calvin, 103

corn: and ethanol, 260, 370; and herbicides, 151, 155; hybridization of, 149, 150, 155

Cornwallis, Charles, 33

corporatization, 60–62

Council on Environmental Quality, 174

Crocker, Stephen, 169

Croly, Herbert, 83

Crompton, Samuel, 28

Cronkite, Walter, 213
Crude Oil Windfall Profits Tax, 196
Cuba, 80, 290
Curtiss, Glenn, 99
Curtiss Aeroplane, 99
Curzon, Lord George Nathaniel, 108
Cutting, Bronson, 139
Cycles of American History, The (Schlesinger), 14
Cynamon, Barry, 236

dams, 248–49
Daschle, Tom, 302
Davies, Donald, 169
Davis, Bancroft, 61
debt: household, 225–26, 230–31, 232, 236–37; public,
    35, 39, 200, 225–26
Declaration of Independence, 325
defense spending, 200–201
deforestation, 265–66
democracy, 215–16, 217–18, 274, 300, 333
Democracy Amendment, 301–6, 378–79
Democratic Party, 51, 55, 196–97, 225, 242; and
    climate security, 204–5; and election of 2008,
    275, 280–83; and election of 2016, 296–98;
    generational support for, 276, 277; and New Deal
    Coalition, 110, 114, 131, 137, 173; and South, 110,
    125–26, 173–74. See also elections
Democratic-Republicans, 35, 38–41
Denver, CO, 133–34, 154
Department of Agriculture, 60, 94, 96
Department of Education, 241
Department of Energy, 176–77, 346
Department of Housing and Urban Development,
    126
deregulation, 198–99, 202; of financial industry, 202,
    204, 227–28, 229–30, 234, 324
desertification, 248–49, 259, 266, 370
developing nations, 159–60, 316–17
Devine, Joseph, 142
Dewey, George, 80
Diamond, Jared, 25
Diesel, Rudolf, 95
diet, 24, 25–26, 143; and meat consumption, 260,
    372–74
Digital Revolution, 160–73; and computers, 161–66,
    170–71; government research in, 166–68; and
    Internet, 168–70, 171–73
Dilts, James, 50

directional drilling, 294
Dodd, Christopher, 287, 330
Dodd-Frank financial regulations, 324
Doggett, Lloyd, 304–5
Doherty, Harry, 109
Douglas, Stephen, 56
Douglass, Frederick, 54
Dow Chemical, 293
Down to Earth (Steinberg), 11
Dred Scott v. Sandford, 56–57
Drew, Daniel, 70
drones, 290, 323
Duany, Andres, 343
Duffield, John, 186
Duncan, Arne, 289
Dunlop, J. B., 95
du Pont, Francis, 137
Dust Bowl, 153
Dutch empire, 27–28, 143, 270

Earth Day, 174
Eckert, J. Presper, 162, 163
Economic Opportunity Act, 126
Edison, Thomas, 74–76, 89, 90
education, 273, 289, 375; inequity in, 239–44; need
    for investment in, 326–28
Edwards, Donna, 223
Edwards, George C., III, 304
Edwards, John, 281
Egypt, 185, 250–51
Eisenhower, Dwight, 93–94, 128, 137
elections: of 1828, 13, 51; of 1860, 13, 57; of 1896, 13,
    78; of 1932, 13, 112–13, 114; of 1968, 13, 173–74; of
    1980, 13, 196–97, 200; of 2008, 275, 280–83; of
    2012, 217, 277; of 2016, 296–98, 378; managing
    national, 305–6; primary, 222–23
Electoral College, 217–18, 298; need for elimination
    of, 303–5
electoral realignment, 12–13
electric cars, 352–53, 356
electricity: and coal, 86–87, 134, 375; direct vs. alter-
    nating current, 74–76, 77; distribution of, 89–90;
    and energy alternatives, 352–53; and energy
    storage, 355–56; and gas, 351; grid for, 90–91, 249,
    354, 356–57, 362; harnessing power of, 72–73; and
    hydrocarbons, 18–19, 261, 262; and hydroelec-
    tricity, 76–78, 130, 248–49, 355, 361; and mass
    production, 92–93; power stations for, 74, 86–87,

88–90; and rail, 76, 364–65; rural electrification, 129–34; and steam turbine, 87–88

electric lighting, 73–74

Electrification Revolution, 18–19, 72–78, 86–91, 129–34, 262, 265, 375

Elementary and Secondary Education Act of 1965, 126, 241, 242

Ellison, Keith, 305

Emancipation Proclamation, 58

Embargo Act of 1807, 40–41

Endangered American Wilderness Act, 177

Endangered Species Act of 1973, 174

Endangered Species Preservation Act of 1966, 126

energy security, 18, 19, 236, 291; Carter efforts for, 175–77, 191–96, 254, 284; and economic development, 42, 81, 235; and energy insecurity, 225, 232, 234, 243, 246–47, 249, 260, 272, 274, 283, 295–96, 299, 313; from gas, 295–96, 340, 350–51; hydrocarbons' providing of, 16–18, 235, 291, 324, 376, 379; and political dysfunction, 208, 210–11, 223–24, 225, 234, 313, 377–78; relative loss of, 19–20, 184, 249–50, 252, 260, 273, 291, 313; and sustainability revolution, 350–62

energy storage, 355–56

Engelbart, Douglas, 170

Engler, Robert, 10, 62, 68, 205

ENIAC (Electrical Numerical Integrator and Computer), 162–63

environmental laws and regulations, 126–27, 174, 177, 334, 335, 337. *See also* Clean Air Act

environmental movement, 9, 151, 188, 295, 371; backlash against, 259, 339; and energy taxes, 336, 342

Environmental Protection Agency (EPA), 174, 334–35, 337, 338

Erie Canal, 48, 50

EROI (energy return on investment), 182–83

ethanol, 260, 370

Evans, Lloyd, 156

executive power hypothesis, 234

extreme weather events, 266, 369, 385

extremism and extremists, 218, 254, 306; in Republican Party, 209–10, 211, 212, 216, 299

Exxon Valdez oil spill, 1–7

Fagan, Brian, 255

Fagg, Fred, 140

Fairchild, Sherman, 164–65

Faisal, King, 108

Fall, Albert B., 103

Falwell, Jerry, 198

Fannie Mae, 226

Faraday, Michael, 72

Far East, 80, 105, 318–19

Farr, Douglas, 343

Farris, Floyd, 293

Fast, Bob, 293

fast breeder reactors, 359

Fazzari, Steven, 236

Federal-Aid Highway Act of 1944, 136–38

Federal Aid Road Act, 96

Federal Deposit Insurance Corporation (FDIC), 115, 325

Federal Election Campaign Act, 214

Federal Emergency Relief Administration, 115

Federal Housing Authority, 226

federalism, 154–55, 218–19, 243–44

Federalist Party, 36, 37, 38

Federal Railroad Administration, 367

Federal Reserve, 84, 85–86, 228, 376, 377

Federal Trade Commission, 84, 376

Feeding the Ten Billion (Evans), 156

Ferguson, Niall, 30–31, 144, 270, 271–72

fertilizers, 147–49, 158–59

filibusters, Senate, 220–22, 308–9, 312

finance and financialization, 224–32; deregulation of, 202, 204, 227–28, 229–30, 234, 324; and Great Recession, 202, 230–31, 325; and hostile takeovers, 228–29; and macroeconomic security, 324–25; New Deal regulation of, 115, 377; Obama reform program for, 287, 296–97, 324, 378; and oil market, 231–32; and public debt, 225–26

fire harnessing, 23

firewood, 29, 67

First Political Age (1789–1860), 19, 56, 208–9, 374–75; Conservative Cycle, 51–55; Progressive Cycle, 31–42

Flom, Joe, 229

Flowers, Tommy, 162

food security, 369–70

Forbath, William, 238

Ford, Gerald, 175

Ford, Henry, 89, 92–94, 95, 135, 146

Fort Mansfield, 383–84

Fourth Political Age (2009–), 19, 272, 378–79; Progressive Cycle, 274–99

Fox News Network, 213–14

Fox-Penner, Peter, 90, 91

fracking, 293–95

France, 37, 44, 118, 219; and US, 38, 40, 253, 254

Frank, Barney, 287, 304–5

Franklin, Benjamin, 72

Freese, Barbara, 10, 65, 67–68, 86

Freight Rail Infrastructure Capacity Expansion Act, 364

Friedman, Milton, 123, 227–28

Friedman, Thomas, 156, 216, 242–43

Fuerth, Leon, 254–55

Fukuyama, Francis, 216

Fulton, Robert, 47

fungicides, 151

fusion reactors, 360–61

Gaddafi, Muammar, 290

Galbraith, John Kenneth, 16, 17, 122–23

Gale, Leonard, 73

Gallatin, Albert, 40

Garner, Jay, 207

Garrison, William Lloyd, 54

gas, 291–95; as bridge fuel, 340, 350–51

gasoline taxes, 96, 138, 349

Gates, Bill, 170, 171

Gates, Robert, 291

General Agreement on Tariffs and Trade, 128

General Electric Company, 76, 77–78, 89

generational cycle theory, 14, 15

generational dynamics, 260–61; in Fourth Political Age, 275–79; generational divides, 212, 280; generation types, 14–15; realignment in, 15–16, 114; theory of, 13–14

generations: Baby Boomer, 15, 197–98, 260–61, 280, 282, 298; GI, 114; Gilded, 57; Greatest, 197, 212; Lost, 114, 212; Millennial, 15, 275–79, 297, 298, 339, 385; Missionary, 114; Silent, 197, 260–61, 297, 298; Transcendental, 57; X, 15, 275–77, 280, 298, 330; Z, 15

Generations: The History of America's Future (Strauss and Howe), 14–15

Generation X, 15, 280, 298, 339; as political factor, 275–77

Generation Z, 15

genetically modified crops (GMOs), 270, 371–72, 380

George, Henry, 62

geothermal energy, 346, 352, 355, 361

Germany, 116–18, 119–20, 250–51

gerrymandering, 216–17

Gibbons, Thomas, 47

GI Generation, 114

Gilded Age, 60, 70, 80–81, 232, 235, 375

Gilded (Nomadic) Generation, 57

Gilens, Martin, 215–16

Gingrich, Newt, 209, 210

Ginnie Mae, 226

Ginsburg, Ruth Bader, 215

glacier melting, 266, 370

Glaeser, Edward, 343

Glass, Henry, 80

Glass-Steagall Act, 115, 204, 287, 324

globalization, 27, 48, 78, 161, 273, 369; and US geopolitics, 127–28, 315–16

global warming: and carbon dioxide, 191, 257–58, 261–62, 265–66, 334; consequences of, 259–60, 370; measures to confront, 248–49, 352. *See also* climate change

Global War on Terrorism, 186, 242

Gold, Russell, 293, 295

Goldin, Claudia, 121

Goldwater, Barry, 244

Good Roads Movement, 94–95

Goodwin, Doris Kearns, 82

Google, 173

Gordiner, Jeff, 276

Gore, Al, 172, 205

Gould, Jay, 70

government regulation, 198–99, 202, 225; of airline safety, 139–40; of energy industry, 110, 176–77; of financial industry, 115, 287, 296–97, 324, 377, 378; during Progressive Era, 82, 83, 84, 102; of railroads, 70–71. *See also* deregulation

Graham, Lindsey, 336

Gramm-Leach-Bliley Act of 1999, 230

Grant, Ulysses, 58, 60

Great Compression, 191, 211, 232, 321; conservative backlash against, 199, 208, 378; as golden age, 122, 208, 235; stability during, 225, 377; term, 121

Great Depression, 15, 166, 238, 330; government policy during, 132, 134–35, 147, 235; and stock-market crash, 103, 233; and World War II, 116, 117

Greatest Generation, 197, 212

Great Recession, 225, 268, 274–75, 295, 330; financialization as cause of, 231–32, 325, 336; Obama stimulus package for, 231, 247, 286–87, 336, 367, 378

Great Society, 124–25, 126, 377

Greece, ancient, 29, 72–73

Greene, David, 187

greenhouse effect, 257, 261

greenhouse gas emissions, 139, 160, 265–66, 339, 340–42, 369–70; Kyoto Protocol on, 337–38; from methane and nitrous oxide, 261–62, 266, 351; Supreme Court on, 334–35. *See also* carbon dioxide; carbon tax

green industries, 363

Green Revolution, 146

Green River, 181

Greenspan, Alan, 229–30

green urbanism, 344–47, 356, 363–64

Griffith, Morgan, 223

Grove, Andy, 165

Gulbenkian, Calouste, 108

Haber, Fritz, 148

Hagel, Chuck, 330

Hall, John H., 45

Hamilton, Alexander, 35–37, 59, 221, 306

Hanauer, Nick, 238–39

Harding, Warren, 102–3, 108

Hartmann, Thom, 29

Hastert, Dennis, 210

Haupt, Herman, 68

Hayes, Denis, 346

Hazard, Erskine, 63

Hazelwood, Joseph, 2

Head Start program, 126

health care, 332; Clinton and, 203–4, 210, 287; Obama and, 287–89, 378

Heart, Frank, 169

Heather, Peter, 268–69

Henry, Patrick, 307

Heppenheimer, T. A., 100

herbicides, 151, 155

Heritage Foundation, 288

Hero of Alexandria, 87–88

Herring, Augustus, 99

Hickel, Wally, 1

high-speed rail, 365–69, 380

high-voltage direct current (HVDC), 357

highways. *See* roads and highways

Hitler, Adolf, 117–18, 119

Hoerni, Jean, 164

Hoff, Ted, 165

Holland, Hezekiah Russel, 4, 5–6

Hollerith, Herman, 161

Holmes Frank, 110–11

Holocene Epoch, 24–25, 255–56

homeownership, 226, 231

Homestead Act of 1862, 60, 152

Hoover, Herbert, 103, 108, 111–12, 131

Hopper, Grace, 162

horizontal drilling, 181–82, 294, 295

horse collar, 26, 157

hostile takeovers, 228–29

Hounshell, David, 45, 92, 166

House of Representatives: constitutional reform of, 303, 309–11; gerrymandering of districts in, 216–17

housing bubble, 231

Howe, Daniel Walker, 52, 114

Howe, Neil, 14–15, 57, 277

Hubbert, M. King, 184

Huckabee, Mike, 279

Humphrey, Hubert, 173–74

Hunter, Louis, 47

hunter-gatherers, 24–25, 152, 157, 256

hybrid seeds, 149–50, 155

hydraulic fracturing, 181–82

hydroelectric energy, 76–78, 130, 248–49, 355, 361

Ibn Husayn, Faisal, 250

Ibn Saud, 111, 251

ice age, 382–83

Ickes, Harold, 110, 119–20

immigrants and immigration, 69, 88, 143, 328–30

imperialism, 78–80, 129, 134, 272

Imperial Valley, 154

Improving America's Schools Act, 242

income tax, progressive, 59–60, 84, 228, 332, 375

India, 155, 316–17, 338

individualism, 101, 111–12, 113, 123, 199

Industrial Revolution, 18–19, 261, 262

Industrial Revolution, First, 17–18, 65, 158, 160; American system of manufacture in, 42–46, 93; in Britain, 28–31; steamboats and canals in, 46–49

Industrial Revolution, Second: bicycle's importance to, 94; coal and steel in, 62–68, 93, 160, 375; and mass production, 91–93

Industrial Revolution, Third, 160–73, 191, 325. *See also* Digital Revolution inequality, income and wealth, 123, 232–39; consequences of, 236–38; growth of, 233–34, 236; reducing, 235–36, 331–33

inflation, 143, 186, 191

Inflation Bill of 1874, 60

infrastructure, 56, 88, 330; decay of, 244–50; public investment in, 245–46, 362; railroad, 364–65, 367–68

INNATE revolutions. *See* Agricultural Revolution; Electrification Revolution; Industrial Revolution; Transportation Revolution

insider trading, 229

Insull, Samuel, 89–91

interchangeable manufacture, 44–45

intercity transport, 365–69

internal-combustion engine, 105, 138–39

International Business Machines Corporation (IBM), 161, 171

International Energy Agency (IEA), 183

International Living Future Institute, 346

International Monetary Fund, 128

international prestige, 250–55

Internet, 214; invention of, 168–70; public access to, 172–73; and World Wide Web, 171–72

Interstate Commerce Commission (ICC), 71

interstate highway system, 93–94, 136–39

Iran, 110, 191, 252, 290

Iran-Iraq War, 202, 252

Iraq, 108, 203, 250, 251, 252

Iraq War, 186, 206–7, 253, 323

irrigation, 25–26, 151–55

Isaacson, Walter, 163, 169, 171, 172

isolationism, 78, 102–3, 128, 272

Israel, 185, 253, 314

Issacharoff, Samuel, 216–17

Jackson, Andrew, 51–52

Jackson, Jesse, Jr., 304–5

Jackson, Lisa, 337

James, Edwin, 152

Japan, 117–18, 119, 318, 365–66

Jay, John, 37, 44

Jefferson, Thomas, 34, 40–41, 44, 48; and American expansion, 39–40, 374; and Federalist policies, 35–36, 38; rural orientation of, 36, 88

Jennings, Jean, 162

jet propulsion, 140–41

Jobs, Steve, 170, 171

Johnson, Lyndon, 126–27, 241; Great Society of, 124–25, 126, 377

Johnson, Stephen, 335

Joiner, Columbus "Dad," 109

Jones, Bryan, 13

junk bonds, 229

Kadak, Andrew, 359

Kahn, Robert, 169–70, 172

Kansas-Nebraska Act, 56

Kay, Alan, 170

Keeling Curve, 258

Kelvin, Lord, 161

Kennan, George, 127

Kennedy, John, 123–24, 125

Kennedy, Paul, 79, 272–73, 313

Kennedy, Robert, 173

Kennedy, Ted, 192, 196, 242, 253

kerosene, 86, 105

Kessler, Jean Baptiste August, 105

Key, V. O., Jr., 12

Keynesianism: Obama and, 231, 286–87, 336, 378; Roosevelt and, 115

Khomeini, Ayatollah, 191

Kilby, Jack, 164–65

King, Angus, 222–23, 324–25

King, Martin Luther, Jr., 125, 173

Kingdon, John, 341

Kinzinger, Adam, 223

Klare, Michael, 250

Klein, Maury, 72, 74, 77, 88

Kleinrock, Leonard, 169

Kleppner, Paul, 12

Kline, Ronald, 132

Knox, Henry, 44

Korean War, 129

Krueger, Alan, 237

Krugman, Paul, 121, 234, 235–36, 286–87

Kunstler, James Howard, 133, 138

Kuwait, 111, 203, 250, 252

Kyoto Protocol, 159–60, 204, 337–38

labor unions, 67–68, 115, 121, 235

laissez-faire economics, 52, 102–3, 112–13, 130, 199, 202

Lamme, Benjamin Carver, 76

Lampson, Butler, 170

Land and Water Conservation Fund Act, 126

Langley, Samuel Pierpont, 97
Latin America, 318
Lavassor, Émile, 95
Lawes, John Bennet, 158–59
Lazare, Daniel, 308
League of Nations, 85
Leahy, Patrick, 223
Lechterman, Ruth, 162
Lee, Robert E., 58, 59
Lee, Roswell, 45
Lend-Lease Act, 116–17, 251
Leonhardt, David, 225
Levinson, Sanford, 220, 299–300
Lewis, John, 222
liberty, positive and negative, 57, 111, 112, 285–86
Libya, 290
Licklider, J. C. R. "Lick," 168
Liebeg, Justus von, 158
Lilienthal, Otto, 97
Limits to Growth, The, 293
Lincoln, Abraham, 57–59, 68, 69; as economic
    modernizer, 59–60, 375
Lincoln Laboratory, 168
Lind, Michael, 42, 43, 59
Linklater, Richard, 276
Little Ice Age, 256
livestock, 144–45, 155
Living Building Challenge, 346
Livingston, Robert, 40
Livingstone, Robert, 47
local-foods movement, 370
Long Depression, 60, 62, 76
Lopes, Carl, 160
Los Angeles, CA, 105, 154, 344
Lost Generation, 114, 212
Lott, Trent, 302
Louisiana Purchase, 39–40
Lowell, Francis Cabot, 43

MacDonald, Thomas, 96, 135, 137
Macintosh computer, 171
MacKay, David, 356, 361
macroeconomic security, 16, 18; and First Political
    Age, 19, 35, 42, 51, 55, 59–60; and Second Political
    Age, 19, 57, 59, 82, 84, 85–86, 102, 209; and Third
    Political Age, 113, 114, 121–22, 126, 132, 138, 183,
    219, 244; and Fourth Political Age, 286, 323–31,
    374–76

Madison, James, 34, 35–36, 41, 221, 306, 310
Madrick, Jeff, 227–28, 230, 231
Mahan, Alfred Thayer, 78
major realignment elections, 13, 16, 34, 55, 57, 280
Malthus, Thomas, 16
Mandelbaum, Michael, 242–43
Manheim, Karl, 13–14
Manifest Destiny, 53
Manly, Charles, 97
Mann, Horace, 240
Mann, Thomas, 209–10, 214–15, 224
Manning, Richard, 24, 26, 145, 150
Margo, Robert, 121
Marías, Julian, 14
Marine Mammal Protection Act, 174
Marshall, John, 47–48
Marshall Plan, 128, 272
Martin, Albro, 71
Martin, Richard, 360
Massachusetts vs. Environmental Protection Agency,
    334–35, 337
mass production, 65, 66, 89, 91–93
Matthews, Scott, 373
Mattson, Kevin, 191
Mauchly, John, 162, 163
Maxwell, James Clerk, 72
Mayhew, David, 12–13
Mazewski, Matt, 238
McCain, John, 279, 282–83, 324–25
McCallum, Daniel, 68
McCarthy, Gina, 337
McCarthy, John, 168
McClellan, Peter, 145
McClendon, Aubrey, 295
McConnell, Mitch, 298
McDonald, Forrest, 91
McGuinn, Patrick, 240
McKiernan, David D., 207
McKinley, William, 78–80
McNeill, J. R., 11
McNulty, Kay, 162
meat, 369–70; reducing consumption of, 372–74, 380
media, 212–14, 258–59
Medicaid, 126, 288, 377
Medicare, 126, 377
Medieval Warm Period, 256
Mentré, Françoise, 14
Merchant, Carolyn, 11

Meriwether, John, 230

Mesopotamia, 29, 157

Mesozoic Era, 255

methane, 261–62, 266, 351

Mexican War, 53

Mexico, 109, 110, 317

Meyer, Frank, 198

microchip, 164–66

microeconomic security, 16, 18; and Second Political
Age, 19, 81–82, 84, 86, 375; and Third Political
Age, 19, 111, 112–16, 121–22, 123–25, 126, 203–4,
209, 375–76, 377; and Fourth Political Age, 283,
287, 289, 331–33, 380

Microsoft, 171

Middle East: Obama policy toward, 290–91; and
Oil Embargo, 2, 174–75, 185–86, 188; and US
petroleum interests, 108–9, 110–11, 185, 186–87,
201–2, 250–54, 315, 377; during World Wars,
107–8, 250–51

Milankovitch cycles, 257, 261

military forces, US, 201–2, 231–32, 321–23

military-industrial complex, 128–29

Milkis, Sidney, 103, 200

Millennial Generation, 15, 339, 385; as political
factor, 275–76, 277–79, 297, 298

Millken, Michael, 229, 230

minor realignment elections, 13, 16, 51, 78, 174, 200

Miocene Epoch, 255

Misa, Thomas, 66, 67

Missionary Generation, 114

Missouri Compromise, 56

Mitchell, George, 293–94

Mitchell, John, 67

Mitchell, Timothy, 10–11, 316

mobility, social-economic, 237, 325

Mokyr, Joel, 65

Monroe, James, 41

Monroe Doctrine, 85

Moore, Gordon, 164, 165

Moral Majority, 198

Morgan, J. P., 77, 89

Morgenthau, Henry, 188

Morrill Land-Grant Colleges Act, 60

Morris, Ian, 17

Morse, Samuel, 73

mortgage-backed securities, 230–31

Mosaddegh, Mohammad, 191

Motor Vehicle Air Pollution Control Act of 1965, 126

Mozilo, Angelo, 230

Murphy, David, 183

Murrow, Edward R., 213

Muscle Shoals, 148–49

Muskie, Edmund, 174

Myth of Rugged American Individualism, The
(Beard), 112

Naím, Moisés, 216

Napatree Point, 381–86

Napoleonic Wars, 40

National Advisory Committee for Aeronautics
(NACA), 99, 166–67

National Arctic Wildlife Refuge, 5

National Commission on Excellence in Education,
241

National Council on Public Works Improvement,
245

National Currency Acts, 59

National Defense Education Act, 240

National Defense Research Council (NDRC),
166–67

National Energy Security Act, 176–77

National Environmental Policy Act, 126

National Environmental Protection Act, 174

National Highway Traffic Safety Administration, 337

National Historic Preservation Act, 126

National Information Infrastructure Act, 172

National Institutes of Health, 167

nationalism: and multipolarity, 273–74; New
Nationalism, 83–84

National Labor Relations Act, 115

National Research Council, 166

National Resources Board, 166

National Science Foundation, 167

national security, 78, 203, 313–16, 374–75, 377; and
American Geopolitical Cross, 317–19; Obama
and, 284, 289–91, 319; and soft power, 319–21; and
US military, 321–23

National Trails System Act, 126

Nation at Risk, A, 241, 289

Naval Consulting Board, 166

Navy, US, 322

Nelson, Michael, 103, 200

neoconservatives, 268, 314

Neolithic Revolution, 24, 26, 157, 255–56

Newcomen, Thomas, 30

New Deal, 114–16, 131, 134–35, 148–49, 377

New Freedom, 84

New Frontier, 123–24

Newman, Max, 162

New Nationalism, 83–84

newspapers, 212–13, 214

Niagara Falls Power Company, 77, 86

nitrous oxide, 261–62

Nixon, Richard, 173, 185, 287, 365; energy policies of, 1, 174–75

Nobel, Ludwig, 105

No Child Left Behind Act, 242, 289

Nolan, Rick, 301–2

Noriega, Manuel, 203

North, Simeon, 45

North American Free Trade Agreement, 203, 317

North Atlantic Treaty Organization (NATO), 128

Northeast Corridor, 248, 365, 367, 368

Noyce, Robert, 164

nuclear fusion, 360–61

nuclear power, 191, 358–60

Nye, David, 130, 131, 133

Nye, Joseph, 320, 321, 322

Oak Ridge National Laboratory Buildings Technology Center, 346

Obama, Barack, 283–84, 305–6, 333; clean energy efforts of, 335–36, 351–52; and climate security, 207, 285–86, 324, 336–38, 342; educational policy of, 289; elections of, 275, 277, 278, 279–83; and financial reform, 287, 296–97, 324, 378; and health care reform, 287–89, 378; on infrastructure, 245, 247; Keynesian measures of, 231, 247, 286–87, 336, 367, 378; and national security, 284, 289–91, 319

Occupy Wall Street, 238

ocean acidity, 265–66

ocean liners, 101, 141, 363

Office of Energy Efficiency and Renewable Energy (EERE), 352, 353, 354

Office of Scientific Research and Development (OSRD), 167

Ogden, Aaron, 47

oil: American industry established, 104–5; becomes dominant energy source, 135, 160; and climate security, 183, 339–40; conventional vs. unconventional, 181–82; dependence on imported, 184, 186–87, 253; *Exxon Valdez* spill of, 1–7; and financialization, 231–32; and gas, 292–93; government subsidizing of, 188–89; and Middle East, 107–9, 110–11, 174–75, 185, 186–87, 188, 201–2, 250–54, 315, 377; peaking of domestic supplies of, 181–90, 377; price of, 110, 182, 183, 185–86, 187–88, 191, 200–201, 211; production of, 2, 105–6, 109–10, 182, 183–84; and Third Political Age, 376; and US foreign policy, 108–11, 127, 184–85, 250–54; and World Wars, 107–8, 118–20, 250–51

Oil Embargo of 1973, 2, 174–75, 185–86, 188

oil pipelines, 249; in Alaska, 1–2, 4–5, 174

oil sands, 181, 182

O'Neill, Tip, 176, 210

OPEC, 2, 174–75, 184, 185–86, 188, 200–201

Open Door trade policy, 80

Opie, John, 153

Oregon Territory, 53

organic agriculture, 370–71

Ornstein, Norman, 209–10, 214–15, 224

Ørsted, Hans-Christiaan, 72

Ortega y Gasset, José, 14

Otto, Nikolaus, 95

Owen, David, 343, 349

Pacific Railway Acts, 60, 69

Packer, George, 280

Page, Benjamin, 215–16

Page, Larry, 172–73

Page, Logan, 95–96

Pahlavi, Mohammad Reza Shah, 191

Paine, Thomas, 32

Painter, David, 127, 184, 185

Paleogene Period, 255

Paleozoic Era, 255

Palin, Sarah, 283

Panama Canal, 246

Panama invasion, 203

Paris Agreement, 338

parliamentary system, 206–7

peak oil theory, 181

Peason, Sir Weetman, 109

pebble-bed reactors, 359

Persian Gulf War, 186, 203, 252

personal computer, 165–66, 170–71

pesticides, 150–51

Peters, Tom, 229

Phillips, Kevin, 198, 231, 232, 270–71

Pierce, Franklin, 56

Piketty, Thomas, 233, 234, 235, 237, 238

Pittsburgh, PA, 47, 63, 65

Plass, Gilbert, 257–58

Pleistocene Epoch, 252

Pliocene Epoch, 255

Plouffe, David, 280–81

Pocan, Mark, 305

Podobnik, Bruce, 68

political dysfunction, 208–24, 320; and energy insecurity, 210–11, 223–24, 225, 234, 313, 377–78

political partisanship, 211, 216–17

political polarization, 209, 238

Politics of Oil, The (Engler), 10

Polk, James, 52–53, 374

Pollan, Michael, 372

population growth, 156–60, 256

poverty, 60, 62, 116, 131, 243–44

Powell, Colin, 207

Precambrian Era, 255

presidential nominations, 312, 379

Priest, Tyler, 110, 128, 225, 251–52, 295, 313

Prince William Sound, 1–7

Progressive Movement, 71, 83, 106

Project Independence, 174–75

Prout, Henry, 75

Prudhoe Bay, 1, 2

public-private relationship, 50–51, 122–23, 306

Public Roads Administration, 137

public transit, 135, 248, 265, 349, 350, 376; need for, 344, 350

Public Works Administration, 115–16

public-works projects, 46, 115, 124, 138

pumped storage, 355–56

punctuated-equilibrium model, 13

Putnam, Robert, 348

Quality Growth Act, 345

Quest, The (Yergin), 181

railroads: and Civil War, 58–59; decline of, 71, 141, 142; and electricity, 76, 364–65; and First Transportation Revolution, 49–51, 246, 374–75; and high-speed trains, 365–69, 380; and infrastructure, 248, 380; and Second Transportation Revolution, 68–71, 246; transcontinental, 69–70, 375

Rand Corporation, 168, 169

Rapid Deployment Joint Task Force, 201

Ratner, Steven, 288

Rawls, John, 197

Reagan, Ronald, 196–97, 199–200, 202, 225, 241, 246

Reagan Corollary, 202, 252

Reardon, Sean, 243

redistricting, 216–17, 222–23, 302–3

Reichley, James, 12–13

Reinhart, Carmen M., 224, 225

Reisner, Marc, 153–54

renewable and alternative energy sources, 46, 313; Carter call for, 194–95, 254; and energy storage, 355–56; research and technology development in, 351–55, 380; solar, 340, 346, 354–55, 356, 361–62; wind, 340, 354, 355, 356, 361

Renewable and Appropriate Energy Laboratory, 352

Republican Amendment, 306–12, 379

Republican Party, 37, 111–12, 217, 225, 288, 339; and educational policy, 241, 244; and election of 2008, 275, 280–81, 282–83; energy policies of, 110, 232, 352; extremism in, 209–10, 211, 212, 216, 299; generational support for, 275–76, 278; laissez-faire approach of, 102–3, 199; and Lincoln, 57, 59–60; and Reagan, 196–97, 198; and South, 125–26, 173–74, 198; and Trump, 296–99. *See also* elections

research: in clean and alternative energy, 351–53, 380; government support to, 166–68; in macroeconomic security, 320–21

Resource Triangle, 291

Revelle, Roger, 258

Revenue Act of 1862, 59–60

Revolution, American, 31–33

Ricardo, David, 16

Richardson, Bill, 186–87, 254

Rickenbacker, Eddie, 107

rising sea levels, 266, 369, 370, 385

roads and highways, 48–49, 93–97, 134–39, 246; and cities, 135–36, 138; interstate highway system, 93–94, 136–39; rural, 96–97

Roberts, Edward A. L., 292–93

Roberts, Larry, 168–69

Roberts, Paul, 29, 183, 251, 351

Robinson, James A., 55, 133

Rock, Arthur, 165

Rockefeller, John D., 104

Rockne, Knute, 139

Roe, Joseph Wickham, 44–45

Roe v. Wade, 198

Rogoff, Kenneth, 224, 225

Rohatyn, Felix, 245–46, 249–50

Rolfe, John, 144

Roman Empire, 29, 268–70

Romney, Mitt, 277, 279

Roosevelt, Franklin, 91, 113, 131, 166–67; agricultural policy of, 148–49; economic rights program of, 120–21; election of, 112–13, 114; and financial regulation, 115, 377; New Deal agenda of, 114–16; and oil, 110, 251; transportation policies of, 134–35, 136, 140; and World War II, 116–18, 120, 251

Roosevelt, Theodore, 67, 81–82, 83–84, 106, 376

Roosevelt Corollary, 85

Rosenberg, Chaim, 43

Royal Dutch Shell, 108, 109

Rubin, Robert, 230

rural electrification, 129–34

Rural Electrification Act of 1935, 129

Rural Electrification Administration (REA), 132, 133, 134

Russia, 105, 203, 290, 297. *See also* Soviet Union

Ryan, John, 83

Ryan, Paul, 298

Sabato, Larry, 310

Saddam Hussein, 203, 206, 207, 252

Safe Drinking Water Act of 1974, 174

Sampson, William T., 80

Samuel, Marcus, 105

Sanders, Bernie, 238, 297, 305

San Remo Agreement of 1920, 108

Santa Clara County v. Southern Pacific Railroad, 61–62, 214

Saudi Arabia, 111, 181, 183, 185, 186, 187; and US, 128, 202, 251–52

savings-and-loan crisis, 202

Schattschneider, E. E., 12

Scher, Bill, 290

Schlesinger, Arthur M., Jr., 14, 80

Schlesinger, James, 175

Schley, Winfield Scott, 80

Schock, Aaron, 222

school buildings, 327

Science: The Endless Frontier (Bush), 167, 168

Scott, Bruce R., 202

Scotti, R. A., 384

Seattle, Wash., 345, 347

Second Bank of the United States, 41, 51–52

Second Political Age (1861–1932), 19, 54, 209, 374–75; Conservative Cycle, 101–3; First Progressive Cycle, 57–62; Second Progressive Cycle, 78–86

Securities and Exchange Commission, 115

Seely, Bruce, 96, 135

Senate: abolition of, 308–9; filibusters in, 220–22, 308–9, 312

shale gas, 292

shale oil, 181–82

Shannon, Claude, 161

Shelby County v. Holder, 215

Shell Oil Company, 105, 184

Sherman, William T., 58–59

Sherman Antitrust Act, 106

Shewe, Phillip, 91

Shinseki, Eric, 206

Shockley, William, 163–64

Silent Generation, 197, 260–61, 297, 298

Silent Spring (Carson), 151

Silicon Valley, 165

Silver, Nate, 217

Singer, Issac, 46

Slacker, 276

Slater, Samuel, 43

slavery, 53–55, 56–57, 58, 144, 216, 303

Smil, Vaclav, 69, 134, 138

Smith, Adam, 16

Snyder, Betty, 162

Social Security, 116, 377

soft power, 319–21

Soil and Water Conservation Act, 177

solar energy, 340, 346, 356, 361–62; techniques and uses of, 354–55

Solid Waste Disposal Act of 1965, 126

Something New Under the Sun (McNeill), 11

Soros, George, 230

Souter, David, 5

South, 60, 144–45; electrification in, 133–34; and political party alignment, 110, 125–26, 173–74

Southeast Asian Treaty Organization (SEATO), 128

South Korea, 318

Southwest region, 133, 152, 356

Soviet Union, 119, 127, 185, 200–201, 250

Spanish-American War, 80

Special Forces, 323

Sprague, Frank, 76

Square Deal, 82

Standage, Tom, 24, 25, 158

Standard Oil, 104–5, 106

Stanford Research Institute (SRI), 168

Stark, Fortney Pete, 304–5

Starobin, Paul, 129, 273

state governments, 51, 96, 126, 218–19, 307; and national elections, 215, 302–3, 305

steamboats, 46–48, 246, 374

steam power, 30–31, 63

steam turbines, 87–88, 89–90, 375

steel, 375; and Second Industrial Revolution, 66–67, 68, 93, 160

Steel, Carolyn, 26

Stein, Howard F., 202

Steinberg, Ted, 11

Steinsberger, Nick, 294

Sterling, Dorothy, 382–83

Stern, Roger, 187

Stevens, John Paul, 214, 301, 335

Stevens, Ted, 2

Stibitz, George, 161

Stone, Roy, 95

Strategic Arms Reduction Treaty (START), 290; START II, 203

Strauss, Rebecca, 244

Strauss, Richard, 277

Strauss, William, 14–15, 57, 114

Strong, Josiah, 78

suburbs, 133, 265, 345, 376

Summers, Larry, 230

Sundquist, James, 12

Superfund Act of 1980, 177

Surface Mining Control and Reclamation Act of 1977, 177

sustainability revolution: and carbon pricing, 339–42; and energy security, 350–62; food and diet for, 369–74; and intercity travel, 362–69; organizing framework for, 374–80; and urban policy, 342–50

Swift, Earl, 96, 97, 136

Syria, 185

Taft, William Howard, 82

Taney, Roger, 56–57

Tarbell, Ida, 106

taxes, 176, 198–99, 336; carbon, 340–42, 350, 353–54, 367, 373, 380; gasoline, 96, 138, 349; progressive income, 59–60, 84, 228, 332, 375; value-added, 332–33

Taylor, Bob, 168

Taylor, Charlie, 98

Taylor, Paul, 274–75

teachers, 327–28

Teagle, Walter, 108

Tea Party movement, 238, 285, 336

Teapot Dome scandal, 103

telegraph, 56, 72–73

television, 212–14

Tennessee Electric Power Company et al. v. TVA, 131–32

Tennessee Valley Authority (TVA), 116, 129, 131–32, 133, 134

terrorism, 252–53, 314

Terry, Eli, 45

Tesla, Nikola, 72, 74–75, 77

textile industry, 28, 43–44

Thacker, Chuck, 170

Theory of Justice, A (Rawls), 197

thermal cracking, 106

Third Political Age (1933–2008), 19, 209, 240, 325, 376–77; Conservative Cycle, 197–207, 272, 377–78; First Progressive Cycle, 111–29; Second Progressive Cycle, 173–77

three-field farming, 157

Three Mile Island, 191, 358

tobacco, 144

Tocqueville, Alexis de, 55

Todd, Chuck, 280

tokamaks, 360

Toll Roads and Free Roads, 135–36

torture, 284, 290

Tousard, Louis de, 44

tractors, 146–47, 376

trade, 26–27, 37, 40–41, 313–14

Trans-Alaskan Pipeline Authorization Act, 2, 4–5

Transcendental (Prophet) Generation, 57

transcontinental railroad, 69–70, 375

transistors, 163–64

Transmission Control Protocol (TCP), 169–70

Transportation Revolution: airlines, 97–101, 139–42, 363–64; and alternative energy sources, 248–49, 352–53; and greenhouse gas emissions, 261, 262, 265; as hydrocarbon-fueled, 18–19, 46, 376; infrastructure for, 246, 247–48, 380; intercity travel, 362–69; railroads, 49–51, 68–71, 365–69, 374–75; roads and highways, 93–97, 134–39; steamboats and canals, 46–48. *See also* aviation and airlines; railroads; roads and highways

Treaty of Ghent, 41

Treaty of Guadalupe Hidalgo, 53

trucking industry, 362–64

Truman, Harry, 110, 118, 127–28, 251–52, 287
Truman Doctrine, 127–28, 251–52
Trump, Donald, 272, 274, 298–99, 338; election of, 218, 238, 296–98, 378
Turing, Alan, 161, 162
Turkey, 108, 251, 254
Twain, Mark, 14
Twitter, 214

Udall, Tom, 301
United Mine Workers (UMW), 67
United Nations Intergovernmental Panel on Climate Change, 258
United States v. Ballin, 221–22
urbanism, 343–44, 353; green, 344–47, 356, 363–64
urbanization, 29, 64–65, 66–67, 88, 261; and agriculture, 26, 44, 132–33, 260, 270. *See also* cities
U.S. v. Standard Oil decision, 106

value-added tax (VAT), 332–33
Van Buren, Martin, 48, 51, 52
Vanderbilt, Cornelius, 70
Veblen, Thorstein, 81
Venezuela, 109
Veterans Loan Program, 226
Vietnam War, 129, 250, 272
Villard, Henry, 89
Volta, Alessandro, 72
von Neumann, John, 163
von Ohain, Hans, 140–41
von Richthofen, Manfred, 107
voter registration, 305–6
Voting Rights Act, 125–26, 173, 215

Wadsworth, Decius, 44
Walsh, Albert, 103
Ward, Lester, 81
Wargames, 172
War of 1812, 41, 63
War on Poverty, 126
Warren, Earl, 240
Warren, Elizabeth, 324–25
Washington, George, 32–33, 34, 36, 37–38
wastewater systems, 248–49
Waterman, Robert, 229
water power, 43–44, 63, 68, 76–77
water scarcity, 266
Watt, James, 30

Waxman-Markey bill, 336, 337
web browsers, 172–73
Weber, Christopher, 373
Welfare Reform Act of 1996, 204
Wells, Christopher, 138
Wescoff, Maylyn, 162
Westinghouse, George, 75–76, 77
Whig Party, 37, 52–53, 55
White, Josiah, 63–64
White, Theodore, 124
Whitney, Eli, 45
Whittle, Frank, 140
Why Nations Fail (Acemoglu and Robinson), 55
Why the West Rules—For Now (Morris), 17
Wigmore, John, 140
Wild and Scenic Rivers Act of 1968, 126
Wilderness Act of 1964, 126
Wilson, Woodrow, 84, 85, 115, 166, 376
wind energy, 340, 354, 355, 356, 361
Winthrop, John, 143
Wolmar, Christian, 49, 68–69, 70, 71
Works Progress Administration, 115–16
World War I, 85, 166; and oil, 107–8, 251
World War II, 116–18, 140, 166–67; and oil, 118–20, 250–51
World Wide Web, 171–72
Worley, Robert, 289
Wozniak, Steve, 170, 171
Wrangham, Richard, 23
Wright, Wilbur and Orville, 97–99
Wriston, Walter, 228
Wurts, William and Maurice, 63–64

Xerox Alto, 170
Xerox PARC, 168, 170

Yalta Conference, 251
Yamamoto, Isoroku, 119
Yellen, Janet, 237–38
Yeltsin, Boris, 203
Yergin, Daniel, 104–5, 108, 109–10, 181, 184; on Middle East, 111, 252; on oil production, 120, 182
Yom Kippur War, 185
Younger Dryas, 24

Zakaria, Fareed, 273, 274, 315–16
Zijlker, Aeilko Jans, 105
zoning, urban, 347–48